Practical Panarchy for Adaptive Water Governance

Barbara Cosens • Lance Gunderson
Editors

Practical Panarchy for Adaptive Water Governance

Linking Law to Social-Ecological Resilience

Springer

Editors
Barbara Cosens
College of Law and Waters of the West Program
University of Idaho
Moscow, ID, USA

Lance Gunderson
Department of Environmental Sciences
Emory University
Atlanta, GA, USA

ISBN 978-3-319-72470-6 ISBN 978-3-319-72472-0 (eBook)
https://doi.org/10.1007/978-3-319-72472-0

Library of Congress Control Number: 2018934386

Printed on acid-free paper

This Springer imprint is published by the registered company Springer International Publishing AG part of Springer Nature.
The registered company address is: Gewerbestrasse 11, 6330 Cham, Switzerland

Preface

The authors appearing in this volume came together with the Adaptive Water Governance Project, inspired by a shared interest in helping inform the deliberate adaptation and transformation of water-dependent social-ecological systems in the Anthropocene. The project brought together ecologists, geographers, political scientists, and legal scholars, all expressing the concern that although science has existed for decades showing that human development and increasing use of water pose a threat to ecosystems and the social systems reliant on them, the uptake of this information has been slow to enter law and policy and, as a result, slow to lead to actual change. Dialogue surrounding these issues tends to view future scenarios as solely based on tradeoffs within a zero-sum game. Further, while science has made great strides in exploring how climate change may affect specific water systems, environmental law, the tool relied on to institutionalize collective choice on the mediation between social and ecological systems in western democracies, has its foundation in an understanding of systems as linear and its primary goal as facilitating social and economic stability. Thus, at first glance, law leaves society woefully unprepared to manage the dynamic and nonlinear nature of water-based ecological systems that will evolve in unpredictable and complex ways. Yet history has shown that legal systems incapable of evolution in the face of change are themselves destabilizing and that modern societies have addressed this need for evolution and the desire for a broad societal voice in how it plays out through the development of democratic institutions and more recently the emergence of new governance that involves public and private actors. In the context of social-ecological systems, much of the literature and hope focuses on the emergence of adaptive governance. The inspiration for the Adaptive Water Governance Project lays not only in the desire to remove legal barriers to adaptation but to consider how law in democratic systems might be a tool in the facilitation of the emergence of adaptive water governance.

The National Socio-Environmental Synthesis Center (SESYNC), funded by the US National Science Foundation, provided the venue for the Adaptive Water Governance team to meet in a series of five workshops over a three-year period. Despite disciplinary differences, team members had all turned to resilience theory to guide their understanding of social-ecological systems. The ecologists on the team came out of the C.S. Holling school of ecological resilience which views resilience as a system property describing both the capacity of a system to adapt and its ability to exist in alternative states, yet lacking a normative overlay. This perspective formed the foundation for a common lexicon that would form the bridge among members. The team also began with a dialogue developed by the Toolbox Project developed by faculty and Michigan State University and the University of Idaho, to facilitate team-based research through an understanding of value-based differences across disciplines. It was, however, the initial effort to gather information on North American water basins that led to the use of the most powerful boundary object for cross-disciplinary dialogue and integration: place.

Teams led by members with policy and ecology backgrounds engaged with external participants in resilience assessment of six North American water basins. This was, in effect, the data gathering stage of the project. The focus on place was key to the ability of the team to integrate across disciplines. Theoretical and visionary concepts in the abstract are difficult to grasp from divergent disciplinary perspectives. Yet, how they play out in a particular place, within a particular ecosystem, inhabited by a particular society, brings abstract concepts to life, allowing deep understanding across disciplinary divides. The second step of cross-team sharing and reviewing of basin assessments for publication allowed teams to move beyond the contextual and begin to identify potential areas of synthesis and transferrable knowledge. The basins chosen reflected the heavily regulated water systems in which the economic systems had grown dependent on water development. From the viewpoint of the project, this allowed consideration of the intersection of resilience, governance, and law and the potential for the emergence of adaptive processes in the context of social-ecological systems with substantial barriers to change.

The six assessments, and a seventh which tested the approach in Australia to move outside both the legal and social settings of North America, provided the context for understanding how the complex interaction among law, governance, society, and the environment plays out and allowed the development of a deeper understanding of the conditions under which adaptive governance emerges and the barriers to that emergence. The assessments and the common threads that flow from them form the basis for this volume.

The Adaptive Water Governance Project is merely a first step in bringing policy disciplines to the understanding of the dynamic nature of ecological systems. In doing so, the authors are aware both of the hubris of such an endeavor and its limitations. The effort has not yet engaged those with deep knowledge of political ecology and the role of power and agency in the adaptation of social systems and would welcome that addition as a next step. At the same time, the project is aware that in a democracy, law is a manifestation of agency and the exercise of political power. As a result, the readers will find that regardless of their relation to an objective definition of resilience, issues of fairness, equity, transparency, accountability, legitimacy, and the balance between stability and flexibility are essential components of any framework for the role of law in adaptation and transformation. Without attention to these aspects, this would be little more than an academic exercise with little prospects for change. In the end, the common value spanning the Adaptive Water Governance team is the desire to play a role in advancing the capacity of social-ecological systems to adapt and transform, hence the need for a reality-based framework.

Moscow, ID, USA Barbara Cosens
Atlanta, GA, USA Lance Gunderson

Acknowledgments

As the editors of this volume, we would like to acknowledge that it has been an honor and a privilege to work with the Adaptive Water Governance team whose effort is reflected in the chapters in this volume, and without whom, this endeavor would not have been possible. This work would also not have been possible without the funding from the US National Socio-Environmental Synthesis Center (SESYNC) under the funding from the US National Science Foundation, NSF DBI-1052875. The authors would like to thank Margaret Palmer and Jonathan Kramer of SESYNC for their belief in and support of this project. Finally, we would like to thank University of Idaho JD student Christopher Murray and JD/MS Water Resources student Jon Rezabek for their long hours of citation editing. We hope this work helps to inspire others to realize that simply bringing together people from different backgrounds and perspectives can advance knowledge and begin to address some of the more difficult problems faced in a time when humans have become not only an integral part of natural systems but dominate those systems. Engaging in this work is not only a privilege but an obligation as we find ourselves living in the Anthropocene.

The views set forth by the contributors to this volume represent their own and do not represent the views of any public or private entity the contributors are affiliated with.

Barbara Cosens
Lance Gunderson

Contents

Contributors

Craig R. Allen U.S. Geological Survey, Nebraska Cooperative Fish and Wildlife Research Unit, School of Natural Resources, University of Nebraska, Lincoln, NE, USA

David G. Angeler Swedish University of Agricultural Sciences, Uppsala, Sweden

Craig Anthony (Tony) Arnold Brandeis School of Law, Department of Urban and Public Affairs, University of Louisville, Louisville, KY, USA

Angela H. Arthington Australian Rivers Institute, Griffith University, Nathan, Australia

Melinda Harm Benson Haub School of Environment and Natural Resources, University of Wyoming, Laramie, WYO, USA

Hannah E. Birgé Nebraska Cooperative Fish and Wildlife Research Unit, School of Natural Resources, University of Nebraska, Lincoln, NE, USA

The Nature Conservancy, Omaha, NE, USA

Emmanuel Frimpong Boamah Department of Urban and Regional Planning, University of Buffalo, Bufallo, NY, USA

Brian C. Chaffin W. A. Frank College of Forestry and Conservation, University of Montana, Missoula, MT, USA

Alexandra Chase National Sea Grant Law Center, Oxford, MS, USA

Barbara Cosens College of Law and Waters of the West Program, University of Idaho, Moscow, ID, USA

Robin Kundis Craig S.J. Quinney College of Law, University of Utah, Salt Lake City, UT, USA

Daniel A. DeCaro Department of Urban and Public Affairs, Department of Psychological and Brain Sciences, University of Louisville, Louisville, KY, USA

Alan Devenish College of Earth, Ocean, and Atmospheric Sciences, Oregon State University, Corvallis, OR, USA

Jennifer Grace Ewa Office of the Colorado State Public Defender, Denver, CO, USA

Alex Fremier School of the Environment, Washington State University, Pullman, WA, USA

Ahjond S. Garmestani School of Natural Resources, University of Nebraska, Lincoln, NE, USA

Hannah Gosnell College of Earth, Ocean, and Atmospheric Sciences, Oregon State University, Corvallis, OR, USA

Olivia Odom Green Atlantic States Legal Foundation, Syracuse, NY, USA

Lance H. Gunderson Department of Environmental Sciences, Emory University, Atlanta, GA, USA

Shana Hirsch University of Idaho, Moscow, ID, USA

Alfred R. Light St. Thomas University School of Law, Miami Gardens, FL, USA

Dagmar Llewellyn Department of Civil Engineering, University of New Mexico, Albuquerque, NM, USA

Ryan R. Morrison Department of Civil and Environmental Engineering, Colorado State University, Fort Collins, CO, USA

Keith W. Rizzardi St. Thomas University School of Law, Miami Gardens, FL, USA

J. B. Ruhl Vanderbilt University Law School, Nashville, TN, USA

Edella Schlager School of Government and Public Policy, The University of Arizona, Tucson, AZ, USA

Mark Stone Department of Civil Engineering, University of New Mexico, Albuquerque, NM, USA

Dirac Twidwell Department of Agronomy & Horticulture, University of Nebraska, Lincoln, NE, USA

Chapter 1
An Introduction to Practical Panarchy: Linking Law, Resilience, and Adaptive Water Governance of Regional Scale Social-Ecological Systems

Barbara Cosens and Lance Gunderson

Abstract This chapter introduces the volume on Practical Panarchy: Linking Law, Resilience, and Adaptive Water Governance of Regional Scale Social-Ecological Systems. It begins by defining the terminology and theoretical concepts to present the bridging framework among ecological resilience, governance, and law relied on throughout this volume and then introduces this three-part volume. Part I presents the effort to assess resilience and transformation in riverine and wetland social-ecological systems in six US watersheds (the Anacostia River, Columbia River, Everglades wetlands, Klamath River, Middle Rio Grand River, and central Platte River) and one Australian system (the Lake Eyre and Great Artesian basin). Part II focuses on the legal dimensions of watershed governance that directly relate to ecological resilience and transformability of the social-ecological systems and synthesizes the results of the basin assessments to advance the understanding of the role of law and governance as a trigger, facilitator, or barrier to adaptation and transformation in the face of rapid environmental change, including shifting climate. Part III looks at the broader relation between social-ecological resilience and governance through synthesis of the basin assessments and resort to the broader literature on institutions and governance. As a whole, this volume presents the results of a 3-year pursuit on the cross-scale interactions among law, ecosystem dynamics, and governance to address the adaptive capacity of regional scale watersheds as they respond to accelerating environmental change.

Keywords Adaptive law · Adaptive water governance · Climate change · Resilience · Social-ecological systems

B. Cosens (✉)
College of Law and Waters of the West Program, University of Idaho, Moscow, ID, USA
e-mail: bcosens@uidaho.edu

L. Gunderson
Department of Environmental Sciences, Emory University, Atlanta, GA, USA
e-mail: lgunder@emory.edu

B. Cosens, L. Gunderson (eds.), *Practical Panarchy for Adaptive Water Governance*, https://doi.org/10.1007/978-3-319-72472-0_1

1.1 Introduction

With society having entered the epoch in which humans dominate earth systems (Steffen et al. 2007, 2011), the prior paradigms of nature separate from society, of environmental governance focused on mediating human use and exploitation of nature as if the effect of one on the other flows only from humans to nature, and of a stationary and predictable response to implementation of the laws developed as a tool for that mediation are no longer relevant. Social-ecological systems interact in complex, nonlinear ways, with multiple avenues for feedback among systems. The system-wide perturbation caused by human domination places those alive today on the leading edge of a no analog future (Milly et al. 2008; Ruhl 2008). Attempting to understand patterns of complexity has led to integrative scholarship that directs attention to linking the social and ecological components of these systems (Delmas and Young 2009; Berkes et al. 2013). This volume represents the results of an integrative pursuit across ecology, law, and governance over a 3-year time frame. The social-ecological systems studied are defined by water basin boundaries; they include linkages beyond those boundaries to larger scales of society and the biophysical system.

During this period of acceleration in the change of social-ecological systems, the human dimensions (including institutions made up of laws, rules, social norms, and patterns of management) have increased in complexity. The pursuit on which this volume is based, therefore, limited its scope of study to North American water basins within a single federal nation-state (the United States) and one pilot test in Australia to understand the transferability of concepts within a legal system with a similar history. This focus also led to the consideration of resilience and governance within basins that are highly developed, and though the development has led to substantial wealth and thus greater capacity to adapt, the interdependence of society on the development of the rivers and the legacy of massive built infrastructure renders this one of the more complex settings for consideration of the capacity of social-ecological systems to adapt.

The synthesis project resulting in this volume is called the Adaptive Water Governance (AWG) pursuit and took place at the US National Socio-Environmental Synthesis Center (SESYNC), funded by the US National Science Foundation. The AWG project explored the role of law in achieving water governance that is capable of facilitating management, adaptation, and transformation in the face of climate change. This chapter introduces the results of this effort after a discussion of the bridging terminology used by the pursuit.

1.2 Ecological Resilience and Adaptive Environmental Governance: Developing a Common Platform

Governance of natural resources addresses issues of access, use, protection, and management of common pool resources including water (Delmas and Young 2009). Historically in North America, we have relied on the mechanisms of markets (i.e., through private property rights) and government regulation to curb overexploitation (Hardin 1968). Markets are highly adaptive in response to change, but they also exhibit negative environmental externalities. Governmental regulation in the environmental area has grown dramatically since the 1960s to address these issues (Lemos and Agrawal 2006; Lyon 2009). Yet government regulation presents problems with adaptability and, when applied to complex large-scale systems, requires massive information. There is a growing recognition that both in the absence of regulation and within highly regulated systems, self-organization of communities dependent on common pool resources is common and provides a path forward beyond the constraints of markets and regulation alone (Ostrom 1990, 2009). Self-organization is of particular importance in bridging governance systems to ecological resilience.

The study of ecological resilience grew from the recognition by applied ecologists of the uncertain, surprising, and nonlinear behavior across a range of resource systems in the second half of the twentieth century (Holling 1973, 1978). The technical solution they proposed for resource management in the face of uncertainty – adaptive management (Holling 1978; Walters 1986) – has run into barriers with respect to legal and governance issues (Dietz et al. 2003; Brunner et al. 2005; Folke et al. 2005; Gunderson and Light 2006). To be effective, tools for flexible management must be imbedded within systems of law and governance that will address not just the feedback from ecological systems responding to change but from complex social-ecological systems. Governance itself must be adaptive.

Governance approaches that appear to support management in the face of uncertainty have been identified and referred to in the literature under terms ranging from new governance to collaborative governance/co-management and adaptive governance (Dietz et al. 2003; Karkkainen 2004; Brunner et al. 2005; Folke et al. 2005; Gunderson and Light 2006; Chaffin et al. 2014). The phenomena to which these terms have been applied are observed to be emergent and, thus, self-organizing (Dietz et al. 2003; Chaffin et al. 2014). The AWG project uses the phrase adaptive governance as an umbrella term to capture emergent, collaborative, and learning-based types of environmental governance (Chaffin et al. 2014).

Cross-disciplinary research requires an initial effort to reach agreement on the definitions of terms (Repko 2011), particularly contested terms such as "resilience" (Davidson 2010; Olsson et al. 2015). This volume does not seek to put new definitions forward or to resolve disciplinary disputes over differing opinions on a particular term or phrase. Instead, the following paragraphs are intended to communicate the scholarly lineage of the definitions agreed upon and used throughout this special issue. The inventory of bridging terminology focuses on four phrases that arise in the study of complex social-ecological systems that were found to be suited to bridging the disciplines involved and to providing meaning within each discipline: (1) resilience thinking, (2) panarchy, (3) environmental governance, and (4) adaptive governance.

1.2.1 Resilience Thinking

Resilience thinking is a family of concepts used to describe the dynamic and often surprising behaviors observed in complex systems and documented in many ecosystems (Gunderson and Pritchard 2002). Resilience thinking posits that coupled social-ecological systems undergoing change may not do so in a predictable linear fashion. Instead, systems may cross a threshold and reorganize into an alternative, metastable regime or state with different structure and function. Once a threshold is crossed, it is likely to be difficult to return to the prior state (Holling 1973; Gunderson and Holling 2002; Walker et al. 2004; Walker and Salt 2006). Although the term was coined to refer to this entire phenomenon of systems, *resilience* is more often used to describe the degree to which the system avoids regime shift either because it is quite adaptable (latitude) or quite resistant to change (resistance) (Walker et al. 2004). Resilience thinking allows society to approach complex systems from the viewpoint of their ability to continue to provide key functions and maintain supporting structure in the face of change (Holling 1973; Gunderson and Holling 2002; Walker et al. 2004; Walker and Salt 2006). It provides a framework for environmental management of nonstationary systems by focusing attention on achieving societal goals without simplification of the processes that support ecosystem function and, in fact, by supporting those processes rather than a single stationary state (e.g., maximum sustainable yield) (Walker and Salt 2006). It ties society's response to change in an ecological system to the complex feedbacks between the social and ecological system (Gunderson and Holling 2002), improving the prospects for achieving societal goals such as sustainability.

Resilience is often discussed in terms of the system society seeks to maintain; thus, the term resilience is often confused with the goal itself. Resilience as used in this volume is a system property rather than a societal goal and may be used to describe systems that society values and those it does not value (Holling 1973; Gunderson and Holling 2002; Walker et al. 2004; Walker and Salt 2006). The terminology relied on by the AWG pursuit brings values into the discussion in the context of governance. It is within governance that the desire for a particular alternative state is expressed; it is resilience – i.e., the system properties – that informs how to get there within a complex system.

1.2.2 Panarchy

Panarchy is a concept related to resilience thinking and provides a lens through which the cross-scale interactions of both social-ecological systems and their systems of governance can be viewed (Gunderson and Holling 2002). Panarchy captures the observation that within any social-ecological system at the landscape scale are multiple nested smaller scale systems down to the microscopic in the ecological realm and down to the individual in the societal realm. Panarchy also describes the cycle used to describe ecological systems observed to be in various stages of growth, collapse, innovation, and reorganization. Panarchy expands the concept of resilience by recognizing that (a) resilience of a system declines as a system matures or develops, (b) larger (slower) and smaller (faster) scale processes interact and can both foster and erode resilience, and (c) cross-scale interactions may play a role in stabilizing large-scale systems while providing opportunity for innovation and adaptation at smaller scales (Gunderson and Holling 2002). These aspects of panarchy can also be used metaphorically to describe cross-scale interactions in systems of governance (Chaffin and Gunderson 2016). Understanding the current trajectory of the system can aid in intentional adaptation and transformation from a less desirable to a more desirable system state (Chaffin et al. 2016; Gunderson et al. 2017).

1.2.3 Environmental Governance

Governance refers to the means through which collective goals are chosen, decisions are made, and action is taken to achieve the chosen goals (Rogers and Hall 2003; Delmas and Young 2009; UN System Task Team 2012). The term governance encompasses not only government but the relationship between government and society, including the means through which private actors, markets, and even interest-based networks influence policy decisions and self-organize to mediate their own behavior (Rogers and Hall 2003; Folke et al. 2005; Lemos and Agrawal 2006; Huitema et al. 2009). Environmental governance is that subset of collective action that mediates societal interaction with nature (Rogers and Hall 2003; Delmas and Young 2009; UN System Task Team 2012) and is particularly difficult in the context of common pool resources – i.e., those resource systems, like a watershed, that are "sufficiently large as to make it costly (but not impossible) to exclude potential beneficiaries from obtaining benefits from its use" (Ostrom 1990). Governance capable of emerging at the biophysical scale through both formal and informal networks may be particularly effective in the environmental context, and it is this type of governance that the AWG pursuit refers to as adaptive governance.

1.2.4 Adaptive Governance

Adaptive governance is a collective action that emerges at the biophysical scale, is capable of facilitating response to change and surprise, and has the capacity to learn and evolve (Dietz et al. 2003; Folke et al. 2005; Pahl-Wostl et al. 2007; Chaffin et al. 2014; Cosens et al. 2014; Green et al. 2015; Schultz et al. 2015). Dietz et al. (2003), building on the work of coauthor and Nobel Laureate Elinor Ostrom (1990), introduced the term adaptive governance in recognizing self-organization as an alternative path to governing the commons in contrast to Hardin's focus on property rights and government regulation as the only options (Hardin 1968). The AWG pursuit uses the term "adaptive governance" to include "new" environmental governance, adaptive co-management, and collaborative governance (Karkkainen 2004; Dietz et al. 2003; Brunner et al. 2005; Folke et al. 2005; Gunderson and Light 2006; Lockwood et al. 2010; Chaffin et al. 2014). In particular, the AWG pursuit relies on the efforts to relate adaptive or collaborative governance to resilience (Holling 1978; Ostrom 1990; Lee 1993; Dietz et al. 2003; Folke et al. 2005; Lebel et al. 2006; Pahl-Wostl et al. 2007; Huitema et al. 2009; Emerson et al. 2011; Plummer et al. 2013; McGinnis and Ostrom 2014; Chaffin et al. 2014; Green et al. 2015; Schultz et al. 2015). The key aspects of the various definitional approaches that we focus on are that adaptive governance (1) is capable of self-organizing through formal and informal networks, (2) generally arises at the biophysical (or problem) scale in response to change in an interdependent social-ecological system, (3) facilitates an ability to learn, and (4) has the capacity to evolve. Yet, no single definition is adequate to describe the components of an emergent phenomena that is contextual – that is, deeply dependent on the complex results of interaction among society and nature in a specifically defined system. Instead the following paragraphs set forth the key literature that forms the basis for the understanding of adaptive governance by the AWG pursuit and the basis for the inquiry into the role of law. These foundational articles do not provide a narrowly constructed list of criteria that must be present to label the approach adaptive governance. Instead, they reveal a quest to identify those aspects of governance that seem most likely to lead to emergence and acceptance of adaptive processes and solutions.

Ostrom (1990) documented that self-organizing community-based common property institutions evolve to become capable of managing common pool resources as effectively as formal government regulation or the division of resources into privately held property (McGinnis and Ostrom 2014). Ostrom's work included community-based governance with little governmental overlay as well as North American irrigation communities organizing within a highly developed hierarchical governmental framework (Ostrom 1990, 2009). It is this self-organizing behavior that must take place within a system of governmental management of the environment and regulation that the AWG pursuit seeks to understand. The focus is on emergence of adaptive governance nested within and interacting with highly developed systems of government and how government, acting through law, might step aside and even facilitate this emergence.

Folke et al. (2005) stressed the need for adaptive governance with focus on the social aspects of governance as a means to gain acceptance of adaptive management and to assure the organizational learning and ability to navigate competing values and interests to facilitate its implementation. The authors include in their criteria: (1) social capital including trust, common rules, leadership, and experience; (2) networks and bridging organizations within a polycentric governance structure; and (3) a devolution of management rights and power sharing that promotes participation. Huitema et al. (2009) synthesize the views on adaptive governance and attempt a list of components needed in the water governance context: polycentricity, public participation, experimentation, and a bioregional perspective.

While the AWG pursuit relies on this work, we go beyond those who would define adaptive governance as simply the means to implement adaptive management. The AWG pursuit is influenced by the work of Craig and Ruhl (2014) who propose that adaptive management is appropriate when goals are set, uncertainty is high, and the ability to control experimentation is also high. In most instances, the management of major river basins is characterized by competing interests, jurisdictional complexity, and multiple drivers of change, and thus the ability to identify single management goals and to control experimentation is limited. In these more complex cases, adaptive governance is needed to assure accountability in goal setting and experimentation, and adaptive management may be only one of many tools relied on for management in the face of uncertainty.

The importance of this distinction is captured in the work of Lebel et al. (2006) who describe the type of governance needed to manage resilience in social-ecological systems. Their attention not only to the adaptive aspects of governance but to what is required for society to adopt a particular approach to governance comes closest to capturing an understanding of adaptive governance that bridges law and resilience. Thus, Lebel et al. (2006) view adaptive governance as requiring (1) certain attributes of good governance (participation and deliberation in process, polycentric and multilayered institutions, and accountable and just authority) and (2) capacity to manage resilience (ability to function at multiple scales and across scales, ability to anticipate and cope with uncertainty, ability to design institutions to fit social and ecological contexts, ability to detect and navigate thresholds, ability to integrate different forms of knowledge, and the ability to maintain ecological and social diversity). While admitting that their findings are "tentative," the authors find evidence to link the attributes of good governance to the capacity to manage resilience in several case studies, documenting that (1) "[p]articipation builds the trust, and deliberation leads to the shared understanding needed to mobilize and self-organize"; (2) "[p]olycentric and multilayered institutions improve the fit between knowledge, action, and socio-ecological contexts in ways that allow societies to respond more adaptively at appropriate levels"; and (3) "[a]ccountable authorities who also pursue just distributions of benefits and involuntary risks enhance the adaptive capacity of vulnerable groups and society as a whole" (Lebel et al. 2006).

In the broader governance literature, the move to distributed, collaborative governance is referred to as "new governance" (Lockwood et al. 2010; Karkkainen 2004; Lee 2003), with many similarities to adaptive governance and thus useful to the understanding of the concept. Lockwood et al. (2010) discuss this move to new governance as a product of the rise of informal, nongovernmental aspects of governance due to the increase in interdependency and interaction and the pressure for a greater citizen voice in natural resource management (Lockwood et al. 2010). Similar to adaptive governance, Lockwood et al. (2010) characterize new governance as capable of addressing problems "characterized by complexity, uncertainty, interdependency, and deficiencies in resources, expertise, and knowledge" (Lockwood et al. 2010).

The AWG pursuit does not view adaptive governance as a panacea for natural resource management. Bridging the concepts of resilience and adaptive governance leads to the conclusion that adaptive governance, acting in concert with other forms of natural resource management, is appropriate when the system is complex (e.g., lies within multiple jurisdictions), the system faces change with a high degree of uncertainty (e.g., climate and other types of environmental change), and the system is approaching a potential threshold or regime shift as evidenced by increasing conflict over resources, increasing scarcity, or actual identification of an approaching threshold (e.g., listing of species, exceeding water quality standards, or insufficient water for basic needs). In addition, as will be apparent in the chapters in this volume, the specific approach to adaptive governance must be tailored to the contextual setting in which it is employed.

1.3 Integrating Law, Resilience, and Adaptive Governance: An Exercise in Practical Panarchy

The chapters in this volume are organized around the three questions that formed the core inquiry of the AWG pursuit. Part I explores the question of whether the social-ecological resilience and adaptive capacity of regional scale water systems to changing climate may be qualitatively assessed for the purpose of understanding the complex interaction between governance and system resilience. Part II relies on the seven basin assessments to explore the question of whether legal and institutional components of society contribute to adaptive governance and specifically seeks to consider the role of law in preventing, triggering, and facilitating adaptive governance. Part III synthesizes aspects of the six North American basin assessments to consider the role of environmental governance in developing the adaptive capacity of these systems to respond to climate change. The following sections introduce each part through an analysis of the contributions to improving water governance in this time of accelerating change.

1.3.1 Part I. Assessing the Resilience of Regional Scale Social-Ecological Systems Defined by Watershed Boundaries

Human management of water resources has taken place for millennia. Reservoirs, levees, and canals manipulated water resources in dry Mesopotamia 4000 years ago, and the code of Hammurabi was developed to allocate water (Cech 2010). This practice of engineering rivers and developing formal institutions for governance continues today. Water systems in North America have been modified and managed to meet a variety of societal goals including water supply, flood control, energy, agricultural and other economic production, as well as a growing environmental demand. Water governance includes multilevel formal water management through government at the local, state, tribal, regional, and national scale. Water governance also includes informal self-organization at the community and watershed scale, as well as local to national nongovernmental entities. Understanding the complex interaction between governance and system resilience is a critical step in the development of knowledge that will aid society in navigating the effects of climate change on its vital water resources.

Part I presents the results of resilience assessment (Resilience Alliance 2007), modified to highlight the legacy effect of water development on the resilience of the social-ecological system and the role of governance and law in that process for the Anacostia River (Chap. 3), Columbia River (Chap. 4), Everglades wetland (Chap. 5), Klamath River (Chap. 6), Middle Rio Grande River (Chap. 7), and Platte River (Chap. 8) of North America (Fig. 1.1) and the Lake Eyre and Great Artesian basins

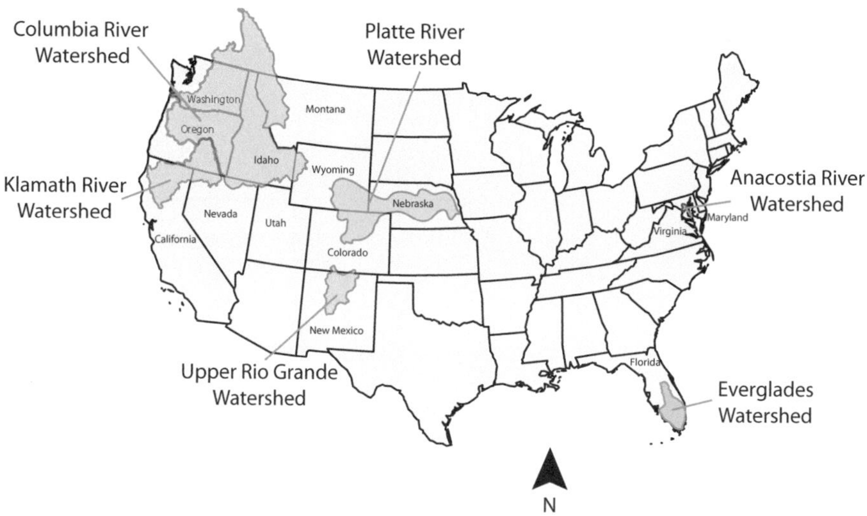

Fig. 1.1 Location of riverine and wetland social-ecological systems in the United States used to study interaction of ecological resilience and adaptive governance (Base map from public domain image, http://www.wikiwand.com/en/List_of_rivers_of_the_United_States)

of Australia (Chap. 9) (Fig. 1.2). These systems range from the undeveloped Lake Eyre basin, an internally drained and sparsely populated basin in Australia, to the highly urbanized Anacostia River, a tributary to the Potomac River in the vicinity of Baltimore and Washington DC in the United States, with over one million residents within its boundaries. Hydrologically, the systems range from the snowmelt-fed rivers of the western United States including the Columbia and Klamath Rivers to the rain-fed cycles of the Everglades and the Lake Eyre basins.

Development of the North American rivers has occurred with substantial federal assistance. While management of water occurs at all scales from the individual to the local, state, and national in all seven basins, it is dominated by a top-down regulatory approach, and water flow controlled by developed infrastructure is generally coordinated at the federal level throughout the North American systems, whereas water allocation is controlled at the state level in both North America and Australia. Nevertheless, self-organization and government-assisted organization at the biophysical scale are occurring to some degree in all the systems studied. While the need for growth in governance capacity and recognition of marginalized indigenous populations in water management is most apparent in the western United States and Australia, all basins illustrate the struggle of those whose needs are not reflected in

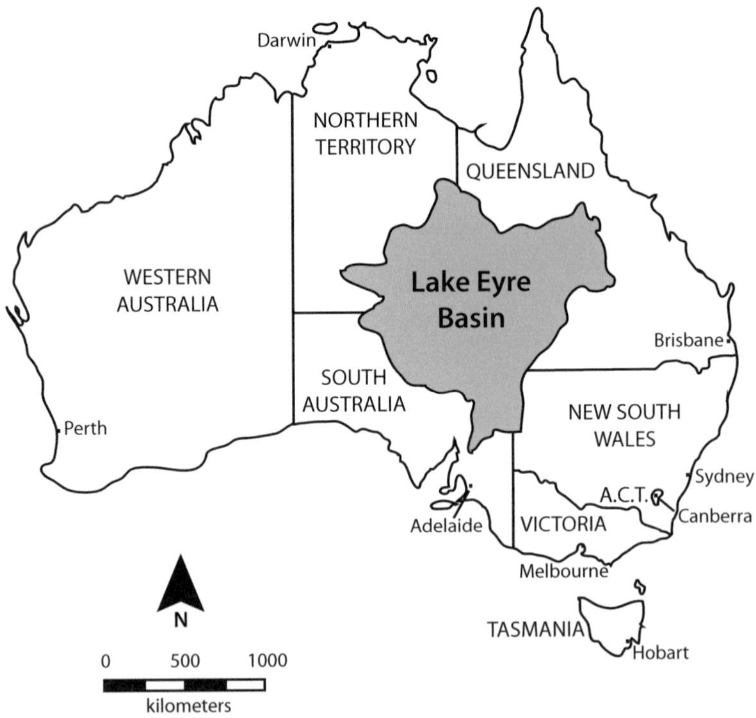

Fig. 1.2 Location of Lake Eyre basin used to study interaction of ecological resilience and adaptive governance in Australia

current river development. In the Anacostia watershed where African-American communities are still struggling from the legacy of slavery and its manifestation in racism, the seed of watershed-focused leadership and activism has been sown, in part, facilitated by the federal Clean Water Act.

While these aspects provide a snapshot of the current status of the basins assessed, it is the process of analyzing social-ecological change in each system through time and assessing the role of law and governance in each major transformation that proved the most useful approach to understanding system resilience and the law-governance-resilience relation in each system. The historic analysis of social-ecological change in the water systems studied provided a means for a contextual assessment of the capacity to respond to change. While that capacity may also change over time, the basic governmental structure supporting it has remained relatively constant in both North America and Australia for more than a century, thus providing a means to address access to higher levels of support and interaction among basin and national levels of governance during periods of rapid change. The approach also allowed qualitative assessment of the complex feedbacks between governance and the social-ecological system during periods of change.

All the basins have been studied to identify the potential changes in both the timing and supply of water through climate change. The basin assessments illustrate that with the onset of the impacts from climate change, some of the water supplies relied on in North America may be close to irreversible thresholds. Once thresholds are crossed, the services provided by altered ecosystems may threaten the adequacy of engineered infrastructure potentially impairing existing water-based economies. At the same time, the ecological system of the undeveloped Lake Eyre basin may be so highly adapted to historic extremes, that it too may be vulnerable to climate change. Parts II and III contribute to the understanding of how society may use and alter its governance of these systems to navigate change.

1.3.2 Part II. The Role of Law in Adaptive Governance and Resilience

That adaptive governance is emergent and law is intentional is a key starting point for inquiry into the role of law. Despite this apparent dichotomy, much of the scholarship on adaptive governance discussed above recognizes that emergence of adaptive processes occurs through both formal and informal networks and through interaction of governmental and nongovernmental entities (Pelling et al. 2007). In their conceptual model, Dietz et al. (2003) provide a set of governance principles that are almost entirely composed of actions that may be facilitated by law. These principles include congruence of rules with ecological conditions, avenues for analytical deliberation/participation, clear boundaries and defined rights, enforced sanctions, mechanisms for dispute resolution, institutional variety, accountability, and nesting (Dietz et al. 2003, Fig. 3).

Part II begins with analysis of the law that provides the authority, sets the boundaries and structure, and dictates the process for government agency action (Chap. 10). While law presents barriers to adaptive governance, it also provides avenues through which government may facilitate and even participate in adaptive governance, and importantly, the tools for doing so already exist. This work reveals the importance of informal networks for increased integration of governmental activity related to the same biophysical system, and this aspect was given specific attention in the context of endangered species (Chap. 12).

Of significance in this part is the exploration of one of the greatest challenges to governmental facilitation of adaptive governance: the need to balance flexibility and stability (Chap. 11). Government and law are designed to provide social stability, and the introduction of flexibility can be viewed as a risky measure and one unlikely to gain traction. However, as accelerating change threatens that highly valued stability, society must find means to adapt. Again, the tools for doing so are already available. Attention to process and procedure through adjustments to administrative law, as well as increased use of substantive standards for the achievement of societal goals (in contrast to rules dictating how to achieve those standards), may allow an increased level of substantive flexibility while maintaining legitimacy. This in turn may provide the requisite levels of psychological, social, and economic stability needed for communities to successfully adapt.

A textual analysis of the basin studies presented in Part II provides both a methodology for synthesis of qualitative resilience assessment and a window on common themes related to drivers of change and features of adaptive governance within the AWG pursuit assessments (Chap. 13). This methodology reveals that economic development and population growth were the initial drivers of change through infrastructure development in the North American basins. As corresponding changes in the ecological systems fed back to the social systems, social movement fed by both growing environmental values and the rise of marginalized communities arose in most of the basins. This in turn has given rise to efforts to restore ecological function. Although activity to date has been primarily around the margins of the legacy of existing infrastructure, the seeds for emergence of adaptive governance and some of the legal changes necessary to facilitate that emergence are already evident. Understanding that emergence and its role in resilience is the subject of Part III.

1.3.3 Part III. Social-Ecological Resilience and Adaptive Governance

The basin studies reveal that understanding the trajectory of the system is an important first step in facilitating its capacity to adapt or transform (Chap. 14). The trajectory describes the status of the system with respect to growth and development, crisis, recovery, and reorganization. During phases of development and stabilization of the North American systems, adaptive capacity was determined by infrastructure that was designed and constructed within specific biophysical limits. That is, these systems were designed and built to control extremes of hydrological input (floods and droughts) up to specific design limits based on historic patterns of precipitation and flow and provided the space and the means for substantial economic development within the basins. The high level of optimization of built infrastructure based on the historic hydrograph places some of these systems in positions to cross thresholds as climate change introduces new extremes and alters long-term trends. Yet all of the systems have experienced periods of crisis, recovery, and regime shift that led to new structures and rules as society learned about the limits of the extant system. Cross-scale interactions among levels and sectors of government and society played both positive and negative roles in adaptive capacity during these periods.

One of the foremost challenges to intentional facilitation of adaptation or transformation is the inability to quickly access system resilience and to manage uncertainty in that assessment. Outreach to experts and stakeholders within the basin studies allowed exploration of techniques to rapidly assess social-ecological resilience (Chap. 15). The approach used an online survey instrument and data/information from that survey to more explicitly understand trade-offs and uncertainty measures in resilience assessments. Rather than seek an absolute measure of resilience, the authors focused on relative resilience across the watershed systems. This approach reveals trade-offs among different sector groups with respect to social, economic, and ecologic components.

Returning to the literature on governance and resilience on which the AWG pursuit builds, Chap. 16 bridges the work of institutional scholars with the efforts to analyze the role of law and reveals broader social requirements for adaptive environmental governance including need to (1) encourage collaborative problem solving, (2) garner social acceptance and commitment, and (3) cultivate trust and tolerance for change and uncertainty. This part then turns to the language of institutional scholars to reframe the design principles for facilitation of adaptive governance through legal systems.

Just as the basin studies reveal the legacy effect of built infrastructure, Part III concludes with the role of social cognition in adaptation (Chap. 17). Because governance systems are socially constructed, the biases, power relations, and mental models held by society may constrain or facilitate adaptation. Understanding these underlying factors will play an important role in the development of policy alternatives to build the adaptive capacity of social-ecological systems.

1.4 Conclusion

The final synthesis of the AWG pursuit is presented in a concluding chapter. The members of the AWG pursuit invite the reader to pursue the arguments presented in this volume, with an eye toward lessons from these highly developed water management systems which paradoxically are perhaps the most technologically advanced yet least resilient social-ecological systems in the face of changing conditions. It is hoped that the volume will spur discussion and discourse to increase society's collective adaptive capacity as it faces the challenges of accelerating global environmental change.

Acknowledgments This work was developed in part under the Adaptive Water Governance project, funded by the US National Socio-Environmental Synthesis Center (SESYNC) under funding from the US National Science Foundation, NSF DBI-1052875.

References

Berkes F, Colding J, Folke C (2013) Navigating social-ecological systems: building resilience for complexity and change. Cambridge University Press, Cambridge

Brunner RD, Steelman TA, Coe-Juell L, Cromley CM, Edwards CA, Tucker DW (2005) Adaptive governance: integrating science, policy, and decision making. Columbia University Press, New York

Cech T (2010) Principles of water resources: history, development, management, and policy, 3rd edn. Wiley, Hoboken

Chaffin BC, Gunderson LH (2016) Emergence, institutionalization and renewal: rhythms of adaptive governance in complex social-ecological systems. J Environ Manag 165:81–87

Chaffin BC, Gosnell H, Cosens B (2014) A decade of adaptive governance scholarship: synthesis and future directions. Ecol Soc 19(3):56. https://doi.org/10.5751/ES-06824-190356

Chaffin BC, Garmestani AS, Gunderson LH, Benson MH, Angeler DG, Arnold CA, Cosens BA, Craig RK, Ruhl JB, Allen CR (2016) Transformative environmental governance. Annu Rev Environ Resour 41:399–423. https://doi.org/10.1146/annurev-environ-110615-085817

Cosens B, Gunderson L, Chaffin B (2014) The adaptive water governance project: assessing law, resilience and governance in regional socio-ecological water systems facing a changing climate. Idaho Law Rev 51:1–27

Craig RK, Ruhl JB (2014) Designing administrative law for adaptive management. Vanderbilt Law Rev 67:1

Davidson DJ (2010) The applicability of the concept of resilience to social systems: some 319 sources of optimism and nagging doubts. Soc Nat Resour 23:1135–1149

Delmas MA, Young OR (eds) (2009) Governance for the environment: new perspectives. Cambridge University Press, Cambridge

Dietz T, Ostrom E, Stern PC (2003) The struggle to govern the commons. Science 302:1907–1912. https://doi.org/10.1126/science.1091015

Emerson K, Nabatchi T, Balogh S (2011) An integrative framework for collaborative governance. J Public Adm Res Theory 22:1–29

Folke C, Hahn T, Olsson P, Norberg J (2005) Adaptive governance of social-ecological systems. Annu Rev Environ Resour 30:441–473. https://doi.org/10.1146/annurev.energy.30.050504.144511

Green OO, Garmestani AS, Allen CR, Gunderson LH, Ruhl JB, Arnold CA, Graham NAJ, Cosens BA, Angeler DG, Chaffin BC, Holling CS (2015) Barriers and bridges to the integration of social–ecological resilience and law. Front Ecol Environ 13(6):332–337. https://doi.org/10.1890/140294

Gunderson L, Holling CS (eds) (2002) Panarchy: understanding transformations in human and natural systems. Island Press, Washington, DC

Gunderson LH, Light SS (2006) Adaptive management and adaptive governance in the everglades ecosystem. Policy Sci 39:323–334. https://doi.org/10.1007/s11077-006-9027-2

Gunderson LH, Pritchard L (2002) Resilience and the behavior of large scale systems. Island Press, Washington, DC

Gunderson L, Cosens BA, Chaffin BC, Arnold CA, Fremier AK, Garmestani AS, Craig RK, Gosnell H, Birge HE, Allen CR, Benson MH, Morrison RR, Stone MC, Hamm JA, Nemec K, Schlager E, Llewellyn D (2017) Regime shifts and panarchies in regional scale social-ecological water systems. Ecol Soc 22(1):31. https://doi.org/10.5751/ES-08879-220131
Hardin G (1968) The tragedy of the commons. Science 162:1243–1250
Holling CS (1973) Resilience and stability of ecological systems. Annu Rev Ecol Syst 4:1–24. https://doi.org/10.1146/annurev.es.04.110173.000245
Holling CS (1978) Adaptive environmental assessment and management. Wiley, London
Huitema D, Mostert E, Egas W, Moellenkamp S, Pahl-Wostl C, Yalcin R (2009) Adaptive water governance: assessing the institutional prescriptions of adaptive (co-)management from a governance perspective and defining a research agenda. Ecol Soc 14(1):26. http://www.ecologyandsociety.org/vol14/iss1/art26/
Karkkainen BC (2004) "New Governance" in legal thought and in the world: some splitting as antidote to overzealous lumping. Minnesota Law Rev 89:471–497
Lebel L, Anderies JM, Cambell B, Folke C, Hatfield-Dodds S, Hughes TP, Wilson J (2006) Governance and the capacity to manage resilience in regional social-ecological systems. Ecol Soc 11(1):19. http://www.ecologyandsociety.org/vol11/iss1/art19/
Lee KN (1993) Compass and gyroscope. Island Press, Washington, DC
Lee M (2003) Conceptualizing the new governance: a new institution of social coordination. Paper presented at the Institutional analysis and development mini-conference, Workshop in political theory and policy analysis. Indiana University, Bloomington, 3–5 May 2003
Lemos MC, Agrawal A (2006) Environmental governance. Annu Rev Environ Resour 31:297–325. https://doi.org/10.1146/annurev.energy.31.042605.135621
Lockwood M, Davidson J, Curtis A, Stratford E, Griffith R (2010) Governance principles for natural resource management. Soc Nat Resour 23(10):986–1001. https://doi.org/10.1080/08941920802178214
Lyon TP (2009) Environmental governance: an economic perspective. In: Delmas MA, Young OR (eds) Governance for the environment: new perspectives. Cambridge University Press, Cambridge
McGinnis MD, Ostrom E (2014) Social-ecological system framework: initial changes and continuing challenges. Ecol Soc 19(2):30. https://doi.org/10.5751/ES-06387-190230
Milly PC, Betancourt J, Falkenmark M, Hirsch RM, Kundzewicz ZM, Lettenmaier DP, Stouffer RJ (2008) Stationarity is dead: whither water management? Science 319:573–574
Olsson L, Jerneck A, Thoren H, Persson J, O'Byrne D (2015) Why resilience is unappealing to social science: theoretical and empirical investigations of the scientific use of resilience. Sci Adv 1(4):e1400217
Ostrom E (1990) Governing the commons: the evolution of institutions for collective action. Cambridge University Press, Cambridge
Ostrom E (2009) A general framework for analyzing sustainability of social-ecological systems. Science 325:419–422
Pahl-Wostl C, Sendzimir J, Jeffrey P, Aerts J, Berkamp G, Cross K (2007) Managing change toward adaptive water management through social learning. Ecol Soc 12(2):30. http://www.ecologyandsociety.org/vol12/iss2/art30/
Pelling M, High C, Dearing J, Smith D (2007) Shadow spaces for social learning: a relational understanding of adaptive capacity to climate change within organisations. Environ Plan A 40(4):867–884
Plummer R, Armitage DR, de Loë RC (2013) Adaptive co-management and its relationship to environmental governance. Ecol Soc 18:21
Repko AF (2011) Interdisciplinary research: process and theory, 2nd edn. Sage, New York
Resilience Alliance (2007) Assessing resilience in social-ecological systems: a workbook for scientists, Version 1.1
Rogers P, Hall A (2003) Effective water governance. Global water partnership technical committee (TEC) background paper 7. Global Water Partnership, Stockholm. http://www.gwp.org/Global/ToolBox/Publications/Background%20papers/07%20Effective%20Water%20Governance%20%282003%29%20English.pdf. Accessed 1 Oct 2017

Ruhl JB (2008) Climate change and the Endangered Species Act – building bridges to the no-analog future. Boston Univ Law Rev 88:1–62

Schultz L, Folke C, Österblom H, Olsson P (2015) Adaptive governance, ecosystem management, and natural capital. PNAS 112(24):7369–7374

Steffen W, Crutzen PJ, McNeill JR (2007) The Anthropocene: are humans now overwhelming the great forces of nature? Ambio 36:614–621

Steffen W, Persson Å, Deutsch L, Williams M, Richardson K, Crumley C, Crutzen P, Folke C, Gordon L, Molina M, Ramanathan V, Rockström J, Scheffer M, Schellnhuber HJ, Svedin U (2011) The anthropocene: from global change to planetary stewardship. Ambio 40:739–761

UN System Task Team (2012) UN system task team on the Post-2015 UN development agenda: governance and development. http://www.un.org/millenniumgoals/pdf/Think%20Pieces/7_governance.pdf. Accessed 1 Oct 2017

Walker B, Salt D (2006) Resilience thinking: sustaining ecosystems and people in a changing world. Island Press, Washington, DC

Walker B, Holling CS, Carpenter SR, Kinzig A (2004) Resilience, adaptability and transformability in social–ecological systems. Ecol Soc 9(2):5. http://www.ecologyandsociety.org/vol9/iss2/art5/

Walters CJ (1986) Adaptive management of renewable resources. McGraw Hill, New York

Part I
Assessing Resilience of Regional Scale Social-Ecological Systems

Chapter 2
Case Studies in Adaptation and Transformation of Ecosystems, Legal Systems, and Governance Systems

Lance Gunderson and Barbara Cosens

Abstract Seven water basins, six in North America and one in Australia, were chosen as case studies to investigate the dynamic interactions among ecosystems, society, and legal systems. These cases were chosen because of local knowledge and expertise of a group of interdisciplinary scholars. The teams were asked to describe the structures and processes that contribute to resilience, adaptations, and transformations in both the ecological and social components of the linked system. The role of law in triggering or hindering change in governance and institutional reformation was explored. The cases also describe conditions for emergence of adaptive governance in heavily regulated and developed water systems nested within a hierarchical governmental system.

Keywords Climate change · Panarchy · Resilience assessment · Adaptive governance · Water management

2.1 Introduction

Human engineered control of water dates as early as 1200 BC in North America but has accelerated at an unprecedented rate and scale over the past century (Cech 2003). From the arid southwestern region of the continent to the relatively water-rich east, humans have sought to control surface waters and tap vast groundwater resources for human use. In the ultimate effort to modify basic ecosystem functions for the service of humans, dams have been built in the mighty snowmelt-driven

L. Gunderson (✉)
Department of Environmental Sciences, Emory University, Atlanta, GA, USA
e-mail: lgunder@emory.edu

B. Cosens
College of Law and Waters of the West Program, University of Idaho, Moscow, ID, USA
e-mail: bcosens@uidaho.edu

B. Cosens, L. Gunderson (eds.), *Practical Panarchy for Adaptive Water Governance*, https://doi.org/10.1007/978-3-319-72472-0_2

rivers of western North America – the Columbia, the Colorado, the Platte, and the Missouri – to control floodwaters, produce electricity, and divert water resources for agricultural irrigation and human consumption. This substantial investment in infrastructure required collective action and government subsidy. Today that infrastructure is aging, and it is not clear whether the political will to modernize it exists. We are just beginning to realize the full extent of the costs to ecosystem services. Loss of biodiversity, as indicated by the large number of endangered and threatened species and cultures, and loss of ecosystem services provided by wetlands, floodplains, and a connected river and riparian corridor continue, but in ways that are difficult to quantify. Both the planned and inadvertent ecological changes associated with the development of water resources have led to unforeseen system shifts characterized as the erosion of ecological resilience (Holling and Meffe 1996; Folke et al. 2004; Gunderson and Pritchard 2002). Moreover, the property approach to water allocation, particularly in the western United States, and the command and control approach to environmental regulation including water quality and species protection have increased social inequity. Marginalized populations do not share equally in the benefits of the twentieth-century development adversely affecting overall social resilience.

As changes to the water resources of North America have occurred, the ways in which humans have valued and governed these systems have also changed (Gunderson et al. 1995). Once viewed merely as a resource to be controlled to serve humans, water is now seen to supply a wide variety of ecosystem goods and services many of which society now values in their unaltered state. The rise of the social objective of sustainability extended the time horizon for social goals and outcomes by considering the needs of future generations (Brundtland Commission 1987). During this period of development and changing values, the human dimensions (including institutions made up of laws, rules, social norms, and patterns of management) have increased in complexity. Attempting to understand such patterns of complexity has led to integrative scholarship that directs attention to linking the social and ecological components of these systems (Delmas and Young 2009; Berkes et al. 2013).

Integrative scholarship at the intersection between social and ecological system shows that across the planet, natural resource systems are being influenced by larger-scale environmental changes. The global environment has been relatively stable over the past 10,000 years, but beginning with the Industrial Revolution, manifold changes are occurring at the scale of the planet (Vitousek et al. 1997; Rockström et al. 2009). These changes include loss of biodiversity, accelerated cycling rates of nitrogen and phosphorus, ocean acidification, atmospheric pollution, and conversion of land use to agricultural and other purposes (Rockström et al. 2009). While many of these changes influence water resource systems, perhaps the greatest threat or problem to regional scale water systems is posed by changing climate.

Broadly defined, climate is the long-term (decades to centuries) pattern of precipitation and temperature in a particular area (Intergovernmental Panel on Climate Change [IPCC] 2007). In regional scale water systems, climatic patterns have been central to the design and management of such systems, and infrastructure and use allocation have been optimized on an assumption that the historic climate will persist. The climatic zones vary widely across the cases (Table 2.1). The Everglades has a subtropical savanna climate that is characterized by little seasonal change in temperature (rare freezing), with pronounced wet and dry seasons (Hela 1952), and the management system has evolved according to this annual cycle to control flooding during the wet season and supply water to agriculture, urban interests, and conservation areas during the dry season. Water basins in western North America experience substantial seasonal variability characterized by spring runoff from snowmelt (Mote et al. 2005), and water infrastructure and management is designed to even out the hydrologic cycle for flood control, hydropower, and irrigation (Cosens and Fremier 2014). These managed systems in the western United States are heavily reliant on natural storage of water in snowpack (Mote et al. 2005). Yet a growing body of literature indicates that long-term changes in the hydrologic processes controlling these patterns in both the east and west are occurring, calling into question fundamental assumptions on which design and management have been based (Milly et al. 2008). At the same time, the compromise of ecosystem functions through narrow-purposed engineering has reduced the latitude within which these water systems may adapt without human intervention. The types of events associated with climate change including greater extremes in water supply will continue to test the resilience of the coupled social-ecological system to respond and adapt to these broadscale changes. Understanding the dynamics of these complex social-ecological systems is urgent as climate change upsets the assumptions on which water law, infrastructure, allocation, and protection have been based. Attempting to understand these complex dynamics and provide some guidance for increasing the adaptive capacity of regional water systems to changing climate was a major reason for undertaking the study that led to this volume.

2.2 Assessing Resilience, Law, and Adaptive Governance in Social-Ecological Water Systems

This volume is the result of a 3-year study called the Adaptive Water Governance (AWG) project of the US National Socio-Environmental Synthesis Center (SESYNC) that was funded by the US National Science Foundation. The project was aimed at seeking a better understanding of the complex ecological and social dynamics around water and resource management by focusing on three themes: resilience of the linked system, relationship between governance and adaptive capacity in the system, and the relationship between law and adaptive governance. The AWG project explored the role of law in achieving water governance that is capable of facilitating management, adaptation, and transformation in the face of climate change. Key questions that were addressed in the project include:

- Could ecological resilience and adaptive capacity of regional scale water systems to changing climate be qualitatively assessed to a useful degree?
- What is the role of environmental governance in developing the adaptive capacity of these systems to respond to climate change?
- What legal and institutional components contribute to adaptive governance?
- What is the role of law in preventing, triggering, and facilitating adaptive governance?

An investigation into these questions was done by bringing together interdisciplinary teams of legal, biophysical, and social science scholars who were embedded in and had in-depth knowledge of six regional social-ecological water systems (Fig. 2.1) (Cosens et al. 2014). Case studies were done in six North American water basins: the Anacostia River (Arnold et al. 2014), Columbia River (Cosens and Fremier 2014), Klamath River (Chaffin et al. 2014), Middle Rio Grande River (Benson et al. 2014), Platte River (Birge et al. 2014), and the Everglades (Gunderson et al. 2014). Since publication of these case studies, associated concepts on the role of law have been applied to the Lake Eyre and Great Artesian basins in Australia (Fig. 2.2) and the lessons from this additional non-US perspective (Cosens 2015).

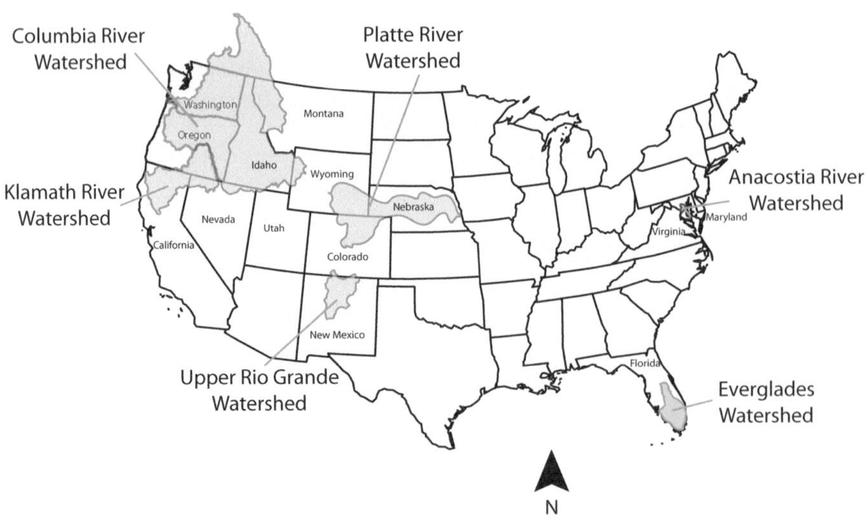

Fig. 2.1 Location of riverine and wetland social-ecological systems in the United States used to study interaction of ecological resilience and adaptive governance (Base map from public domain image, http://www.wikiwand.com/en/List_of_rivers_of_the_United_States)

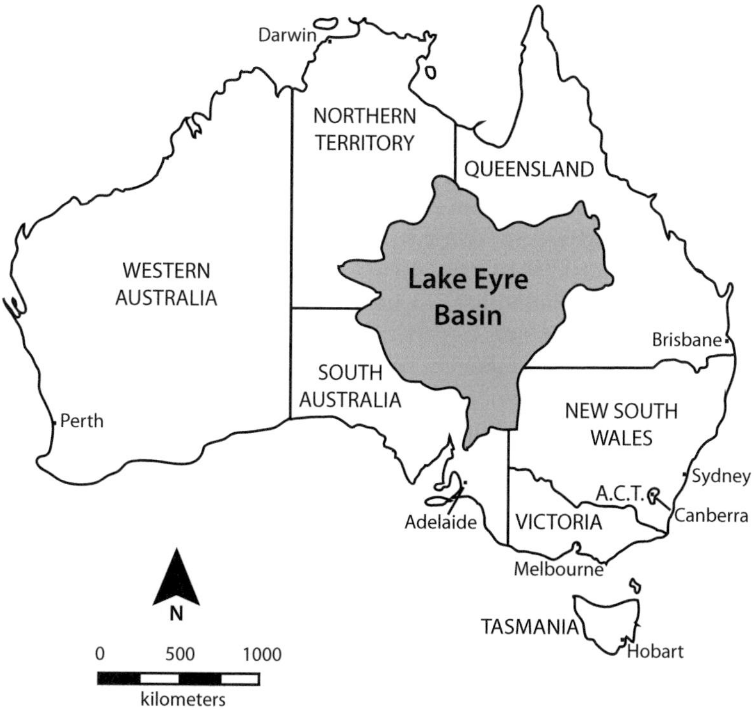

Fig. 2.2 Location of Lake Eyre Basin used to study interaction of ecological resilience and adaptive governance in Australia

The six basin teams have used a variety of approaches that build off earlier approaches to resilience assessment (Walker et al. 2004; Resilience Alliance 2010; Walker and Salt 2006; Nemec et al. 2014), by adding assessment of governance and the role of law. By testing different approaches to evaluate basin governance, these assessments will form the basis for the development of a governance assessment method. In each case study, the question was posed as to the resilience of the basin's social-ecological system to broader-scale environmental changes.

Many of the case studies take a historical approach to assessment to understand both the change in resilience and governance attributes through time and their legacy effect today, including the key role of governance in both the facilitation of and barriers to adaptation. Second, the cases have examined the role of law and identified its importance (a) in establishing boundaries that, once crossed, signal approaching thresholds, (b) in creating conditions for establishment of rights that alter expectations sufficiently to open a window to new and sometimes collaborative approaches to water governance, (c) in providing an avenue for the development of new process tools to facilitate emergence of adaptive forms of governance, and (d) in presenting barriers to adaptation as a result of rigid and fragmented authority (Garmestani and Allen 2014; Cosens et al. 2017).

These assessments illustrate that with the onset of water balance impacts from climate change, some of the water supplies relied on in North America are close to irreversible thresholds that, once crossed, will alter natural ecosystem services and the adequacy of engineered infrastructure, potentially impairing existing water-based economies. Moreover, such assessments of the capacities to govern in the face of change should help identify the conditions under which adaptive governance may emerge and, in particular, the role of law to either prevent or facilitate such forms of governance. We hope to identify the legal tools necessary for adaptive governance, which, along with other factors such as major investments in conservation, green infrastructure, ecological restoration, and reoperation of dams (Richter and Thomas 2007), will be necessary to increase the adaptability of water-based economies in the face of climate change. Achieving this will require governance that is capable of navigating change as well as itself evolving.

All of the basin assessments illustrate an increasing attention to public input and participation in resource decision-making. The recognition of treaty-based water and fishing rights of Native Americans in both the Klamath and Columbia River has led to increased participatory capacity from formerly marginalized populations. The emergent collaborative process among irrigators and Native American tribes in the Klamath basin illustrates both the change in power distribution and participatory capacity resulting from litigation and thus its role in opening a window to collaborative processes. This in turn has led to consideration of changes in basin management that may enhance general resilience in the face of climate change by focusing the attention on the restoration of impaired ecosystem services.

2.3 Case Studies

The following paragraphs briefly describe the seven basins presented in Chaps. 3, 4, 5, 6, 7, 8 and 9 and the key points in their assessment found at the intersection of their social-ecological systems with governance and law. The locations of the six North American basins studied are shown in Fig. 2.1, and the Australian basin is shown in Fig. 2.2. Table 2.1 provides the key hydrologic characteristics of the basins.

Table 2.1 Characteristics of hydrologic basins in the United States used as case studies in assessing adaptive capacity and ecological resilience to global environmental change

Name	Basin area (km^2)	Average flow (m^3/s)	Maximum flow (m^3/s)	Climate zone(s)	Political units
Anacostia River	456	1.5	51	Humid subtropical climate	United States, State of Maryland, Washington, DC
Columbia River	668,000	7500	35,100	Semiarid steppe, alpine, marine west coast	United States, States of Washington, Oregon, Montana, Idaho, Wyoming, Nevada, Utah, Canada, Province of British Columbia,
Everglades basin	28,205	12	80	Humid subtropical climate, tropical wet/dry	United States, State of Florida
Klamath River	40,790	484	15,777	Semidesert, dry alpine, temperate rainforest	United States, States of Oregon and California
Middle Rio Grande river	72,000	41	707	Semiarid steppe	United States, Colorado, New Mexico
Central Platte River	219,916	199	4530	Semiarid steppe, humid continental	United States, State of Nebraska, Colorado, and Wyoming
Lake Eyre Basin	1.14 million	Known for having the highest variability of any river system in the world ranging from 0 flow to flood stage	Highest recorded flow of Diamantina tributary: 4,700	Monsoonal north and arid south, internally drained basin	Australia states of Queensland, New South Wales, South Australia, and Northern Territory

2.3.1 Anacostia River Basin

The Anacostia River is located in the central Atlantic region in eastern North America (Fig. 2.1). The river runs through Washington, DC, and then enters the Potomac River (Chap. 3). The Anacostia has transitioned from a natural to an urban watershed in which restoration efforts will require intensive human intervention. The watershed is home to over one million people. Changes in land use and other pollution sources have led to highly degraded waters. Implementation of the Clean Water Act and subsequent litigation has led to the emergence of local, watershed organizations and adaptive efforts to restore aesthetic and recreational qualities in the watershed. The Anacostia governance structures are multi-scalar across space including formal local, state, and federal governments as well as nongovernmental actors at every level and are influenced by larger-scale restoration programs (Chesapeake Bay). Thus, the federal and regional levels provide much of the knowledge and funding necessary for local capacity building and response. Increased resources for the emerging local organizations will be necessary to enhance adaptive capacity as the watershed responds to climate change (Arnold et al. 2014).

2.3.2 *Columbia River Basin*

The Columbia River, located in the Pacific Northwest of the United States and Canada (Fig. 2.1), is the largest producer of hydropower in North America. Federal investment in the early twentieth century led to the development of major dam infrastructure to achieve the social objectives of flood control, navigation, irrigation, and hydropower in a system largely driven by snowmelt from the Rocky Mountains of Canada and the northern United States (Chap. 4). Almost eight million acres of crop land are irrigated from the Columbia and its tributaries. The basin is also a major site of spawning runs for populations of Pacific salmon and steelhead relied on by commercial and sport fishing and of economic, subsistence, and spiritual importance to indigenous peoples. Development has contributed to the precipitous decline in salmon populations. By the latter half of the century, the assertion of rights by Native American tribes led to their engagement in governance of fisheries. This major capacity building by formerly marginalized communities was made possible by the recognition of rights in federal court and funding for salmon recovery as a result of the listing of numerous populations under the US Endangered Species Act. Although the economic goal of river development has been largely successful, its achievement through optimization for hydropower has left the basin with limited room for adaptation and thus vulnerable to changing climate. In the Columbia River, the scale of governance extends to the international level. Current review of the treaty between the United States and Canada may be an opportunity for increasing management and infrastructure flexibility as well as reconciling certain ecosystem functions with river development (Cosens and Fremier 2014).

2.3.3 *Everglades*

The Everglades is a biologically rich, subtropical wetland that supplies water to about eight million people, a multibillion dollar agriculture enterprise, and the conservation of biodiversity. The system covers about half of the Florida peninsula, in the southeastern United States (Fig. 2.1). Over the past century, the system has successfully promoted economic and social development (Chap. 5). But like the Columbia River, this has come at an environmental cost measured in the listing of a dozen endangered species and the imperiled Everglades National Park. The Everglades Restoration Act of 2000 called for implementation of adaptive management to recover this vast ecosystem. The Everglades system has many of the attributes necessary for adaptive governance such as identified thresholds, the authority to experiment (e.g., adaptive management), and a diversity of institutions. Nevertheless, adaptive governance is hindered by overly prescribed planning and litigation, leaving the social-ecological system of the Florida Everglades constrained in its capacity to adapt to climate change. In both the Columbia River Basin and the Florida Everglades, rigid management at higher levels and failure to balance stability of economic investment with flexibility to adjust management measures have formed impediments to implementation of a more flexible adaptive governance (Gunderson et al. 2014).

2.3.4 *Klamath River Basin*

The Klamath River Basin in south central Oregon and Northern California in the western United States (Fig. 2.1) have been the stage for a classic water conflict among Native American tribes, commercial and recreational fishing interests, environmental groups, and irrigators served by a federal reclamation project (Chap. 6). In a reversal of normal water basin geography, the Klamath has an arid upper basin with rivers, lakes, and marshes supporting irrigated agriculture and a lower basin that drains mountainous, forested terrain and provides significant salmon spawning habitat. Between the upper and lower basin, the drainage constricts as it passes through the Cascade Mountain Range. This constriction at the Oregon/California border provided the ideal sites for development of four hydroelectric dams in the mid-twentieth century, blocking fish passage to the upper basin. Although conflict over water and fish management in the Klamath basin reached a stage of near-violent public protest in 2001, the continued role of law, in particular the Endangered Species Act and the assertion of Native American reserved water rights, ultimately served as the catalyst for emergence of collaborative processes and local adaptive solutions. These solutions are precarious if not formally institutionalized and currently await federal approval (Chaffin et al. 2014).

2.3.5 *Middle Rio Grande Watershed*

The Middle Rio Grande in central New Mexico (Fig. 2.1) is defined as the portion of the river that runs from Cochiti Dam near Santa Fe to Elephant Butte Reservoir south of Albuquerque in the arid US southwest (Chap. 7). Native American Pueblos, communities that date to Spanish settlement, and Anglo-Americans hold irrigation water rights. The river is regulated to provide water downstream to both Texas and Mexico. Management has been modified to protect endangered aquatic species. The system is very close to a threshold due to a combination of: over allocation of water pursuant to the prior appropriation doctrine, lax management including lack of definition and enforcement of water rights, urban development of groundwater hydrologically connected to the river despite an absence of consideration of groundwater lag times in conjunctive management, separation of the river from the floodplain, and extended drought due to climate change that is not only reducing water supply but altering the upland forest ecosystem and fire regime. Rigid political adherence and economic dependency on the existing development place the watershed's society in a vulnerable position. Transition without economic dislocation will require local leadership and capacity building as well as federal investment to restore some of the watershed's ecologic capacity to adapt and to reduce the degree of water dependency (Benson et al. 2014).

2.3.6 *Platte River Basin*

The Platte River basin in the central US drains east from the Rocky Mountains of Colorado and Wyoming to meet the Missouri River after traversing the state of Nebraska (Fig. 2.1). The water laws, policies, and infrastructure of the central Platte River basin in south central Nebraska have evolved during post-European settlement to optimize the needs of irrigation and flood control (Chap. 8). Development has come at a high ecological cost to the system including aquatic and riverine habitat degradation and the listing of several endangered species. Listing has triggered responses to ecological degradation that include a tri-state and federal collaborative Platte River Recovery and Implementation Program with the capacity to coordinate an adaptive approach to system-wide ecological restoration. The Platte River Recovery Program is a first step toward applying an adaptive management approach to restoration at the social-ecological system scale (Birge et al. 2014).

2.3.7 *Lake Eyre and Great Artesian Basins: Australia*

The assessment of the Lake Eyre Basin and its connections to the Great Artesian Basin in Australia (Fig. 2.2) provided an opportunity to apply the results of the initial phase of the AWG project and was used to test the legal guidelines presented in this volume (Chap. 10). The internally draining Lake Eyre Basin covers 1.14 million square kilometers or roughly 15% of Australia, including much of Australia's outback. The basin encompasses parts of New South Wales, Queensland, and the

Northern Territory, and its terminal lake, Lake Eyre, or Kati Thanda, as it is known to the traditional owners of the land, the Arabana (or Arabunna or Urabunna) people, is in South Australia. The Lake Eyre Basin is sparsely populated and its highly variable rivers remain free-flowing. The primary legacy effect of the human development of water in the basin is the thousands of bores developed in the late 1800s and early 1900s in the Great Artesian Basin (the groundwater basin extending under and beyond the Lake Eyre surface water basin) for pastoral use. Efforts are underway to cap and control bore flows as pressures within the Great Artesian Basin aquifers decline, but many remain free-flowing. The impact of colonization and the lack of recognition of Native Title to land and waters until recent years has had a lasting impact on the capacity of Aboriginal communities in the basin to participate in water management. Recent studies indicate that climate change may reduce precipitation and increase temperatures in the southern portion of the basin, while the northern portion of the basin, which supplies the runoff from monsoonal rains to the basin, may experience increased precipitation and greater extremes. The Lake Eyre and Great Artesian basins are currently managed separately. Lake Eyre Basin is subject to an intergovernmental agreement between the Commonwealth, the states of Queensland and South Australia, and the Northern Territory, which only addresses the avoidance of cross-border impacts and, despite policy statements aspiring to a whole-of-basin management approach, does not provide the framework or authority for basin-wide management; rather, intrastate water management is the subject of state law. The fragile and highly adapted nature of the basin ecosystem leaves it vulnerable in the face of climate change (Cosens 2015).

2.4 Conclusion

The remaining chapters in this Part (I) are comprised of the case study descriptions. The case studies, in varying degrees, examine details of the interactions among the structures and dynamics of the social-ecological system, law, and adaptive governance. The system dynamics highlight historical trajectories and patterns to indicate how the system evolved to its current state. The case studies describe how legal, ecological, or social drivers of change have led to ecological and social structures undergoing regime shifts once thresholds have been crossed. The cases depict the role of law in adaptation and transformation of governance and institutions. All of the cases discuss the presence of adaptive governance and the factors that facilitate or obstruct emergence of this type of environmental governance that is a key source of the system capacity to respond to broadscale environmental change.

Acknowledgments This work was developed in part under the Adaptive Water Governance project, funded by the US National Socio-Environmental Synthesis Center (SESYNC) under funding from the US National Science Foundation, NSF DBI-1052875.

References

Arnold CA, Green OO, Decaro DA, Chase A, Ewa JG (2014) The social-ecological resilience of an eastern urban-suburban watershed: the anacostia river basin. Idaho Law Rev 51:29–90

Benson MH, Llewellyn D, Morrison R, Stone M (2014) Water governance challenges in New Mexico's Middle Rio Grande Valley: a resilience assessment. Idaho Law Rev 51:195–228. https://doi.org/10.2139/ssrn.2464387

Berkes F, Colding J, Folke C (2013) Navigating social-ecological systems: building resilience for complexity and change. Cambridge University Press, Cambridge

Birge H, Allen CR, Craig RK, Garmestani AS, Hamm JA, Babbitt C, Nemec K, Schlager E (2014) Social-ecological resilience and law in the Platte River Basin Platte River. Idaho Law Rev 51:229–256

Brundtland Commission (1987) Report of the world commission on environment and development: our common future. http://www.un-documents.net/our-common-future.pdf

Cech TV (2003) Principles of water resources: history, development, management, and policy. Wiley, Hoboken

Chaffin B, Craig RK, Gosnell H (2014) Resilience, adaptation, and transformation in the Klamath River Basin social-ecological system. Idaho Law Rev 51:157–193

Cosens B (2015) Application of the adaptive water governance project to the management of the Lake Eyre Basin and it connections to the Great Artesian Basin. In: Report to the Goyder Institute, Adelaide

Cosens B, Fremier A (2014) Assessing system resilience and ecosystem services in large river basins: a case study of the Columbia River Basin. Idaho Law Rev 51:91–126

Cosens BA, Gunderson L, Chaffin BC (2014) The adaptive water governance project: assessing law, resilience and governance in regional socio-ecological water systems facing a changing climate. Idaho Law Rev 51:1–27

Cosens BA, Craig RK, Hirsch S, Arnold CA, Benson MH, DeCaro DA, Garmestani AS, Gosnell H, Ruhl JB, Schlager E (2017) The role of law in adaptive governance. Ecol Soc 22(1):30. https://doi.org/10.5751/es-08731-220130

Delmas MA, Young OR (eds) (2009) Governance for the environment: new perspectives. Cambridge University Press, Cambridge

Folke C, Carpenter S, Walker B, Scheffer M, Elmqvist T, Gunderson L, Holling CS (2004) Regime shifts, resilience, and biodiversity in ecosystem management. Annu Rev Ecol Evol S 35:557–581. https://doi.org/10.1146/annurev.ecolsys.35.021103.105711

Garmestani AS, Allen CR (2014) Social-ecological resilience and law. Columbia University Press, New York. https://doi.org/10.7312/garm16058

Gunderson L, Pritchard L (eds) (2002) Resilience and the behavior of large-scale systems. Island Press, Washington, DC

Gunderson LH, Holling CS, Light SS (1995) Barriers and bridges to the renewal of ecosystems and institutions. Columbia University Press, New York

Gunderson L, Garmestani A, Rizzardi KW, Ruhl JB, Light A (2014) Escaping a rigidity trap: governance and adaptive capacity to climate change in the everglades social ecological system. Idaho Law Rev 51:127–156

Hela I (1952) Remarks on the climate of South Florida. B Mar Sci 2(2):438

Holling CS, Meffe GK (1996) Command and control and the pathology of natural resource management. Conserv Biol 10:328–337. https://doi.org/10.1046/j.1523-1739.1996.10020328.x

Intergovernmental Panel on Climate Change (2007) Appendix 1: glossary A-D. In: Climate change 2007: impacts, adaptation and vulnerability. Cambridge University Press, New York. http://www.ipcc.ch/publications_and_data/ar4/wg2/en/annexessglossary-a-d.html. Accessed 28 July 2017

Milly PCD, Betancourt J, Falkenmark M, Hirsch RM, Kundzewicz ZW, Lettenmaier DP, Stouffer RJ (2008) Stationarity is dead: whither water management? Science 319:573–574. https://doi.org/10.1126/science.1151915

Mote PW, Hamlet AF, Clark MP, Lettenmaier DP (2005) Declining mountain snowpack in western North America. B Am Meteorol Soc 86:39–49. https://doi.org/10.1175/BAMS-86-1-39
Nemec KT, Chan J, Hoffman C, Spanbauer TL, Hamm JA, Allen C, Hefley T, Pan D, Shrestha P (2014) Assessing resilience in stressed watershed. Ecol Soc 19(1):34. https://doi.org/10.5751/ES-06156-190134
Resilience Alliance (2010) Assessing resilience in social-ecological systems: workbook for practitioners, version 2.0. http://www.resalliance.org/files/ResilienceAssessmentV2_2.pdf
Richter BD, Thomas GA (2007) Restoring environmental flows by modifying dam operations. Ecol Soc 12(1):12. https://doi.org/10.5751/es-02014-120112
Rockström J, Steffen W, Noone K, Persson Å, Chapin FS III, Lambin E, Lenton TM, Scheffer M, Folke C, Schellnhuber H, Nykvist B, De Wit CA, Hughes T, van der Leeuw S, Rodhe H, Sörlin S, Snyder PK, Costanza R, Svedin U, Falkenmark M, Karlberg L, Corell RW, Fabry VJ, Hansen J, Walker B, Liverman D, Richardson K, Crutzen P, Foley J (2009) Planetary boundaries: exploring the safe operating space for humanity. Ecol Soc 14(2):32. http://www.ecologyandsociety.org/vol14/iss2/art32/. Accessed 28 July 2017
Vitousek PM, Mooney HA, Lubchenco J, Melillo JM (1997) Human domination of earth's ecosystems. Science 277:494–499
Walker B, Salt D (2006) Resilience thinking: sustaining ecosystems and people in a changing world. Island Press, Washington, DC
Walker B, Holling CS, Carpenter SR, Kinzig A (2004) Resilience, adaptability and transformability in social–ecological systems. Ecol Soc 9(2):5. http://www.ecologyandsociety.org/vol9/iss2/art5/. Accessed 28 July 2017

Chapter 3
Resilience of the Anacostia River Basin: Institutional, Social, and Ecological Dynamics

Check for updates

Craig Anthony (Tony) Arnold, Olivia Odom Green, Daniel DeCaro, Alexandra Chase, and Jennifer-Grace Ewa

Abstract The Anacostia watershed traverses the urban-suburban areas around Washington, D.C., and Maryland. Historically, the Anacostia River basin has transitioned from a biologically rich natural ecology prior to European settlement through three periods of ecosystem degradation due to agriculture and navigation, industrialization, and urbanization. The current regime is dominated by restoration and green-infrastructure activities yet is still influenced by previous regimes' legacy effects and continued urban-development pressures. The major drivers of regime shifts from presettlement to the present are (1) societal treatment of the basin's waters, lands, vegetation, and wildlife as exploitable goods and services for short-term economic benefit (even in the current regime in which improved water quality and restored lands are public goods and services); (2) shifts from weak to strong environmentalist values and activism; (3) changing ways that humans psychologically relate to the basin and its functions; (4) patterns of structural inequality, oppression, discrimination, and movements to seek social and environmental justice; and (5) changes in governance institutions, including laws, to support and

C. A. Arnold (✉)
Brandeis School of Law and Department of Urban and Public Affairs, University of Louisville, Louisville, KY, USA
e-mail: tony.arnold@louisville.edu

O. O. Green
Atlantic States Legal Foundation, Syracuse, NY, USA
e-mail: ogreen@aslf.org

D. DeCaro
Department of Urban and Public Affairs, Department of Psychological and Brain Sciences, University of Louisville, Louisville, KY, USA
e-mail: daniel.decaro@louisville.edu

A. Chase
National Sea Grant Law Center, Oxford, MS, USA

J.-G. Ewa
Office of the Municipal Public Defender, Denver, CO, USA

B. Cosens, L. Gunderson (eds.), *Practical Panarchy for Adaptive Water Governance*, https://doi.org/10.1007/978-3-319-72472-0_3

facilitate the dominant social values and policies of the time. Institutions have played strong and pervasive roles in both the watershed's declining ecological resilience and potential for improving social-ecological resilience. The greatest opportunities for a more resilient, climate-adaptive Anacostia River watershed require continued and improved changes in watershed governance, restoration and green-infrastructure initiatives, land-use regulation, public engagement, integration of social justice into watershed decision-making, and monitoring and feedback loops.

Keywords Resilience · Adaptive governance · Water management · Institutional analysis · Resilience assessment · Clean Water Act · Water quality

3.1 Introduction

The Anacostia River is a small river and watershed located in the humid, temperate climate of the eastern region of the United States. The watershed area is the smallest of the case studies but is also the subject of the longest period of European development. This chapter begins with a description of the social and ecological components of the watershed and how it has changed over time. It considers the adaptive capacity of the Anacostia River watershed social-ecological system through assessment of the social, ecological, institutional, and governance capacity. This leads to consideration of the potential for adaptive governance institutions to emerge, evolve, or strengthen in order to enhance adaptive capacity and resilience in the system. The chapter concludes with recommendations for continued and new improvements in watershed governance, restoration and green-infrastructure strategies, land-use regulation, public engagement with watershed conditions, integration of social justice principles and processes into governance structures and decisions, and monitoring and feedback loops that contribute to scientific and social learning.

3.2 The Social-Ecological System of the Anacostia Watershed

The Anacostia River basin stretches from rural and suburban Maryland through the heavily urbanized District of Columbia (Fig. 3.1). It consists of 173 square miles of land in Prince George's and Montgomery Counties, Maryland, and Washington, D.C., draining into the Anacostia River (Maryland Department of the Environment 2012; Anacostia Watershed Society 2009a, b). The Anacostia River flows into the Potomac River, which in turn flows into the Chesapeake Bay (Maryland Department of the Environment 2012). Approximately 84% of the watershed is located within Maryland, with the remaining 16% being located within Washington, D.C. (Maryland Department of the Environment 2012). Underground streams and seeps in the upper watershed, as well as stormwater runoff from throughout the watershed, feed the streams and main

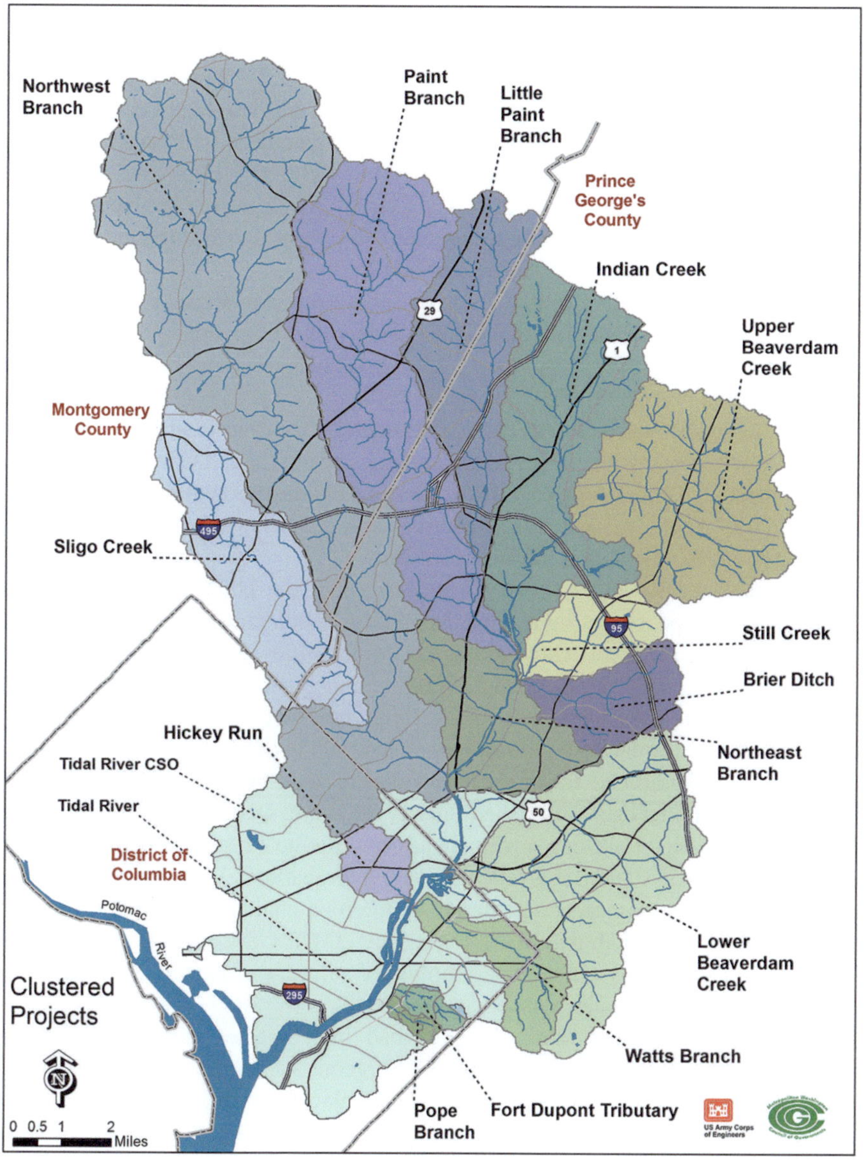

Fig. 3.1 The Anacostia River and watershed, within Maryland and District of Columbia. Shadings depict different subbasins within the watershed, identified by the name of the tributary to the Anacostia (branches and creeks). The Anacostia River joins the Potomac River in the southwestern portion of the figure. The Anacostia is part of the Chesapeake Bay watershed in the mid-Atlantic region of the United States (Figure adopted from USGS, USACOE, and Metropolitan Washington Council of governments used through Creative Commons license)

stem of the Anacostia River (Maryland Department of the Environment 2012). Tides influence the main stem of the Anacostia River, as well as lower portions of some tributaries (Maryland Department of the Environment 2012). The watershed straddles two different ecoregions—the Piedmont and the Coastal Plains. The Piedmont portion of the basin has steep stream valleys, rocky streambeds with steep gradients, well-drained loamy soils, and elevations of 200 to 400 ft above sea level (Maryland Department of the Environment 2012). The Coastal Plains portion has gentle slopes, slowly meandering streams, sandy soils, and elevations of 0 to 200 ft above sea level (Maryland Department of the Environment 2012).

The Anacostia River watershed is heavily urbanized, containing about 1 million residents in the watershed with an average density of 2.66 people per acre, but the population density is not evenly spread throughout the basin (Maryland Department of Natural Resources 2000). The distribution of land uses is 75% urban or suburban, 20% forest, and 5% agricultural, and approximately one-quarter of the watershed's lands are covered with impervious surfaces (Maryland Department of the Environment 2012). Extensive urbanization and industrialization have strongly influenced current watershed characteristics, adversely affecting the basin's biological, hydrological, and physical functions.

The Anacostia River has been designated under the US Clean Water Act as impaired with respect to nutrients (i.e., phosphorus), sediments (i.e., total suspended solids), fecal bacteria, impacts to biological communities in nontidal waters, polychlorinated biphenyls (PCBs), heptachlor epoxide, and trash/debris (Maryland Department of the Environment 2012). At times, portions of the River and its tributaries have been deemed to be too contaminated with toxics or fecal bacteria (often increased from combined sewer overflows during and after storm events) for any human contact (Anacostia Watershed Society 2009a, b; District Department of the Environment 2008). Pollution adversely affected the health of fish and wildlife in the watershed. PCBs and polycyclic aromatic hydrocarbons (PAHs) in stream and river sediment and their bioaccumulation up the food chain have caused cancerous tumors in fish (U.S. Fish and Wildlife Service 2000). Government officials have issued health advisories against consumption of several different species of fish in the Anacostia (Anacostia Watershed Society 2012).

Historically, the Anacostia's fisheries had high species richness and were characterized by thriving populations of "sturgeon, American and hickory shad, white and yellow perch, redbreast sunfish, pickerel, catfish, and herring" (US Army Corps of Engineers et al. 2010). Many fish species have declined or been extirpated from the river due to industrial pollutants, low dissolved oxygen caused by nutrient contamination and associated algae blooms, lower spring water temperatures and fewer floods from snowmelt, and shallow flows resulting from sedimentation and streambed alterations (US Army Corps of Engineers et al. 2010). Major fish kills in warm summer months with low flows have been a problem, although fish kills have been declining over the past two decades (US Army Corps of Engineers et al. 2010).

Feeder streams and the river have been altered both by human channelization of stream segments and the effects of increased runoff from urban impervious surfaces (Maryland Department of the Environment 2012). Development of riparian buffer zones, deforestation, and draining and filling of wetlands have also adversely affected several characteristics of the watershed, including water flow regimes, water quality, waterway structural features, flood management capacity, and the healthy functioning of biological communities (Maryland Department of the Environment 2012). Upstream tributaries have flashy runoff and flood characteristics, while the downstream tidal areas of the basin are sluggish and therefore trap sediment that washes downstream (US Army Corps of Engineers et al. 2010). The basin has approximately 120 to 130 human-created barriers to fish migration, including buried utility lines, road culverts, and weirs from channelization (US Army Corps of Engineers et al. 2010).

In the basin, 93% of the predevelopment tidal wetland acreage and 63% of the predevelopment nontidal wetland acreage have been destroyed or altered (US Army Corps of Engineers et al. 2010). Only 2550 total wetland acres remain in the watershed, often in fragmented segments with impaired flood management capacity, and the watershed's beavers have been extirpated due to development of nontidal wetlands (US Army Corps of Engineers et al. 2010). More than 70% of the watershed has been deforested with the greatest tree loss occurring in riparian areas (US Army Corps of Engineers et al. 2010; National Oceanic & Atmospheric Administration 2010). However, natural succession processes on former farmlands and acquisition and management of public parklands have led to increased tree coverage in six, primarily upstream, watersheds since the 1930s (US Army Corps of Engineers et al. 2010). Mature hardwood stands on public lands are considered to have especially high ecological value (US Army Corps of Engineers et al. 2010).

3.3 Social-Ecological System Dynamics of the Anacostia River Watershed

The Anacostia has played a critical role in supporting and enabling important economic developments in the region and nation. The Anacostia's transformation from a natural watershed system into a heavily polluted and highly engineered urban watershed was driven by powerful interests seeking to use it as an engine for economic growth (Kronthal 1998; Wennersten 2008). At the same time, the Anacostia has been neglected, in part because it flows through low-income and minority urban neighborhoods. The Anacostia has been called America's "Forgotten River," in the shadow of its much more famous neighbor, the Potomac River, which flows through more elite communities.

3.3.1 Drivers of Change

During the past four centuries, the Anacostia River watershed has undergone four major transitions in its social-ecological state: conversion of forests and wetlands to farmland for cash crops and development of commercial navigation, industrialization, and urbanization (Anacostia Watershed Restoration Partnership 2009b).

This has altered the watershed from a watershed of forests, wetlands, and flows to a watershed of agriculture and navigation, to an industrialized watershed, to a heavily urbanized watershed, and to a watershed of restoration and green infrastructure—landscaping and design principles that mimic natural processes for managing urban stormwater. In the past three decades, the Anacostia has received increasing attention, as citizen groups, like the Anacostia Watershed Society, have pressured government agencies to clean up and restore the Anacostia and its tributaries and to create multi-stakeholder partnerships to develop plans to do so (U.S. Environmental Protection Agency 1997). This last transition is only partially underway, and it remains to be seen whether this becomes a stable state or temporary stable state for the watershed or whether the watershed will quickly shift to another regime. In each transition, the fundamental ecological, social, and institutional characteristics of the watershed have changed, and the watershed has crossed a threshold into a new systemic state or regime with reorganized structures, processes, and functions. Each of these social-ecological changes has been driven by larger forces of change in linked institutional-social-ecological dynamics: (1) commodification and consumerism; (2) environmentalism; (3) watershed psychology; (4) inequity, discrimination, and social justice movements; and (5) institutional change. These drivers and their impact on the basin are described in the following paragraphs.

First, land cover, river/stream structure, and water quality have been altered or destroyed for economic gain driven by commodification and consumerism. The land, forests, and wetlands were treated as raw materials to be transformed into agricultural production of marketable crops like tobacco. The river's structural system was treated as a malleable conduit for the transportation of goods in commerce. The river has been and continues to be a sink and drain for waste from industrial production and urban development. Critical watershed lands have been converted into marketable residential and commercial units. Even now, as restoration and green-infrastructure policies dominate watershed governance decisions, the river and the watershed are being treated as aesthetic, recreational, and residential amenities for human consumption.

Second, the rise in environmental values and activism, in the latter half of the twentieth century, has led to civic and political activism for environmental protection, pro-environmental changes in the missions and actions of government agencies and business entities, and changes in individual and group behaviors. Environmentalism is a major driving force behind many of the current efforts to green and restore the watershed and the legal and policy tools that are being used to do so. Nonetheless, the overall social-ecological-institutional system is vulnerable to shifts in public values, political power, and antienvironmental changes in laws and policies.

Third, place-based connections with, or attachments to the watershed have become a significant part of human cognition and emotion, social behavior, and norms, values, and attitudes in the Anacostia watershed (National Wildlife Federation Lake Research Partners 2012; Arnold 2004; Kronthal 1998). Resource issues and governance problems are cognitively framed as watershed issues and problems of watershed governance (Arnold 2014). Collective action is organized around watershed protection and restoration, and people adjust their behaviors based on the effects that they could have on the watershed's health and functioning (Arnold 2014). Social organizations and institutions develop around the watershed as the central organizing element (Arnold 2014).

Fourth, the legacy of slavery is substantial in the watershed. It includes segregation of people and land uses by race and class, environmental injustices that placed low-income people and racial and ethnic minorities in close proximity to pollution and intensive land uses, redevelopment practices that gentrified communities and displaced existing low-income and minority residents, disparities in the distribution of infrastructure and civic and social services, and differences in the amount of attention that the Potomac and the Anacostia received, based in large part on where society's elites live. By the 1980s, when the Anacostia River waterfront in Washington, D.C., was polluted, crime-ridden, physically deteriorating, industrialized, urbanized, unhealthy, and ignored, the strong interconnections between social decline and ecological decline were hard to miss. At the time of this writing, civil rights and social justice movements, including environmental justice movements are nevertheless changing the direction of the watershed's social-ecological conditions and governance processes.

Fifth, a major driver of social-ecological change in the Anacostia River basin has been institutional change—i.e., changes in the systems of rules that shape human behaviors and structure social action (Ostrom 2006; Scott 2008). The US Clean Water Act (CWA) and amendments to the CWA to regulate municipal stormwater systems have stimulated efforts to address and reverse the effects of impervious surface-cover and urban-suburban pollution. Also, the primary mission and focus of the powerful US Army Corps of Engineers has changed from reengineering waterways and wetlands for commercial navigation and flood control to reengineering them for watershed restoration and wetland mitigation. Informal institutions of social norms and self-governance systems have also changed in important ways. In recent years, watershed-focused organizations and governance collaborations have emerged with robust energy and activity to change trends in the basin's land cover and hydrology. Despite institutional change, the legacy of past rules, norms, and beliefs often remains in a new system in coexistent or hybridized ways. Thus, urban growth and industrial pollution persist in or alongside the new green-infrastructure regimes and the structural elements of racial bias and discrimination persist (Bonilla-Silva 2014).

3.3.2 *Transformations of the Watershed*

The five broad forces or drivers of systemic change have shaped the Anacostia River basin throughout four major transitions. First, the transition from a watershed dominated by forests, wetlands, and clear natural streams (Natural Resources Defense Council 2010), inhabited by the Necostan or Anacostan Native Americans, was transformed for agriculture through deforestation, draining, and filling wetlands (Wennersten 2008; Anacostia Watershed Partnership 2009a). Second, transformation for navigation began with a major port established on the upper Anacostia River in 1742 and the US Navy establishing the Washington Navy Yard with the Anacostia River as the southern boundary of the property in 1799 (Wennersten 2008). In the eighteenth and nineteenth centuries, canals and locks were developed (Heine 1953). Third, in the latter half of the nineteenth century, development of industrial activities and the pollution that they generated quickly altered the fundamental characteristics of the watershed from farming-dominated and navigation-dominated functions to industry-dominated functions, particularly in the lower reaches of the basin in and near Washington, D.C. (Wennersten 2008). The advent of the railroad reduced the economic dominance of commercial navigation, and the Industrial Revolution in the American economy generally contributed to this shift (Wennersten 2008). Fourth, urbanization began at the same time as industrialization but would become a more lasting altering force. This included conversion of lands, including riparian lands, forests, and wetlands, to urban development (US Army Corps of Engineers et al. 2010; National Oceanic & Atmospheric Administration 2010). Urbanization has been an increasing and relentless phenomenon in the watershed. Over 70% of the watershed is now developed, and 45% of the watershed is residential development (Metropolitan Washington Council of Governments 2010).

The current structure of the Anacostia watershed's urbanized regime is poorly adapted to the hydrological, ecological, and sociopolitical pressures from urbanization's adverse effects on the watershed's functioning. Just as urbanization has driven the Anacostia watershed toward a threshold that may be categorized as functional severance of the hydrologic cycle, there is a movement toward reversing this trend by reconnecting the hydrologic cycle through green infrastructure. New multi-jurisdiction, multiagency, multi-stakeholder partnerships have been created to govern the watershed with attention to its social-ecological resilience. Over 20 community-based or citizen-based organizations have arisen with a focus on the Anacostia River watershed or one of its sub-watersheds. Many federal, state, and local government agencies and community-based groups are participants in several different partnerships, and virtually all of them are aimed at improving the environmental conditions and social-ecological functions of the degraded Anacostia River and its watershed. Major legal changes have improved environmental protection and management for watershed function in the Anacostia River basin. Federal environmental statutes and regulations have stimulated action to restore the watershed, control runoff and pollution, and increase the use of green infrastructure. Changes in watershed governance have resulted not only from broad statutory standards but also the implementation of these standards through regulations and enforcement actions. Most significantly, state and local governments have had to seek new watershed-oriented solutions due to the terms of (1) stormwater permits under the CWA, (2) settlements of litigation over combined stormwater and sewage overflow from treatment plants that violate the CWA, (3) plans to achieve compliance with pollutant loads based on water quality standards developed pursuant to the CWA, and (4) remediation plans for contaminated sites under federal statutes

(CERCLA) (Urban Waters Federal Partnership 2013). Watershed groups have been active in commenting on proposed terms of these legal instruments, seeking tougher standards, and encouraging green infrastructure (Natural Resources Defense Council 2010).

Another major development in the Anacostia has been the creation and implementation of a wide range of policies, plans, and projects to improve the ecological, hydrological, and social functions of the watershed or at least prevent their continued deterioration. The centerpiece plan is the Anacostia River Watershed Restoration Plan, calling for restoration and developed by the Anacostia Watershed Restoration Partnership (US Army Corps of Engineers et al. 2010). The Metropolitan Washington Council of Governments prepared an Anacostia Watershed Forest Management and Protection Strategy for the Anacostia Watershed Restoration Partnership in 2005 in order to protect the remaining forests and trees in the watershed and to engage in strategic reforestation for watershed restoration and conservation purposes (Levendosky et al. 2005). Additional efforts include the 2008 Washington, D.C., Department of Environment plan that arose out of ongoing restoration efforts and multi-stakeholder collaborations and calls for making the Anacostia River fishable, swimmable, boatable, visually presentable, and supportive of stable fish and wildlife populations by 2032 (District Department of the Environment 2008). Land conservation has received special attention in the Anacostia in the past few decades (Metropolitan Washington Council of Governments 2010; Arnold 2006).

3.4 Adaptive Capacity and Adaptive Governance in the Anacostia River Watershed

Certain features of adaptive watershed governance systems that contribute to their plasticity and incremental changes over time are present in the Anacostia watershed governance system. Watershed governance in the Anacostia has a polycentric structure with loosely linked networks of diverse and engaged stakeholders and government agencies (Walker and Salt 2006; Garmestani et al. 2014). Multiple modes or methods of watershed governance are being used in moderately integrated ways to achieve multiple goals for the long-term ecological and social functioning of the watershed (Arnold and Gunderson 2013). Governance processes are participatory, engaging multiple stakeholders and many area residents. Moreover, cooperation among stakeholders is relatively high, yet conflict, litigation, and formal legislative and regulatory processes are used to hold both government agencies and resource users accountable to watershed-conservation standards and to break down entrenchment of interests that resist collective problem-solving (Sabatier et al. 2005; Huitema et al. 2009). Litigation and political activism lead to collaboration, not merely escalation of the conflict (Karkkainen 2008). Some formal monitoring mechanisms and feedback loops have been built into plans and projects, and other informal feedback loops have emerged; there is capacity for both expert and public learning from implementation of governance decisions and actions (Scholz and Stiftel 2005). The basic foundations of an evolving adaptive governance system appear to exist in the Anacostia River watershed.

Changes within federal environmental regulatory institutions have already played important roles in the current transition to focus on watershed restoration and green infrastructure. The Clean Water Act has been the federal regulatory hammer that gets the relevant stakeholders and governance entities working together to innovate solutions and change degradation trajectories. The Clean Water Act went from being an institutionalized point source pollution control system aimed at industry and wastewater treatment plants to having a much more diverse array of objectives, including control of nonpoint source pollution and runoff through stormwater system permitting and establishing daily pollutant load standards based on water quality goals, funding for watershed planning, and promoting green infrastructure as an alternative to engineered controls (Innovative Stormwater Infrastructure Act 2013). Both top-down command-and-control regulation and rule-enforcing litigation have forced government agencies and polluters to address watershed problems. These legal tools stimulated innovation, cooperation, and problem-solving among many stakeholders at several different scales. Watershed institutions have arisen out of the dynamic and interwoven successes and failures of more formal legal and regulatory regimes.

Emergent watershed governance institutions in the Anacostia River basin should be continued and strengthened, because they show adaptive features and capacity. They are organized around the watershed and thus are scaled to govern ecological and hydrological problems at the ecosystem scale. Yet, they have smaller-scale components, such as plans organized around each sub-watershed, and nesting within larger-scale basin management and governance activities in the Chesapeake Bay basin. Thus, Anacostia River watershed governance is multi-scalar with governance activities appropriately scaled to the relevant problems.

The hydrology and ecology of the Anacostia River watershed affects the opportunities for adaptive governance. While episodic droughts in the Anacostia River basin are likely to become greater stresses on both natural and human systems under conditions of climate change, the primary stressors have to do with too much water, not too little water (Maryland Commission on Climate Change: Adaptation & Response & Scientific & Technical Working Groups 2011). Peak storm events, which are likely to increase in frequency and intensity as the climate changes, and stormwater runoff from impervious surfaces combine to scour streambeds and stream banks, erode soils, and carry sediment and pollutants into waterways. The slow-flowing, shallow downstream tidal portions of the Anacostia have become more sluggish and shallow due to anthropogenic land and waterway alterations and are traps where pollutant-laden sediment collects and remains toxic. Natural forests and wetlands have been eliminated from much of the watershed and are now replaced with built structures, a fact that limits the options of governance systems. Merely improving the ecological management of natural resources—often at least one policy option in large river basins in the western part of the United States—will not do much to improve the ecological resilience of a watershed where most of those natural resources no longer exist.

The Anacostia Watershed Society and other watershed-focused organizations have helped to develop adaptive governance institutions and processes by engaging the participation and voice of low-income people and neighborhoods of color, including children, neighborhood organizations, and social justice groups. However, concerns about gentrification threaten the legitimacy and efficacy of watershed restoration actions and governance systems. The watershed will not be resilient unless governance decisions and actions address past injustices, are fair in their processes and their distribution of environmental harms and benefits, and strengthen, not weaken, the health and resilience of low-income and minority communities, including neighborhoods in the Anacostia River area. Assessments of urban watershed resilience and adaptive governance capacity must give thoughtful and thorough attention to the histories of racism, class discrimination, segregation, and environmental injustices that continue to have influence over the linked ecological, social, and institutional conditions of the watershed.

Adaptive watershed governance requires extensive monitoring and feedback loops in which lessons learned from monitoring the effects of actions and decisions end up shaping and reshaping future decisions and actions (Holling 1978; Lee and Lawrence 1986). Considerable informal feedback loops exist among various government agencies, nongovernmental organizations, and other participants in many of the Anacostia River watershed partnerships and projects. Information and ideas are shared through informal networks, as well as formal and semiformal networks. In addition, some restoration and green-infrastructure projects have monitoring activities built-in and have produced some important lessons to guide decision-makers or managers. However, like most examples of adaptive management or adaptive governance (Camacho 2007, 2009), there has been very little systematic attention to designing and imbedding feedback loops into governance processes in order to ensure monitoring, assessment, learning, and appropriate revisions of plans, policies, and actions. Nonetheless, watershed governance systems in the Anacostia may be flexible enough to provide the space for revisions to rules, policies, plans, and actions if the right variables are carefully and persistently monitored and if the monitoring data are analyzed for lessons that could inform governance institutions.

Based on this resilience assessment of the Anacostia River basin, we identify seven key variables to be studied, monitored in an ongoing, systematic, and thorough manner, and managed adaptively for the overall social-ecological-institutional resilience of the watershed. These variables are (1) land cover and land use; (2) the quantity, velocity, and quality of stormwater runoff; (3) streamflow; (4) pollutant loading, adjusted for changing ecological conditions; (5) performance of green infrastructure; (6) the interactions between social values/norms and political forces, including not only measures of public attitudes and values toward the Anacostia and the environment but also trends and patterns in environmental, natural-resource, and land-use politics at various governance scales (Schlager and Blomquist 2008; Karvonen 2011); and (7) institutional capacity and change.

3.5 Conclusion

The history of the Anacostia River watershed offers several important lessons about adaptive watershed governance and social-ecological resilience in small urban-suburban watersheds. The distinctive characteristics of watersheds like the Anacostia require particular attention in assessing their social-ecological resilience and in developing and supporting adaptive watershed governance systems.

Institutions matter. Throughout this case study of the Anacostia River watershed, we have identified the strong and pervasive roles of institutions in the watershed's declining ecological resilience and potential for improved social-ecological resilience. However, institutions change, often in complex interrelationships with social change and ecological change. The institutional-social-ecological dynamics of the Anacostia River basin over time give us both concerns and optimism about its potential for improved social-ecological resilience.

Resilience assessments must give attention to the major drivers of systemic change that can strengthen or weaken systemic resilience. Analyzing the major drivers of land cover change and alterations of watershed structural features, we have concluded that the greatest opportunities for a more resilient, climate-adaptive Anacostia River watershed require continued and improved changes in watershed governance, restoration and green-infrastructure initiatives, land-use regulation, public engagement, integration of social justice into watershed decision-making, and monitoring and feedback loops.

Acknowledgments This work was developed in part under the Adaptive Water Governance Project, funded by the US National Socio-Environmental Synthesis Center (SESYNC) under funding from the US National Science Foundation, NSF DBI-1052875. The views set forth by contributors to this volume represent their own and do not represent the views of any public or private entity the contributor is affiliated with.

References

Anacostia Watershed Restoration Partnership (2009a) Historical changes of the watershed. http://www.anacostia.net/history/history.html. Accessed 5 Oct 2017

Anacostia Watershed Restoration Partnership (2009b) Upper Beaverdam Creek. http://www.anacostia.net/Subwatershed/Upper_Beaverdam.html. Accessed 5 Oct 2017

Anacostia Watershed Society (2009a) Anacostia River water trail guide: a voyage through time: from Captain John Smith to the modern day. http://www.anacostiaws.org/images/maps/AnacostiaRiverWaterTrailGuide.pdf. Accessed 4 Oct 2017

Anacostia Watershed Society (2009b) Anacostia River watershed map. http://www.anacostiaws.org/images/maps/WatershedProgression.jpg. Accessed 4 Oct 2017

Anacostia Watershed Society (2012) Addressing the risk: understanding and changing anglers' attitudes about the dangers of consuming Anacostia River fish. www.anacostiaws.org/userfiles/file/AWS_angling_FINAL_web.pdf. Accessed 4 Oct 2017

Arnold CA (2004) Working out an environmental ethic: anniversary lessons from Mono Lake. Wyoming Law Rev 4:1–55

Arnold CA (2006) For the sake of water: land conservation and watershed protection. Sustain J Environ Sustain Issues 14:16–31

Arnold CA (2014) Framing watersheds. In: Hirokawa KH (ed) Environmental law and contrasting ideas of nature: a constructivist approach. Cambridge University Press, Cambridge, pp 271–302. https://doi.org/10.1017/cbo9781139519762.015

Arnold CA, Gunderson LH (2013) Adaptive law and resilience. Environ Law Rep 43(5):10426–10443

Bonilla-Silva E (2014) Racism without racists: color-blind racism and the persistence of racial inequality in America, 4th edn. Rowman & Littlefield Publishing, Lanham

Camacho AE (2007) Can regulation evolve?: lessons from a study in maladaptive management. UCLA Law Rev 55:293–358

Camacho AE (2009) Adapting governance to climate change: managing uncertainty through a learning infrastructure. Emory Law J 59:1–78

District Department of the Environment, Government of the District of Columbia (2008) Anacostia 2032: plan for a fishable and swimmable Anacostia River. http://green.dc.gov/sites/default/files/dc/sites/ddoe/publication/attachments/Anacostia2032.pdf. Accessed 4 Oct 2017

Garmestani AS, Allen CR, Arnold CA, Gunderson LH (2014) Introduction: social-ecological resilience and law. In: Garmestani AS, Allen CR (eds) Social-ecological resilience and law. Columbia University Press, New York

Heine CW (1953) The Washington City canal. In: Records of the Columbia Historical Society, 42nd Volume 1953/1956, Washington, DC, pp 1–27. http://www.jstor.org/stable/pdfplus/40067664.pdf. Accessed 5 Oct 2017

Holling CS (ed) (1978) Adaptive environmental assessment and management. Wily, Chichester

Huitema D, Mostert E, Egas W, Moellenkamp S, Pahl-Wostl C, Yalcin R (2009) Adaptive water governance: assessing the institutional prescriptions of adaptive (co) management from a governance perspective and defining a research agenda. Ecol Soc 14(1):26. http://www.ecologyandsociety.org/vol14/iss1/art26/

Karkkainen BC (2008) Getting to "Let's talk": legal and natural destabilizations and the future of regional collaboration. Nevada Law J 8:811–852

Karvonen A (2011) Politics of urban runoff: nature, technology, and the sustainable city. MIT Press, Cambridge, MA

Kronthal ML (1998) Local residents, the Anacostia River, and "Community". https://www.sfaa.net/files/1513/7329/3792/kronthal.pdf. Accessed 4 Oct 2017

Lee KN, Lawrence J (1986) Adaptive management: learning from the Columbia River basin fish and wildlife program. Environ Law 16:431–460

Levendosky K, Galli, J, Trieu, P, Vatovec, C (2005) The Metropolitan Washington Council of Governments, Department of Environmental Programs, Anacostia Watershed Forest Management and Protection Strategy. http://www.anacostia.net/restoration/Reports_and_Data/Final_FMPS.pdf. Accessed 5 Oct 2017

Maryland Commission on Climate Change: Adaptation & Response & Scientific & Technical Working Groups (2011) Comprehensive strategy for reducing Maryland's vulnerability to climate change: phase II: building society, economic, and ecological resilience. http://ian.umces.edu/pdfs/ian_report_299.pdf. Accessed 10 Oct 2017

Maryland Department of Natural Resources (2000) Maryland's surf your watershed, Watershed Profile: Anacostia River. www.dnr.state.md.us/watersheds/surf/prof/pdf/02140205_wp.pdf

Maryland Department of the Environment (2012) Watershed report for biological impairment of the non-tidal Anacostia River Watershed, Prince Georges and Montgomery Counties, Maryland and Washington DC.: Biological stressor identification analysis results and interpretation. http://www.mde.state.md.us/programs/Water/TMDL/Documents/BSID_Reports/Anacostia_River_BSID_Report_020112_final.pdf. Accessed 4 Oct 2017

Metropolitan Washington Council of Governments (2010) Anacostia watershed environmental baseline conditions and restoration report. http://www.anacostia.net/Restoration_Plan/download/PLAN_report_Jan_2010.pdf

National Oceanic & Atmospheric Administration (2010) The Anacostia: partnering to restore an Urban Estuary. http://oceanservice.noaa.gov/news/features/mar10/anacostia.html. Accessed 4 Oct 2017

National Wildlife Federation Lake Research Partners (2012) Restoring the Anacostia River: determining the best path forward and building public will for action. http://studylib.net/doc/13651400/restoring-the-anacostia-river--determining-the-best-path-... Accessed 5 Oct 2017

Natural Resources Defense Council (2010) NRDC, local partners petition DC to use smarter practices to clean up its waterways. https://www.nrdc.org/experts/jon-devine/nrdc-local-partners-petition-dc-use-smarter-practices-clean-its-waterways. Accessed 10 Oct 2017

Ostrom E (2006) Understanding institutional diversity. Princeton University Press, Princeton

Sabatier P, Focht W, Lubell M, Trachtenberg Z, Vedlitz A, Matlock M (eds) (2005) Swimming upstream: collaborative approaches to watershed management. MIT Press, Cambridge, MA

Schlager E, Blomquist W (2008) Embracing watershed politics. University Press of Colorado, Boulder

Scholz JT, Stiftel B (eds) (2005) Adaptive governance and water conflict: new institutions for collaborative planning. Resources for the future, Washington, DC

Scott WR (2008) Institutions and organizations: ideas and interests, 3rd edn. Sage, Thousand Oaks

U.S. Army Corps of Engineers, Metropolitan Washington Council of Governments, Montgomery and Prince George's Counties, the District of Columbia, the Maryland-National Capitol Park and Planning Commission, Maryland Department of the Environment, and Maryland Department of Natural Resources (2010) Anacostia River watershed restoration plan and report. Final Draft. http://www.anacostia.net/Restoration_Plan/download/Anacostia-Report-Web-Quality.pdf. Accessed 4 Oct 2017

U.S. Environmental Protection Agency (1997) Anacostia rebirth: restoration for the river, risk reduction for the Community EPA # 903-F-97-001. https://nepis.epa.gov/Exe/ZyNET.exe/91017OV3.txt?ZyActionD=ZyDocument&Client=EPA&Index=1995%20Thru%201999&Docs=&Query=&Time=&EndTime=&SearchMethod=1&TocRestrict=n&Toc=&TocEntry=&QField=&QFieldYear=&QFieldMonth=&QFieldDay=&UseQField=&IntQFieldOp=0&ExtQFieldOp=0&XmlQuery=&File=D%3A%5CZYFILES%5CINDEX%20DATA%5C95THRU99%5CTXT%5C00000032%5C91017OV3.txt&User=ANONYMOUS&Password=anonymous&SortMethod=h%7C-&MaximumDocuments=1&FuzzyDegree=0&ImageQuality=r75g8/r75g8/x150y150g16/i425&Display=hpfr&DefSeekPage=x&SearchBack=ZyActionL&Back=ZyActionS&BackDesc=Results%20page&MaximumPages=1&ZyEntry=1&slide. Accessed 5 Oct 2017

U.S. Fish & Wildlife Service (2000) Evaluating the health of the Tidal Potomac River watershed: tumor surveys in Brown Bullhead Catfish http://www.fws.gov/chesapeakebay/pdf/bullhead.pdf. Accessed 4 Oct 2017

Urban Waters Federal Partnership (2013) New life for the Anacostia River watershed. https://www.epa.gov/sites/production/files/2014-06/documents/anacostia-river-background-2013.pdf. Accessed 10 Oct 2017

Walker B, Salt D (2006) Resilience thinking: sustaining ecosystems and people in a changing world. Island Press, Washington, DC

Wennersten JR (2008) Anacostia: the death & life of an American River. Chesapeake Book, Chambersburg

Legal Documents

Innovative Stormwater Infrastructure Act, S. 1677, 112th Cong. (2013); H.R. 3449, 111th Cong. (2013)

Chapter 4
Social-Ecological Resilience in the Columbia River Basin: The Role of Law and Governance

Barbara Cosens and Alex Fremier

Abstract The Columbia River is a complex water basin shared by 2 countries, 15 Native American Tribes, 15 First Nations, 7 US states, and 1 Canadian province. Dam construction during the twentieth century has engendered a basin economy that is dependent on low-cost electricity and irrigated agriculture. Yet, these dams are a major factor in the decline of populations of salmon and steelhead species that are critical to the culture of Indigenous peoples. Climate change scenarios predict a transformation from snow- to rain-dominated precipitation in the basin's lower latitudes, greater extremes in flood and drought, and an increasing water deficit as a result of higher rates of evapotranspiration with increasing temperature. Reduced late summer flow may pose challenges for the sustainability of irrigation and fish. The basin provides a unique laboratory to explore resilience of a highly developed social-ecological system to changing climate and rising empowerment of Indigenous peoples. Review of the Columbia River Treaty between the United States and Canada that governs much of the operation of the river presents a window of opportunity for change. This window provides a moment in time to rethink environmental governance and to consider an approach which reflects neither top-down nor bottom-up control of resources but a third path in which each level of government plays a supporting role to a regional vision of the basin's future governance.

Keywords Resilience · Adaptive governance · Adaptive law · Resilience assessment · Climate change · Water management

---

B. Cosens (✉)
College of Law and Waters of the West Program, University of Idaho, Moscow, ID, USA
e-mail: bcosens@uidaho.edu

A. Fremier
School of the Environment, Washington State University, Pullman, WA, USA
e-mail: alex.fremier@wsu.edu

B. Cosens, L. Gunderson (eds.), *Practical Panarchy for Adaptive Water Governance*, https://doi.org/10.1007/978-3-319-72472-0_4

4.1 Introduction

The Columbia River Basin is located in the Pacific Northwest of the United States and Canada and drains an area that is roughly the size of France. Its headwaters lie in the Rocky Mountains of the two countries. Its mid-region is rural and arid with the need for irrigation to achieve its massive agricultural industry. Its coastal region, after cutting through the Cascade Mountains, is relatively wet and includes its largest city of Portland, Oregon. The substantial drop from its headwaters to the Pacific Ocean made the Columbia a prime candidate for the production of hydroelectricity. Today, it is the largest producer of hydropower in North America with its energy transmitted to coastal cities in the Pacific Northwest of the United States and Canada and to the southwestern United States. The river is also the site of 13 runs of Pacific salmon and steelhead populations, substantially impaired by human changes to the river. The river development has fueled the economic growth of the region, and as a result, the region's society is highly dependent on the developed services from the river. Climate change is already altering the timing of runoff as the winter snow that provides the natural storage in high elevation snowpack relied on by the engineered system begins to transition to rain (Cosens and Fremier 2014; Cosens et al. 2016).

The Columbia River Basin provides a living laboratory for consideration of a basin in the growth and development trajectory (Gunderson et al. 2017; Chap. 14) of the adaptive cycle in which the legacy of engineered development to optimize for a narrow range of purposes (primarily hydropower, flood control, and irrigation) has compromised the adaptive capacity of the system (Cosens and Fremier 2014). This chapter will explore the potential window of opportunity to increased adaptive capacity opened by the intersection of the listing of endangered species, review of the Columbia River Treaty between the United States and Canada, and the increased participatory capacity of Indigenous peoples following legal victories in the recognition of rights, the combination of which may have resulted in a paradigm shift within the basin. It will then explore how the combination of built infrastructure, mental models that favor the status quo (DeCaro et al. 2017a; Chap. 17), and a legal framework for international river operation that did not anticipate the evolution in the basin's society and climate (Craig et al. 2017; Chap. 11) may present barriers to change. This chapter will conclude with consideration of avenues for increased capacity for adaptive governance through consideration of adjustments to the legal framework governing the structure, capacity, and processes of governmental actors in basin governance.

4.2 The Social-Ecological Setting of the Columbia River Basin

The Columbia River begins at Columbia Lake and wetlands in the Rocky Mountains of British Columbia, Canada, and flows 1200 miles before it reaches the Pacific Ocean in a rich estuary near Astoria, Oregon, USA. Major tributaries include the Kootenai (Kootenay in Canada), the Pend Oreille/Clark's Fork/Flathead, Spokane, Okanogan,

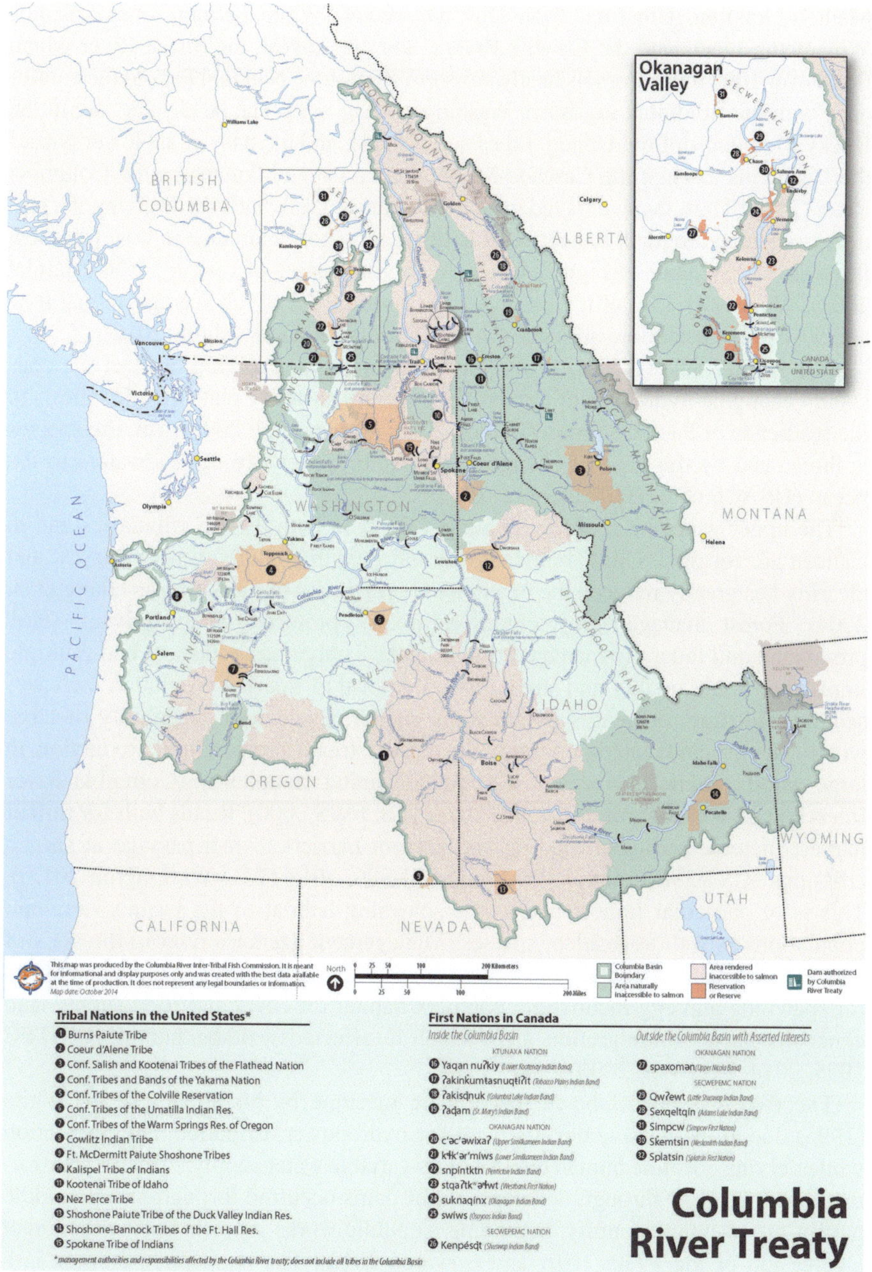

Fig. 4.1 Columbia River Basin shared by the United States and Canada. The figure illustrates the complex jurisdictional boundaries, including international, state, provincial, and tribal boundaries, and First Nation interests. *Green areas* are those historically blocked from anadromous fish migration due to natural barriers. *Salmon colored* areas are those blocked from anadromous fish migration due to human constructed dams. This figure is used with permission of the Columbia River Inter-Tribal Fish Commission

Methow, Yakima, Umatilla, John Day, Deschutes, White Salmon, Wind, Sandy, Willamette, Lewis and the Cowlitz Rivers, and, the largest, the Snake River which flows over 1000 miles from its headwaters in Yellowstone National Park to the confluence with the Columbia in eastern Washington (Fig. 4.1). The headwaters are in the Rocky Mountains of the United States and Canada, and the path in the lower part of the basin cuts through the Cascade Mountains in a stretch known as the Columbia River Gorge. Thus, the river is fed by snow-dominated sub-watersheds giving it a pre-development hydrograph indicative of high spring runoff and range in flow from low flows in late summer to high flows during spring runoff of roughly 1:34 (Hamlet 2003). While only 15% of the basin lies in Canada, 38% of the average annual flow and 50% of the peak flow originates in the Canadian Rockies (Shurts 2012).

The 10-million-year history of salmon in the basin, evolving to meet the constantly changing hydraulics of a river that traverses an active plate margin, attests to the resilience of Pacific salmon (Healey 2009; Waples et al. 2009). Yet, the species is threatened by the pace of change in the Anthropocene from headwaters to the ocean (USACE et al. 2008; 2014).

The upper watersheds of the Columbia River in both the US tributaries and in Canada are forested. Once dominated by old growth conifers, timber harvest and clearing has greatly reduced the extent and biodiversity of these forests (Jiang et al. 2004). Forest management includes a mix of protected areas, private, state/provincial, and federally managed timberlands. Timber harvests have had multiple impacts on water quality and streamflow. Removal of riparian vegetation increases sediment loading; decreases overhanging shelter, shade, and availability of large woody debris; and correspondingly increases stream temperature. Reduction in large woody debris also decreases aquatic habitat complexity (Columbia River Inter-Tribal Fish Commission 1995, 2014; U.S. FWS 2000). Roads built for timber harvest increase sediment loading and present barriers to fish passage at stream crossings (US v. Washington. 20 F.Supp.3d 986, 1013 (W.D Washington 2013)). This is of particular importance to the spawning habitat of the basin's salmonid populations as well as resident species. Management agencies refer to the 4-Cs of native salmon conservation: cold water for spawning and rearing, clean gravel beds for spawning and egg incubation, complex habitat for cover and food supply, and connected habitat for migration, all of which are affected by timber harvest (CRITFC 1995, 2014; US FWS, 2000).

The river has been labeled the organic machine by historian Richard White (1995) due to its massive development for hydropower, irrigation, and navigation while defying absolute human control of its variable water supply. A major transformation of the river through development of dams occurred throughout the middle part of the twentieth century, beginning as public works projects during the Great Depression of the 1930s (Hirt and Sowards 2012). The engineered storage is a smaller volume relative to supply than other western US rivers, with storage capacity only 40% of the average annual flow (Barton and Ketchum 2012). The Columbia River Basin is the largest producer of hydropower on the North American continent (United States Army Corps of Engineers et al. 2001). The United States and Canada jointly operate the river under the Columbia River Treaty (CRT), which provides for coordination of numerous dams for hydropower production and flood control.

The engineered development of the basin also supports roughly 7.8 million acres of irrigated agriculture (Foundation for Water Energy and Education 2016). Irrigation enables a growing wine industry, the largest production of potatoes and hops in North America (USDA 2015), and an additional 2.5 million acres of dryland farming that have transformed the interior uplands to monoculture (Balkwill and Duffin 2008). Irrigated regions have developed both surface and groundwater to the maximum extent possible leading to shortage in dry years. The realization that connected surface and groundwater must be managed as one source if their capacity to deliver services is to be sustained has come late to irrigation communities (Cosens 2006). The reliance on groundwater as a primary source of irrigation water, rather than a supplemental supply available to bridge production during extended drought, reduces the adaptive capacity of the agricultural regions.

Salmon and steelhead runs struggle to survive due to the combined impact of dams on migration, water flow and quality, loss of spawning and rearing habitat, continued commercial and recreation harvests, and hatchery influences (CRITFC 1995, 2014). Anadromous species migrations are blocked from 37% of their former habitat by dams (Harrison 2008). Salmon runs are supplemented by over 200 hatcheries, with hatchery bred fish amounting to 80–90% of the anadromous runs (Peery 2012). Changes in climate influence stream temperature and timing of flows through reduced snowpack and earlier seasonal runoff (Mote et al. 2005; Stewart et al. 2005; Hamlet et al. 2007; Abatzaglou et al. 2014). For example, in 2015 a warming event in combination with low flows led to lethal temperatures for salmon. All 13 runs of salmon and steelhead are listed under the US Endangered Species Act (Code of Federal Regulations 2013; Federal Register: June 14, 2004, Feb. 1, 2006; NOAA Fisheries n.d.; United States Army Corps of Engineers et al. 2008, 2014).

The Columbia River Basin is jurisdictionally complex with transboundary issues at the international, federal/tribal, interstate, and interlocal levels. The basin includes areas claimed as the ancestral lands of 15 First Nations in Canada, 15 Native American Reservations in the United States, 7 US states, 2 countries, and 1 Canadian province. Native American Tribes and First Nations in the basin have a deep connection to salmon as both a primary protein source and cultural icon and to historic fishing sites as a source of trade and intertribal ceremony (Hirt 2008; Landeen and Pinkham 1999). Native American Tribes led the effort to seek listing of anadromous fish under the US Endangered Species Act in the mid-1990s. As an outgrowth of the environmental movement of the 1960s and 1970s, the basin society in general increasingly values the fish, recreation, and aesthetic amenities and has preserved vast stretches of wild river in the headwaters of its tributaries.

4.3 The Social-Ecological Dynamics of the Columbia River Basin

The dynamics and possible trajectories of a social-ecological system – a system in this case defined by river basin boundaries – are a product of the legacy effects of various periods of human interaction with an ecological system that would itself evolve over time if left without human intervention. History plays a role in both constraining the options available to the system and in building its capacity to sustain, adapt, and transform as its future unfolds.

Human contact with the Columbia River Basin is documented as much as 9000 years ago (Bonnichsen v. US 367 F.3d 864), and oral history of today's native people of the Columbia River Basin indicates that Indigenous peoples have had a special relation to the river since pre-European contact. Salmon was not only the primary protein source for Indigenous people in the basin but formed the cornerstone of their religion, culture, and economy (Hines 1999; Pearson 2012). Indigenous people adapted to the natural variation in ecosystem services by taking advantage of river morphology to harvest salmon (Landeen and Pinkham 1999). The life cycles of Columbia Basin fisheries were used to mark time suggesting a strong integration between Indigenous culture and ecosystem services (Landeen and Pinkham 1999). Evidence also exits that Indigenous practices served to regulate fish harvest (Barber 2006).

Initial contact between Indigenous people and Europeans focused on trade, in particular for fur, which benefited the Tribes as much as the white traders (Josephy 1965). But as European settlement along with commercial fishing and development of agriculture and railroads continued, Indigenous peoples experienced direct impacts through war and disease. Negotiations concerning tribal territory in the mid-1800s were driven by the desire to expand settlement opportunities for emigrants and by railroad interests (Josephy 1965). Indigenous people who held out were defeated in battle, and many succumbed to disease (United States v. Washington, 384 F.Supp. 312 (D.Wash., 1974), aff'd by 520 F.2d 676 (9th Cir. 1975), cert denied Washington v. U.S. 423 U.S. 1086 (1976)). The devastating impact of this period to native people is magnified by the fact that it happened predominantly within the time period of a single generation. The legacy of nineteenth century colonization and decimation of Indigenous populations and culture remains in the marginalization of their descendants in western government, science, and culture (Cosens and Fremier 2014).

The next transformation in the social-ecological system of the Columbia River occurred through the massive engineering of the river in the early to mid-twentieth century (Cosens and Fremier 2014). Infusion of federal resources and technology fueled the development of an economic region in the Pacific Northwest with cheap energy, irrigated agriculture, and reduced catastrophic flooding (Vogel 2012; Shurts 2012; Mouat 2012). Nevertheless, this engineered solution muted feedbacks from the ecological to the social system, delaying the recognition of the impact on salmonid species until a threshold beyond which recovery would not be possible was almost reached or possibly crossed (CRITFC 1995, 2014; Bottom et al. 2009).

The legacy effect of optimization of the river for key ecosystem services has substantially narrowed the latitude to adapt to changes in the hydrograph outside historic norms, leaving the system vulnerable to change (Cosens and Fremier 2014). Without extensive investment in reengineering the existing infrastructure to reconcile ecosystem function and the benefits of hydropower, flood control, and irrigation, the system will become increasingly vulnerable as population growth and the projected overall drying associated with climate change unfold (Mote et al. 2005; Stewart et al. 2005; Hamlet et al. 2007, 2010; Abatzoglou et al. 2014; Cosens et al. 2016). In concert with reengineering, the operation and governance of the system must be altered in a way to enhance flexibility and assist society in navigating change.

The slow unfolding of the effects of climate change may be an insufficient trigger for an emergent governance response. However, just as a sudden change in an ecological system brought on by a natural or human-caused disaster may trigger change in the social-ecological system as a whole, the law in its regulatory role is also an agent of change. This is particularly true when the literal application of the law would displace economic interests. In the Columbia River Basin, litigation under Treaty rights and the Endangered Species Act (ESA) has redistributed power to formerly marginalized populations dependent on salmonid runs (United States v. Washington (Boldt Decision), 384 F. Supp. 312, 330 (W. D. Wash. 1974) aff;d 525 F.2d. 676 (9th Cir. 1975), cert. Denied, 423 US 1086 (1975) and Washington v. Washington State Commercial Passenger Fishing Vessel Ass. 443 US 658, 685 (1979)), as will be discussed in greater detail below. This has also led to substantial investment in habitat restoration. Nevertheless, adjustment to river operation has occurred only around the margins with no systemic change to the basin's aging infrastructure or its economic base despite numerous attempts.

Rethinking of basin governance to address declines in salmonid populations began in the 1970s and culminated in the institutionalization of ideas in 1980 with passage of the Northwest Power Act (Pacific Northwest Electric Power Planning and Conservation Act, (Northwest Power Act), Pub. L. No. 96- 501, 94 Stat. 2697). The Northwest Power Act led to the creation of the Northwest Power and Conservation Council, a governmental entity across four US states. The council was one of the first regional scale attempts to consider adaptive management as a means to manage under uncertainty (Lee and Lawrence 1986; Volkman and McConnaha 1993; Lee 1993, 1999).

Although the basin hopes rested on the ability of the Council to balance its competing mandates of power planning and fish and wildlife conservation, the Council did not interpret its mission to be to balance competing interests. Rather, its literal interpretation of its statutory authority is to oversee two independent programs, one for power planning and the other for fish and wildlife, and the hope of implementation of adaptive management to reconcile the two was never realized. Frustration over the lack of progress in this approach led to petitioning under the ESA in the 1990s, with the Shoshone-Bannock Tribe leading a petition for the listing of sockeye (Mann 2004). Twelve additional salmonid listings followed ((Code of Federal Regulations 2013, Federal Register: June 14, 2004, Feb. 1, 2006; NOAA Fisheries n.d.; United States Army Corps of Engineers et al. 2008, 2014).). Listing has led to numerous biological opinions concerning operation of the Federal Columbia River Power System to comply with the ESA and litigation concerning the adequacy of the biological opinions (USACE et al. 1993, 1995, 2000, 2004, 2008, 2014; see generally, Blumm 2002; Thomas-Morse 2012). While this has resulted in some changes to increase flows during out-migration of salmon smolts, modeling indicates that the system continues to be operated as if solely optimized for hydropower production (Cecchini-Beaver 2013). Coincidentally, the capacity built by those seeking restoration of salmon runs, along with the production of knowledge and increased public awareness, prepared the basin for future opportunities. One such opportunity is the review of the CRT.

The United States and Canada have shared the management of the Columbia River under the CRT since 1964. The partnership has provided both countries with significant direct benefits from flood control and power generation and fueled the economic growth of the region (Mouat 2012; Hirt and Sowards 2012; Vogel 2012). The two countries share equally in the benefits of hydropower production in the United States made possible by release of water from storage in Canada behind dams built under the CRT. While the CRT has no expiration date, the United States paid upfront for 60 years of assurance that dams in Canada would result in a very low flood risk. The expiration of this conservative approach to flood risk management and fallback to much heavier reliance on US storage occurs in 2024. By the terms of the CRT, either country may unilaterally terminate the Treaty beginning in 2014, after giving 10 years notice. This looming change in flood control and the power to unilaterally terminate have triggered reviews on both sides of the border. The reviews went well beyond flood risk management, by considering energy demand, water supply, ecosystem function, and other shifts in stakeholder goals since 1964 (Cosens and Williams 2013; Cosens 2016). These changes may simply be too great for the existing CRT to navigate (Shurts 2012).

Formal reviews on both sides of the border have been shadowed by informal cross-border dialogues led by the Universities Consortium on Columbia River Governance (UCCRG n.d.) and by Native American – First Nation cross-border dialogues. In 2010, the 15 Tribal nations on the US side of the Basin cooperated in the development of a set of common views on the future of the Columbia River (Columbia Basin Tribes 2010). Despite both historic and current differences, the Tribes worked together throughout the review process co-led by the Administrator of the Bonneville Power Administration and Division Engineer of the Northwestern Division of the United States Army Corps of Engineers. The US review led to a recommendation to the US Department of State calling for elevation of ecosystem function to a third prong of any modernized treaty (US Army Corps of Engineers and Bonneville Power Administration n.d., 2013). British Columbia transmitted a position to the Commonwealth Cabinet, indicating that while ecosystem issues are important, they were not matters for the CRT (British Columbia 2013).

The US review included substantial public input (US Army Corps of Engineers and Bonneville Power Administration n.d.); however, agreement at such a large scale is fragile until memorialized in law (DeCaro et al. 2017b, Chap. 16). The authority to represent the United States in international negotiations is vested in the executive branch and delegated to the Department of State. Treaties however must go through the advice and consent of the Senate. Almost as soon as the US review recommendation was finalized, hydropower, irrigation, environmental, and Tribal interests began lobbying Congress and the Department of State for a better deal. On October 12, 2016, the US Department of State announced that it was ready to move forward with negotiations along the lines of the regional recommendation. At the time of this writing, both countries have appointed lead negotiators and the United

States has announced that negotiations will begin in 2018. Leadership in both countries and in the Province of British Columbia has changed since finalization of the reviews. It is unclear at this point if the hoped-for modernization of the Columbia River Treaty will prevail or if it is likely to give way to a small fix such as extension of flood control provisions. This presents both a trigger and an opportunity to consider means to improve the basin's general resilience by enhancing its capacity to adapt to change.

4.4 The Role of Law in Governance and Resilience in the Columbia River Basin

Lessons and perhaps solutions relevant to the reduction in the latitude for adaptation in the Columbia River Basin can be found in considering the role of law in governance, particularly its role in the structure, capacity, and process of government (Cosens et al. 2017, Chap. 10). Law determines the level of government that holds authority over the management and regulation of the interaction between society and the environment. With a window of opportunity created by the review of the CRT, the need for modernization of both water management and infrastructure in the basin may lead to consideration of every issue at the scale of the international basin. However, the call for elevation of ecosystem function to a third prong of the Treaty between the United States and Canada raises issues ranging from local habitat restoration to basin-scale operation of dams. A critical step in the process must be to consider which issues related to ecosystem function require coordination at the scale of the basin and which would be better served by placing authority at the local level (Bankes and Cosens 2014; Cosens et al. 2016).

In the Columbia River Basin, control of flow is coordinated via international cooperation on the regulation of dams (Barton and Ketchum 2012). Of the portions of the basin in which reconnection of the river to former spawning habitat for anadromous fish may be possible once new technology is developed and investment in aging infrastructure becomes a reality, only one implicates transboundary issues – the passage of fish beyond Chief Joseph and Grand Coulee Dams located in the United States, serving as a barrier to salmon runs up the river main stem into Canada since the middle of the twentieth century (Columbia Basin Tribes and First Nations 2015). In contrast, the restoration of habitat implicates local land use and requires tailoring of measures to local conditions and societal values (Bankes and Cosens 2014; Cosens et al. 2016). At the same time, habitat degradation has a cumulative effect. The journey anadromous fish must take through multiple jurisdictions to reach locally restored spawning grounds ties even in local measures to a need for broader coordination.

Should revival of ecosystem function become a future goal, whether through the CRT or another mechanism, it presents a unique governance process problem of how to coordinate efforts that range from individual to local land use decisions and planning, to state/provincial allocation of water, to federal/provincial/private operation of dams, and to international coordination of storage and flow. Any single entity at the basin scale attempting to coordinate all this would require perfect knowledge and abundant resources for enforcement (Ostrom 1990). But there is increasing recognition of an alternative path if it is acknowledged that this should not be the work of a single entity but the coordinated effort of many. The alternative path requires resources and a degree of authority over implementation and decision-making focused at the local level, standard or goal setting at the regional level, and resources, technology, and enforcement at the federal and international level (Cosens et al. 2016). Such nested governance is a form of polycentric governance described by panarchy theory (Gunderson and Holling 2002). Scaled nesting, or panarchy, may allow for local innovation and experimentation, while larger scales provide resources and stability.

Karkkainen (2013) points to the Great Lakes Compact and Agreement (Great Lakes–Saint Lawrence River Basin Agreement 2005; Compact 2008) as a model for agreement across an international boundary by governments below the national level. This approach allows for both greater flexibility and greater local control due to its subnational status. The Great Lakes Compact and Agreement provide subnational instruments for coordination among states and provinces across the United States – Canada border. The cooperative nature enhances information exchange while allowing for a maximum level of local discretion. Bankes and Cosens (2014) relied on the work of Karkkainen (2013) to recommend a similar approach for coordination of the habitat restoration aimed at improving ecosystem function in the Columbia River Basin, while also proposing that Tribes and First Nations take an equal seat at the table with states and provinces. This approach would result in the nesting of local discretion and authority within an international setting for coordination.

Institutional nesting is incomplete without a political body at the scale of the basin, something lacking in the current CRT implementation that relies on agency engineers (Barton and Ketchum 2012). It will also require mechanisms for cross-scale coordination. While the current approach allows for a degree of discretion at the operational level, it does not provide for adaptive capacity at the formulation and policy-making levels essential for response during times of change and uncertainty (Ostrom 1990). While this involvement of local, place-based governments in water management is critical to tightening the tie between the social and ecological system, such engagement requires the capacity to participate. The substantial growth in participatory capacity of the Basin's Native American Tribes and First Nations since the 1960s and their current rise to co-managers of the basin fisheries illustrates the impact of local capacity building on adaptive capacity.

In the 1855 Council of Walla Walla, treaties were negotiated among the US federal government and numerous Tribes of the Pacific Northwest. A key position of the Tribes was their right to continue spiritual, cultural, and subsistence reliance on Columbia Basin fisheries both on and off the reservations the negotiations would create (Josephy 1965). In 1905, the US Supreme Court gave meaning to the Treaty language guaranteeing the right to fish off-reservation at what had been the usual and accustomed places by holding that the right carries with it implied rights that are necessary to its fulfillment. In the case in question, the Court recognized an implied right to access across private land (U.S. v. Winans 1905). It would be many years before further progress would be made in defining these rights.

Not until the Civil Rights movement of the 1960s and 1970s in the United States did a similar movement among Native Americans arise and lead to the assertion of treaty rights in various parts of the United States. This was called the American Indian Movement or AIM (Wittstock and Salinas n.d.). As a result of an assertion of treaty fishing rights in the Pacific Northwest, the language recognizing off-reservation fishing was the subject of a lawsuit challenging Tribal fishing free of state control. The federal district court in the case held that the language entitles treaty tribes to up to 50% of the harvestable fish that pass (or would pass absent harvest) the usual and accustomed fishing places (United States v. Washington 1974; Washington v. Washington State Commercial Passenger Fishing Vessel Association 1979). To facilitate division and protection of tribal harvests, the governments of the Nez Perce Tribe, Confederated Bands of the Yakama Nation, Confederated Tribes of the Umatilla Indian Reservation, and Confederated Tribes of the Warm Springs Reservation formed the Columbia River Inter-Tribal Fish Commission (CRITFC). CRITFC is a fisheries science and policy agency that has grown to be recognized as a leading agency in co-management of salmonid fisheries in the US portion of the Columbia River Basin. Thus, law was the tool used to redistribute power and access to basin fisheries, but resources and self- organization made the shift in governance a reality.

This level of self-organization would come later to the Tribes whose fishing grounds lie in areas blocked from salmon runs by dams. Today, the five upper Columbia tribes in the United States have joined together on various resource issues of common concern as the Upper Columbia United Tribes (UCUT), and the Tribes on the Columbia's largest tributary have organized as the Upper Snake River Tribes (USRT) for similar purposes.

Such incremental growth in capacity over a long period of time illustrates the power of the initial step in capacity building to spiral up (Emery and Flora 2006). The level and sophistication of the capacity resulting from the initial steps of a few to assert Treaty fishing rights in the Pacific Northwest are illustrated by the major diplomatic effort of all 15 tribal nations in the US portion of the Columbia River Basin. Despite the tendency of the dominant society to lump Indigenous peoples into a single category such as Native Americans, Pacific Northwest Tribes have a history of war and competition that predates European contact. Nevertheless, all 15 Tribes came together in 2010 to develop The Common Views on the future of the Columbia River Treaty (Columbia Basin Tribes 2010) and have stood behind this document as a united block throughout the review of the CRT. While these changes indicate a complex interaction between law and society as well as avenues for greater flexibility in the governmental environmental management, what is their role in the emergence of adaptive governance? How can new approaches to government structure and the use of law to shift power imbalance and build capacity lead to emergence of adaptive response to surprise that involves both government and nongovernmental actors? The following section explores these questions.

4.5 Pathways to Adaptive Governance in the Columbia River Basin

Moving from a system predominantly managed from the top-down to one that also involves flexible, innovative, and adaptive local management begins with the legacy effect of past actions and its effect on the possible trajectories of collective action. Social decision-making captures the interplay of individual and group psychology with governance (DeCaro et al. 2017a). Of particular interest in the Columbia River Basin is the bias for the status quo reinforced by the development of a substantial economy reliant on continued optimization of the system for hydropower. The cheap non-carbon electricity produced by the Columbia River supplies 60% of the power needs of the region and, through an interconnected grid, supplements summer power needs in the southwestern United States (USACE et al. 2001; Hamlet et al. 2010). From a rural economically weak region that required substantial federal investment to come out of the Great Depression of the 1930s, the Pacific Northwest has become an economic powerhouse with its major cities of Seattle, Portland, Vancouver, and Victoria (Cosens and Fremier 2014). Typing this manuscript in Microsoft Word with Microsoft headquarters in the Pacific Northwest is but a small reminder of this. Though of this list of cities, only Portland is in the basin, all are reliant on its energy and food production.

In addition, the drier inland parts of the basin have used storage on both the main stem and tributaries to develop an agricultural industry worth billions (USDA n.d.). Only since the successful assertion and litigation of Treaty fishing rights in the 1970s, combined with listing of salmonid runs under the Endangered Species Act, has the Tribal voice become a force within the basin (Cosens 2012). Yet the sheer economic power of the hydro-system and reliance on the status quo creates a predominant mental bias against change (Cosens and Fremier 2014). Until the seed for a shift began with the CRT review process, the impossibility of change permeated the prevailing mental model of the basin, and as noted above, it remains to be seen if the emerging mood for change will prevail. It is likely to fail unless careful attention is paid to the balance between stability and flexibility (Craig et al. 2017; Chap. 11).

Functioning legal and economic systems are designed to foster social stability. Stability in governmental, corporate, and personal affairs is essential to confidence in economic investment and peaceful resolution of conflict. It is the legal system through its establishment of governmental structure, its granting of authority that both enables and limits governmental action, its processes for assuring governmental legitimacy and accountability, and its refereeing of social interaction that assures this stability exists (Craig et al. 2017). Yet this system – a system that allows a final decision between two competing views of a dispute even in the face of uncertainty and one that is so beautifully designed to provide a final peaceful resolution to conflict – is poorly suited to mediate social-ecological interaction (Cosens 2008). Resolution of an environmental dispute based on science later found to be wrong has not resolved the underlying issue. In a legal system designed for stability, this is destabilizing. The answer in the mediation of social-ecological interaction, although not entirely clear in its application, is to balance stability with flexibility.

In the Columbia River Basin, this tension plays out in the ongoing effort to recover salmonid populations while continuing to receive the benefits of cheap, carbon-free hydropower, irrigation, and the protection from floods. While legal battles concerning compliance of federal operation of dams with the ESA have served a useful purpose in keeping the issue alive and altering the distribution of power by giving voice to those for whom the endangered species are important, it has not provided a solution. Part of any solution must be engineered – i.e., the redesign of aging dams to better account for their coexistence with salmonids – but part lies in governance. This need, both in the Columbia River Basin and elsewhere, has given rise to considerable literature on collaborative governance, adaptive governance, and co-management (Chaffin et al. 2014). Common to all is the need for restructuring of government to increase local resources and discretion and to provide for inclusive mediation of disputes. Feedback between the ecological system and local entities is much more direct; thus change is detected as it occurs, and given the resources to respond, local communities may adapt in an iterative manner, rather than the more disruptive punctuated change that would have to occur if only done at large scales (Cosens et al. 2017). At the same time, we have a long history of local attention to select sectors of society; in fact, it is this attention to the enormous benefits of cheap, clean power that can be blamed in part for the failure to halt the salmonid decline at an earlier stage. Thus, the nesting of local discretion within regional goal setting and national to international resources and enforcement is essential to achieving the proper balance.

4.6 Discussion and Conclusion

What does this Columbia Basin case study suggest about a way forward either as a future research agenda on adaptive governance or on the ground in the effort of the people of the Pacific Northwest to adapt to climate change? Either path indicates the need to rethink environmental governance. The history of the Columbia Basin illustrates the shortcomings of both unchecked and under-resourced local control and of top-down management that severs feedback from both local communities and local ecosystems to those making decisions. This study suggests a third possible path, one in which local entities have the authority and capacity to respond to feedback from both the social and ecological systems: one in which local control has sufficient oversight that it does not become a proxy for a single economic sector or a means to exclude certain portions of the population from either access to or control over the resource in question; one in which larger scales of governance play the role of connector, facilitating information exchange, dialogue, and goal setting among local entities; and one in which larger scales of governance play the role of local capacity building, knowledge transfer, and development of new technology.

Consider that this third path is not so far from the situation today if one simply sets aside the more polarizing debate of local versus national or international governance. Consider also that a substantial beginning to the research needed already exists in the modern writings of those thinking about governance of the commons and environmental governance (Ostrom 1990; McGinnis 1999; Brunner et al. 2002; Dietz et al. 2003; Rogers and Hall 2003; Folke et al. 2005; Olsson et al. 2004, 2006; Raadgever et al. 2008; Huitema et al. 2009; Pahl-Wostl 2009; Lockwood et al. 2010; Chaffin et al. 2016; etc.). Then the question of environmental governance in our time becomes: how do we get there from here?

Acknowledgments This work was developed in part under the Adaptive Water Governance Project, funded by the US National Socio-Environmental Synthesis Center (SESYNC) under funding from the US National Science Foundation, NSF DBI-1052875.

References

Abatzoglou J, Rupp D, Mote P (2014) Understanding seasonal climate variability and change in the Pacific Northwest of the United States. J Clim 27:2125–2142

Balkwill D, Duffin A (2008) Plowed under: agriculture and environment in the Palouse. University of Washington, Seattle

Bankes N, Cosens B (2014) Protocols for adaptive water governance: the future of the Columbia River Treaty. In: Program on water issues. Munk School of Global Affairs. University of Toronto. http://powi.ca/wp-content/uploads/2014/10/Protocols-for-Adaptive-Water-Governance-Final-October-14-2014.pdf. Accessed 22 July 2017

Barbar K (2006) Indigenous regulations of the harvest, in canneries on the Columbia: a new western history, narrative. In: The Oregon history project. http://www.ohs.org/the-oregon-history-project/narratives/canneries-on-the-columbia/native-fishery-katrine-barber/indigenous-regulations-of-harvest.cfm. Accessed 22 July 2017

Barton J, Ketchum K (2012) Columbia River Treaty: managing for uncertainty. In: Cosens B (ed) The Columbia River Treaty revisited: transboundary river governance in the face of uncertainty. Oregon State University Press, Corvallis

Blumm M (2002) Sacrificing the Salmon: a legal and policy history of the decline of Columbia Basin Salmon. BookWorld, Oshkosh

Bottom, D, Jones K, Simenstad C, Smith C eds. (2009) Pathways to resilient salmon ecosystems. Ecol Soc 14(1):34 [online] URL: http://www.ecologyandsociety.org/issues/view.php?sf=34

British Columbia (2013) Columbia River Treaty Review: B.C. Decision. B.C. Government. http://blog.gov.bc.ca/columbiarivertreaty/files/2012/03/BC_Decision_on_Columbia_River_Treaty.pdf. Accessed 23 July 2017

Brunner R, Colburn C, Cromley C, Klein RA, Olson EA (2002) Finding common ground: governance and natural resources in the American west. Yale University Press, New Haven

Cecchini-Beaver, M (2013) Transboundary Columiba River operational alternative analysis in a collaborative framework. Masters Thesis in Water Resources, University of Idaho. http://www.lib.uidaho.edu/digital/etd/. Accessed 23 July 2017

Chaffin B, Gosnell H, Cosens B (2014) A decade of adaptive governance scholarship: synthesis and future directions. Ecol Soc 19(3):56. https://doi.org/10.5751/ES-06824-190356. Accessed 26 August 2017

Chaffin B, Garmestani A, Gunderson L, Benson MH, Angeler DG, Arnold CA, Cosens BA, Craig RK, Ruhl JB, Allen CR (2016) Transformative environmental governance. Annu Rev Environ Resour 41:399–423

Columbia Basin Tribes (2010) Common views on the future of the Columbia River Treaty. http://www.usea.org/sites/default/files/event-/Common%20Views%20statement%20NQ.pdf. Accessed 23 July 2017

Columbia Basin Tribes and First Nations (2015) Fish passage and reintroduction into the U.S. & Canadian Upper Columbia Basin. http://www.ucut.org/Fish_Passage_and_Reintroduction_into_the_US_And_Canadian_Upper_Columbia_River3.pdf. Accessed 23 July 2017

Columbia River Inter-Tribal Fish Commission (1995) Spirit of the Salmon. http://plan.critfc.org/2013/spirit-of-the-salmon-plan/about-spirit-of-the-salmon/. Accessed 23 July 2017 (1995 and 2014 update)

Cosens B (2006) The role of hydrology in the resolution of water disputes. J Contemp Water Res Educ: River Adjudications 133. doi:https://doi.org/10.1111/j.1936-704X.2006.mp133001004.x/epdf. Available at http://onlinelibrary.wiley.com/doi/10.1111/j.1936-704X.2006.mp133001004.x/epdf. Accessed 26 Aug 2017

Cosens B (2008) Resolving conflict in non-ideal, complex systems: solutions for the law-science breakdown in environmental and natural resource law. Nat Resour J 48:257–301

Cosens B (2012) Changes in empowerment: rising voices in Columbia Basin resource management. In: Cosens B (ed) Columbia River Treaty revisited: transboundary river governance in the face of uncertainty. Oregon State University Press, Corvallis

Cosens B (2016) The Columbia River Treaty: an opportunity for modernization of basin governance. Colorado Nat Resourc Energy Environ Law Rev 27(1): 1–19. Available at http://www.colorado.edu/law/sites/default/files/CNREELR-V27-I1-Cosens%20Final_0.pdf. Accessed 26 August 2017

Cosens B, Fremier A (2014) Assessing system resilience and ecosystem services in large River Basins: a case study of the Columbia River Basin. Idaho Law Rev 51:91–125

Cosens B, Williams M (2013) Resilience and water governance: adaptive governance in the Columbia River Basin. Ecol Soc 17(4) Art. 3. URL: https://doi.org/10.5751/ES-04986-170403. Accessed 26 Aug 2017

Cosens B, Fremier A, Bankes N, Abatzoglou J (2016) The Columbia River Treaty and the dynamics of transboundary water negotiations in a changing environment: how might climate change alter the game? In: Water policy and planning in a variable and changing climate. CRC Press, Boca Raton

Cosens B, Craig R, Hirsch S, Arnold CA, Benson MH, DeCaro DA, Garmestani AS, Gosnell H, Ruhl JB, Schlager E (2017) The role of law in adaptive governance. Ecol Soc 22(1):30. https://doi.org/10.5751/ES-08731-220130. Accessed 26 August 2017

Craig R, Garmestani A, Allen C, Arnold CA, Birgé H, DeCaro DA, Fremier AK, Gosnell H, Schlager E (2017) Balancing stability and flexibility in adaptive governance: an analysis of tools available in U.S. environmental law. Ecol Soc 22(2):3. https://doi.org/10.5751/ES-08983-220203

DeCaro D, Arnold C, Boamah E, Garmestani AS (2017a) Understanding and applying principles of social cognition and decision making in adaptive environmental governance. Ecol Soc 22(1):33. https://doi.org/10.5751/ES-09154-220133

DeCaro D, Chaffin B, Schlager E, Garmestani AS, Ruhl JB (2017b) Legal and institutional foundations of adaptive environmental governance. Ecol Soc 22(1):32. https://doi.org/10.5751/ES-09036-220132. Accessed 26 Aug 2017

Dietz T, Ostrom E, Stern P (2003) The struggle to govern the commons. Science 302:1907–1912. https://doi.org/10.1126/science.1091015. Accessed 26 Aug 2017

Emery M, Flora C (2006) Spiraling-up: mapping community transformation with community capitals framework. Commun Dev 37(1):19–35

Folke C, Hahn T, Olsson P, Norberg J (2005) Adaptive governance of social-ecological systems. Annu Rev Environ Resour 30:441–473. https://doi.org/10.1146/annurev.energy.30.050504.144511. Accessed 26 August 2017

Foundation for Water Energy and Education (2016) What makes the Columbia River Basin unique and how we benefit. http://fwee.org/environment/what-makes-the-columbia-river-basin-unique-and-how-we-benefit/. Accessed 22 July 2017

Gunderson L, Holling C (2002) Panarchy: understanding transformations in human and natural systems. Island Press, Washington, DC

Gunderson L, Cosens B, Chaffin B, Arnold CA, Fremier AK, Garmestani AS, Craig RK, Gosnell H, Birge HE, Allen CR, Benson MH, Morrison RR, Stone MC, Hamm JA, Nemec K, Schlager E, Llewellyn D (2017) Regime shifts and panarchies in regional scale social- ecological water systems. Ecol Soc 22(1):31. https://doi.org/10.5751/ES-08879-220131. Accessed: 26 Aug 2017

Hamlet A (2003) The role of transboundary agreements in the Columbia River basin: an integrated assessment in the context of historic development, climate, and evolving water policy. In: Diaz H, Morehouse B (eds) Climate and water: transboundary challenges in the Americas. Kluwer, Dordrecht, pp 263–289

Hamlet A, Mote P, Clark M, Lettenmaier DP (2007) 20th century trends in runoff, evapotranspiration, and soil moisture in the western U.S. J Clim 20:1468–1486

Hamlet A, Lee S, Mickelson K, Elsner MM (2010) Effects of projected climate change on energy supply and demand in the Pacific Northwest and Washington State. Clim Chang 102:103–128

Harrison J (2008) Fish Passage at Dams, North West Power & Conservation Council. http://www.nwcouncil.org/history/fishpassage. Accessed 23 July 2017

Healey M (2009) Resilient salmon, resilient fisheries for British Columbia, Canada. Ecol Soc 14(1): 2. [online] URL: http://www.ecologyandsociety.org/vol14/iss1/art2/

Hines D (1999) Tales of the Nez Perce. Ye Galleon Press, Fairfield

Hirt P (2008) Developing a plentiful resource: transboundary rivers in the Pacific Northwest. In: Whiteley JM, Ingram H, Perry R (eds) Water, place, and equity. MIT Press, Cambridge, MA, pp 147–188

Hirt P, Sowards A (2012) The past and future of the Columbia River. In: Cosens B (ed) The Columbia River Treaty revisited: transboundary river governance in the face of uncertainty. Oregon State University Press, Corvallis

Huitema D, Mostert E, Egas W, Moellenkamp S, Pahl-Wostl C, Yalcin R (2009) Adaptive water governance: assessing the institutional prescriptions of adaptive (co) management from a governance perspective and defining a research agenda. Ecol Soc 14(1):26

Jiang H, Strittholt J, Frost P, Slosser NC (2004) The classification of late seral forests in the Pacific Northwest, USA using Landsat ETM+ imagery. Remote Sens Environ 91:320–221

Josephy A (1965) The Nez Perce Indians and the opening of the northwest. Yale University, New Haven

Karkkainen B (2013) The Great Lakes water resources compact and agreement: transboundary normativity without international law. William Mitchell Law Rev 39:997

Landeen D, Pinkham A (1999) Salmon and his people: fish and fishing in Nez Perce culture. Confluence Press, Lewiston

Lee K (1993) Compass and gyroscope: integrating science and politics for the environment. Island Press, Washington, DC

Lee K (1999) Appraising adaptive management. Conserv Ecol 3(2):3

Lee K, Lawrence J (1986) Adaptive management: learning from the Columbia River basin fish and wildlife program. Environ Law 16:431–460

Lockwood M, Davidson J, Curtis A, Stratford E, Griffith R (2010) Governance principles for natural resource management. Soc Nat Resour 23(10):986–1001. https://doi.org/10.1080/08941920802178214. Accessed 26 Aug 2017

Mann J (2004) Sacajawea's people: the Lemhi Shoshones and the Salmon River country. University of Nebraska Press, Lincoln

McGinnis M (ed) (1999) Polycentric governance and development: readings from the workshop in political theory and policy analysis. University of Michigan Press, Ann Arbor

Mote P, Hamlet A, Clark M, Lettenmaier DP (2005) Declining mountain snowpack in western North America. Bull Am Meteorol Soc 86:39–49

Mouat M (2012) The Columbia exchange: a Canadian perspective on the negotiation of the Columbia River Treaty. In: Cosens B (ed) The Columbia River Treaty revisited: transboundary river governance in the face of uncertainty. Oregon State University Press, Corvallis

NOAA Fisheries (n.d.) West Coast Salmon & Steelhead Listings. URL: http://www.westcoast.fisheries.noaa.gov/protected_species/salmon_steelhead/salmon_and_steelhead_listings/salmon_and_steelhead_listings.html

Olsson P, Folke C, Berkes F (2004) Adaptive co-management for building resilience in social-ecological systems. Environ Manag 34:75–90

Olsson P, Gunderson L, Carpenter S, Ryan P, Lebel L, Folke C, Holling CS (2006) Shooting the rapids: navigating transitions to adaptive governance of social-ecological systems. Ecol Soc 11(1):18 [online] URL: http://www.ecologyandsociety.org/vol11/iss1/art18/. Accessed 26 Aug 2017

Ostrom E (1990) Governing the commons: the evolution of institutions for collective action. Cambridge University Press, New York
Pahl-Wostl C (2009) A conceptual framework for analysing adaptive capacity and multi-level learning processes in resource governance regimes. Glob Environ Chang 19:354–365
Pearson M (2012) The river people and the importance of Salmon. In: Cosens B (ed) The Columbia River Treaty revisited: transboundary river governance in the face of uncertainty. Oregon State University Press, Corvallis
Peery C (2012) The effects of dams and flow management on Columbia River ecosystem processes. In: Cosens B (ed) The Columbia River Treaty revisited: transboundary river governance in the face of uncertainty. Oregon State University Press, Corvallis
Raadgever G, Mostert E, Kranz N, Interwies E, Timmerman JG (2008) Assessing management regimes in transboundary river basins: do they support adaptive management? Ecol Soc 13(1):14 [online] URL: http://www.ecologyandsociety.org/vol13/iss1/art14/. Accessed 26 Aug 2017
Rogers P, Hall A (2003) Effective water governance. Global Water Partnership Technical Committee (TEC) background paper 7. Global water partnership, Stockholm. http://www.gwp.org/Global/ToolBox/Publications/Background%20papers/07%20Effective%20Water%20Governance%20%282003%29%20English.pdf. Accessed 23 July 2017
Shurts J (2012) Rethinking the Columbia River Treaty. In: Cosens B (ed) The Columbia River Treaty revisited: transboundary river governance in the face of uncertainty. Oregon State University Press, Corvallis
Stewart I, Cayan D, Dettinger M (2005) Changes toward earlier streamflow timing across western North America. J Clim 18:1136–1155
Thomas-Morse C (2012) When courts run regulated rivers: the effects of scientific uncertainty. In: Cosens B (ed) The Columbia River Treaty revisited: transboundary river governance in the face of uncertainty. Oregon State University Press, Corvallis
U.S. Army Corps of Engineers (USACE), Bonneville Power Administration (BPA), and the United States Bureau of Reclamation (BoR) (1993, 1995, 2000, 2004, 2008, 2014) Federal Columbia river power system biological opinion. http://www.westcoast.fisheries.noaa.gov/fish_passage/fcrps_opinion/federal_columbia_river_power_system.html. Accessed 23 July 2017
U.S. Army Corps of Engineers (USACE), the Bonneville Power Administration (BPA) and the United States Bureau of Reclamation (BoR) (2001) The Columbia river system: inside story. https://www.bpa.gov/power/pg/columbia_river_inside_story.pdf. Accessed 23 July 2017
U.S. Army Corps of Engineers and Bonneville Power Administration (n.d.) Columbia River Treaty: 2012/2024 review. https://www.crt2014-2024review.gov/. Accessed 23 July 2017
U.S. Army Corps of Engineers and Bonneville Power Administration (2013) U.S. entity regional recommendation for the future of the Columbia River Treaty after 2014. http://www.crt2014-2024review.gov/Files/Regional%20Recommendation%20Final%2013%20DEC%202013.pdf. Accessed 23 July 2017
U.S. Department of Agriculture (USDA) (2015). National hop report. https://www.nass.usda.gov/Statistics_by_State/Regional_Office/Northwest/includes/Publications/Hops/Nat%20Hop%20Rept%202015.pdf. Accessed 23 July 2017
U.S. Department of Agriculture (USDA). (n.d.) National agricultural statistics service. https://www.nass.usda.gov/. Accessed 23 July 2017
U.S. Fish and Wildlife Service (2000) Findings and recommendations on issuance of an incidental take permit (TE034609-0) to Plum Creek Timber Company for their native fish habitat conservation plan for states of Montana, Idaho, and Washington
Universities Consortium on Columbia River Governance (UCCRG). (n.d.) http://www.columbiarivergovernance.org/. Accessed 23 July 2017
Vogel E (2012) Can an international treaty strengthen a region and further social and environmental inclusion? Lessons from the Columbia River Treaty. In: Cosens B (ed) The Columbia River Treaty revisited: transboundary river governance in the face of uncertainty. Oregon State University Press, Corvallis
Volkman J, McConnaha W (1993) Through a glass, darkly: Columbia River salmon, the Endangered Species Act, and adaptive management. Environ Law 23:1249–1272

Waples R, Beechie T, Press G (2009) Evolutionary history, habitat disturbance regimes, and anthropogenic changes: what do these mean for resilience of Pacific salmon populations? Ecol Soc 14(1):3. [online] URL: http://www.ecologyandsociety.org/vol14/iss1/art3/. Accessed 26 August 2017

White R (1995) The organic machine: the remaking of the Columbia River. Hill and Wang, New York

Wittstock L, Salinas E A (n.d.) Brief history of the American Indian movement. https://www.aimovement.org/ggc/history.html. Accessed 27 July 2017

Legal Documents

Great Lakes-Saint Lawrence River Basin Sustainable Water Resources Agreement. Dec. 13, 2005

Great Lakes-St. Lawrence River Basin Water Resources Compact. Pub. L. No. 110–342, 122 Stat. 3739 (2008)

Treaty Between the United States of America and Canada Relating to Cooperative Development of the Water Resources of The Columbia River Basin, U.S.-Can., Jan. 17, 1961, 15.2 U.S.T 1555 (CRT)

Listings of Anadromous Fish Populations

Code of Federal Regulations 50: § 223.102 (2013) (listing salmon species found in the Columbia Basin: Snake River Sockeye (endangered), Upper Willamette River Chinook (threatened), Lower Columbia River Chi-nook (threatened), Upper Columbia River spring-run Chinook (endangered), Snake River fall-run Chinook (threatened), Snake River spring/summer-run Chinook (threatened), Lower Columbia River Coho (threatened), and Columbia River Chum (threatened))

Endangered and Threatened Species: Final Protective Regulations for Threatened Upper Columbia River Steelhead, Federal Register 71: 5177, 5178 (Feb. 1, 2006);

Note that four evolutionarily significant units (ESUs) of steelhead are also currently listed. See Endangered and Threatened Species: Proposed Listing Determinations for 27 ESUs of West Coast Salmonids, Federal Register 69: 33102, 33105 (June 14, 2004);

See also Status of ESA Listings & Critical Habitat Designations for West Coast Salmon & Steelhead, NAT'L OCEANIC & ATMOSPHERIC ADMIN. FISHERIES (updated Oct. 31, 2012),

Judicial Opinions

Bonnichsen v. U.S. 367 F.3d 864 (9th Cir. 2004)

U.S. v. Washington. 20 F.Supp.3d 986, 1013 (W.D Washington 2013)

U.S. v. Winans, 198 US 371 (1905)

United States v. Washington, 384 F.Supp. 312 (D.Wash., 1974), aff'd by 520 F.2d 676 (9th Cir. 1975), cert denied Washington v. U.S. 423 U.S. 1086 (1976)

Washington v. Washington State Commercial Passenger Fishing Vessel Ass. 443 U.S. 658, 685 (1979)

Laws

Pacific Northwest Electric Power Planning and Conservation Act, (Northwest Power Act), Pub. L. No. 96–501, 94 Stat. 2697

Chapter 5
Social, Legal, and Ecological Capacity for Adaptation and Transformation in the Everglades

Lance Gunderson, Ahjond S. Garmestani, Keith W. Rizzardi, J. B. Ruhl, and Alfred R. Light

Abstract The Florida Everglades is a subtropical wetland in the southeastern USA that has been hydrologically modified to protect urban and agricultural lands from flooding, while supplying water for urban interests and Everglades National Park. The social-ecological system has sought control over the surface water through infrastructure of canals, levees, and pumps to direct water movement, managed by a complex institutional arrangement of federal, state, and local governments. Water control has been largely achieved through adaptation and transformation to unforeseen environmental events, too much or too little rainfall and nutrient movement across the wet landscapes. Law has provided both the resources to foster economic and conservation objectives but also resulted in rigid planning and continuing litigation, constraining the adaptive capacity of the social-ecological system of the Florida Everglades to respond to seen and unforeseen environmental changes. Rigid management at higher levels and failure to balance stability of economic investment and varied stakeholder values of the Everglades with flexibility to adjust management measures have limited the emergence of adaptive governance.

Keywords Resilience · Adaptive governance · Adaptive law · Resilience assessment · Climate change · Water management

---

L. Gunderson (✉)
Department of Environmental Sciences, Emory University, Atlanta, GA, USA
e-mail: lgunder@emory.edu

A. S. Garmestani
School of Natural Resources, University of Nebraska, Lincoln, NE, USA

K. W. Rizzardi · A. R. Light
St. Thomas University School of Law, Miami Gardens, FL, USA
e-mail: krizzardi@stu.edu; alight@stu.edu

J. B. Ruhl
Vanderbilt University Law School, Nashville, TN, USA
e-mail: jb.ruhl@vanderbilt.edu

B. Cosens, L. Gunderson (eds.), *Practical Panarchy for Adaptive Water Governance*, https://doi.org/10.1007/978-3-319-72472-0_5

5.1 Introduction

The Everglades is an internationally recognized, subtropical wetland ecosystem, described as a River of Grass (Douglas 1947). Conservation of the unique biota has been and continues to be a social priority (Gunderson and Loftus 1993). A mild, subtropical climate also contributes to a tourist economy which attracts visitors from all over the world to the unique cultural and ecological systems of South Florida. Other areas of the Everglades support globally linked agricultural production of sugar and vegetables (Snyder and Davidson 1994). Since much of South Florida is a wetland ecosystem, the current social-ecological system is enabled by a collective capacity to control surface waters (Blake 1980).

The large, complex water management system of the Everglades has changed, adapted, and evolved since the late 1800s (Light and Dineen 1994; Light et al. 1995). Driven by the social objectives of draining for purposes of agriculture and human settlement, human activities focused on controlling the surface waters of the wetland. The history of water management did not unfold in a linear, gradual, progressive fashion, but rather happened in abrupt, disjunctive, and unpredictable transitions from one era to another (Gunderson et al. 2002). The quest for water control was driven by the application of human technologies that altered key biophysical processes to facilitate land uses of agriculture and biodiversity conservation. These actions were successful but led to unpredictable ecological changes. The history of the Everglades is also described by infrequent, large-scale events such as storms or extreme precipitation that overwhelmed the system's capacity to control, contain, or manage water, thereby leading to new ecological or institutional regimes.

This chapter attempts to assess the ecological resilience (sensu Holling 1973) and adaptive capacity of the South Florida social-ecological water system. The first section of the chapter describes the social-ecological system of the Everglades, highlighting the ecologic characteristics and governance structures. That is followed by a brief history of the dynamics in the Everglades SES during the twentieth century, indicating the ebb and flow of resilience which led to regime shifts in the ecological and social regimes, triggered by shifts in social values and environmental crises. The role of law in adaptive and transformative changes is then described, followed by an assessment of the adaptive capacity (the ability of the system to manage regime shifts) of the current governance. The final section describes obstacles to and opportunities for transformation to escape the current rigidity trap of the Everglades SES.

5.2 The Everglades Social-Ecological System

Unlike the riverine systems in the other chapters in this volume, the Everglades is a large freshwater wetland system. Wetland ecosystems are defined by sufficient inundation that saturates and modifies soils and supports plant and animal species adapted to intermittent surface flooding. The Everglades wetland is characterized by annual cycles of flooding, when water covers the land surface then recedes leading to extended dry periods. The cycles of flooding and drying are a result of a flat topography and climate. South Florida has a tropical savanna climate (Hela 1952). This type of climate is characterized by temperatures that are mild during the winter months and hot during the summer months. The savanna climate is also marked by distinct dry and wet seasons. Most (80%) of the annual rain falls during the wet summer season (Ali et al. 2000). Tropical storm occurrences (including hurricanes) tend to peak in September. Less precipitation occurs during fall, winter, and spring months, and the wetlands continue to dry until the onset of the summer monsoon (Duever et al. 1994). The area receives an average annual rainfall of over 5 ft, with historic ranges from 3 to 10 ft of precipitation.

The Everglades is the southern part of a larger hydrologic system, which includes the Kissimmee River basin, Lake Okeechobee basin, and the Everglades (Fig. 5.1). The abundant rainfall collects in the Kissimmee River basin and flows south into the lake. Lake Okeechobee is a large, shallow lake and collects and stores freshwater. Historically, above certain levels, the Lake would overflow into the Everglades wetland. Water from upstream and local rainfall would fill up the Everglades annually, resulting in slow, overland flow through the wetlands. The overland, surface water would move into the estuarine regions prior to entering Florida Bay or exit through small rivers along the eastern ridge of the Everglades. Evapotranspiration is a major pathway of water loss from the wetland, nearly equaling the precipitation on an annual basis.

Since the late 1800s, people have attempted to command and control the water resources of South Florida for meeting goals of flood protection and water supply. These goals have been achieved through construction and operation of a massive system of levees, canals, pumps, spillways, and other structures and adopting a complex set of operating rules that are implemented by government agencies at the local, state, and federal level. That system of water control has enabled dramatic development of urban and economic development along the southeastern coast. Currently, about eight million people reside in the watershed and depend upon this large system for water supply and flood control, as do a viable agricultural community and a thirsty environment. The allocation of water among urban, agricultural, and environmental sectors has as rich and disputed a history as the water management system itself.

The social components involved in Everglades SES water management include both formal and informal organizations that cover scales from local to national in a polycentric arrangement (Gunderson et al. 2014). The design and construction of the water management structures have largely been the purview of the US Army Corps of Engineers. Other federal agencies, such as the US National Park Service and Fish and Wildlife Service (FWS), manage Everglades National Park and Arthur Marshall National Wildlife Refuge, respectively. The FWS also administers the Endangered Species Act. The state of Florida manages the hydrologic aspects of the SES, largely through the South Florida Water Management District. Other state agencies, such as the Florida Fish and Wildlife Commission and Department of

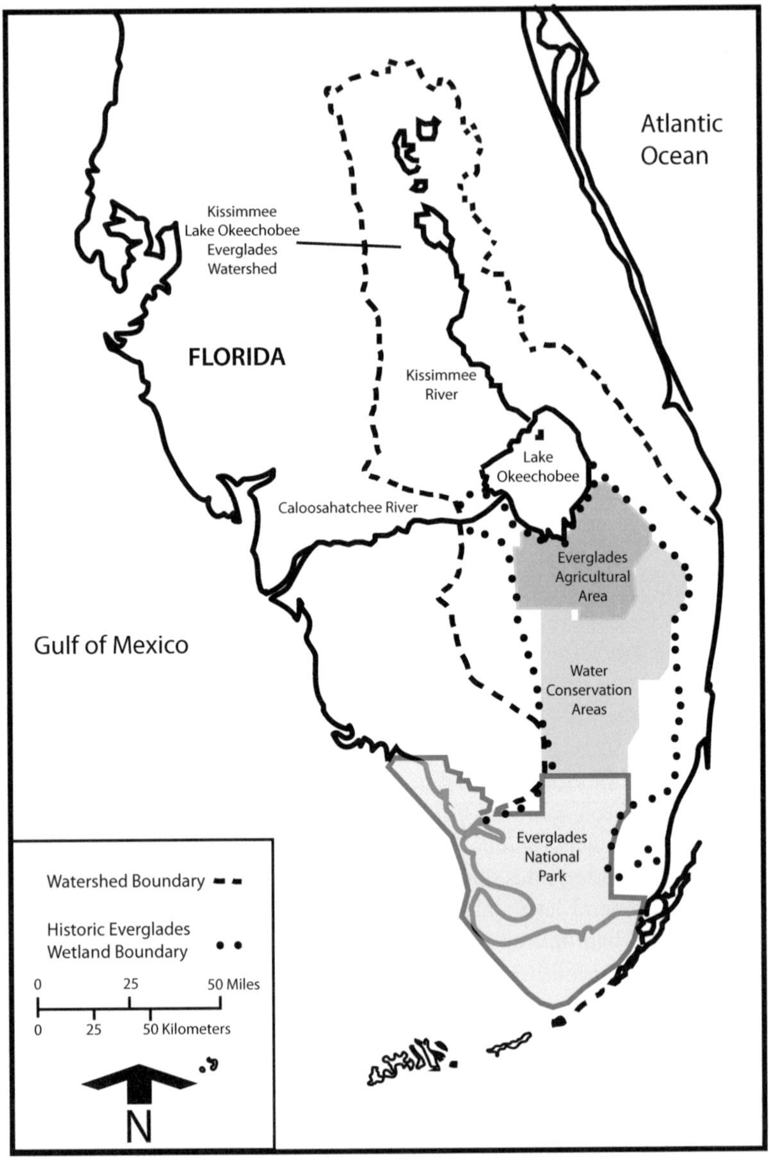

Fig. 5.1 Outline of watershed boundary in southern Florida, showing the Everglades wetlands, Kissimmee River, Lake Okeechobee physiographic regions

Environmental Protection, are active in ongoing management. Local governments include counties and some city-based water supply and conservation groups. The Miccosukee tribe is also engaged in management issues, as are nongovernmental environmental groups, such as the Everglades Foundation, Audubon Society, and many others. There are epistemic groups associated with universities and the National Academy of Sciences.

5.3 System Dynamics: A History of Adaptation and Transformation

Understanding the historical development of the South Florida water management system provides insight into the resilience and adaptive capacity of the SES. Historical shocks or perturbations in the Everglades water management SES include unforeseen floods, droughts, and water pollution. Many of these shocks were viewed as environmental crises, when the ecosystem behaved in a surprising or unexpected manner, and usually signaled a failure of extant policy. Light et al. (1995) presented a pattern of development of the South Florida water management system in which environmental crisis led to shifts in the management regime. That is, the resilience of the system was exceeded, leading to a transformation into an alternative system regime (Table 5.1). The following paragraphs describe how the SES transformed or adapted in response to three categories of events that tested the resilience of the SES: (1) too much rainfall (floods), (2) too little rainfall (droughts), and (3) unclean water (pollution episodes).

During the twentieth century, flood events in 1905, 1928, 1947, and 1983 triggered social-ecological transformations (Table 5.1). Responses to the early flood events involved building a series of canals to remove excess water from the northern and eastern Everglades (Blake 1980). Flooding from hurricanes in 1928 led to federal government funded construction of the Herbert Hoover Dike around Lake Okeechobee. Wet years in 1947–1948 led to passage of the federal Flood Control Act in June 1948 (Flood Control Act 1948). The Act authorized the US Army Corps of Engineers to develop a plan known as the Central and Southern Florida Project for Flood Control and Other Purposes. The plan called for the protection of the east coast and agricultural areas from flooding and to provide recharge of regional aquifers in order to prevent saltwater intrusion. In 1949, the state legislature created the Central and Southern Florida Flood Control District (FCD) to act as local sponsors for the federal project. In 1977, the FCD was renamed the South Florida Water Management District (SFWMD), at which time other objectives of managing water supply and enhancing environmental resources were added to the agency's mission (Hamman 2005).

A series of droughts prompted reexamination of the rules for allocation among users in the system. In 1962, a drought prompted federal scientists to question the security of water deliveries to Everglades National Park (Tarlock 1987). The conflict over allocation was resolved by the passage of the River Basin Monetary Authorization and Miscellaneous Civil Works Amendments Act of 1970 (River Basin Act 1970), which assured the park a minimum flow of water. Droughts of the late 1960s and early 1970s led to sweeping reforms in state legislation with the passage of the Florida Water Resources Act of 1972 (Florida Water Resources Act 1972) codifying a reasonable-beneficial use doctrine that anticipated severe droughts requiring water restrictions (Maloney et al. 1972). In the same act, the Florida Legislature established five water management districts around the state, each based on hydrologic boundaries. A nine-member board governs each district, with each

Table 5.1 A summary of historical regime shifts in the Everglades SES since 1900

Year	Trigger/shock	Policy	Infrastructure	Organizations	Law
1905	Flood 1903	Drain for agriculture and development	Canals	Everglades Drainage District	Unknown
1932–1938	Hurricane 1928	Protect against Lake Okeechobee floods	Hoover Dike	US ACOE Drainage District	River and Harbor Act
1948	Flood 1947	The system was design and built to prevent flooding of agricultural and urban areas. The land uses of Everglades Agricultural Area (EAA) and Water Conservation Areas were defined. The Kissimmee River channelization began	Levees, canals, pumps	Central and Southern Florida Flood Control District USACOE	River and Harbor Act
1970	Drought 1962, 1966	Minimum flow to Everglades Park	Modify S-12 gates, regulation schedule	USACOE	River Basin Monetary Authorization and Miscellaneous Civil Works Amendments Act
1977	Drought 1972	Water supply, water quality	Saltwater dams, water supply to urban aquifers	South Florida Water Management District (SWFMD) created	Water Resources Act
1982	Algal blooms in Lake Okeechobee	Changes to water schedules, rules for delivery	Cessation of use of pumps to move water to Lake Okeechobee	SFWMD/ Florida Department of Environmental Protection	State Permit
1983	ENSO Flood	Open flow to ENP restore hydrology	Adjust delivery rules	SFWMD Everglades Coalition	Supplemental Appropriations Act
1985	ENSO Flood	Rainfall plan, expansion of Everglades National Park	Adjust water delivery	SFWMD USA COE DOI NPS	Everglades National Park Protection and Expansion Act

(continued)

Table 5.1 (continued)

Year	Trigger/shock	Policy	Infrastructure	Organizations	Law
1989	Chronic environmental losses, ENSO Flood, Lawsuit	Seek integrated water quantity and quality. Feasibility of ecosystem restoration	Delivery across Tamiami Trail, rather than through existing S-12 gates.	Epistemic Collaborative Group, AEAM workshops	US v Florida lawsuits
2000	Restudy	Comprehensive Everglades Restoration Plan (CERP)	STA, Aquifer recharge	USACOE/ SFWMD	Water Resources Development Act (1999)

The table is organized by a triggering event or disturbance to the SES, which led to an adaptation (modification of policy) or transformation (change in infrastructure, organizations, and/or laws). For each event, the adaptation is indicated by the year that a change in the rules and norms occurred. A transformation occurred when the perturbation led to changes in physical infrastructure, organizations, and law (Gunderson et al. 2014)

member appointed by the Governor. The districts have authority over almost all aspects of freshwater notwithstanding the estuaries and certain water quality responsibilities (Hamman 2005). They oversee four principal functions: water supply, flood control, water quality, and natural systems. The districts allocate permits to municipalities for withdrawal, treatment, and distribution. All withdrawals for agricultural, residential, industrial, and commercial developments require permits that are managed by the districts. The districts have been characterized as innovative and responsive organizational structures for water management (Scholz and Stifte 2005; Gunderson et al. 2006).

In the 1980s, the degradation of water quality became the major environmental issue in South Florida. Recurring algae blooms in Lake Okeechobee were attributed to nutrient runoff from cattle farms to the north of the lake and from crop and agriculture to the south (Aumen and Wetzel 1995). With the water quality degradation in the lake, water was no longer pumped from agricultural fields to the lake. Instead it was diverted south into the conservation areas, which led to regime shifts in the wetland ecosystems (Davis and Ogden 1994). By 1988, the US federal government filed a lawsuit against the state of Florida alleging that the state had failed to enforce the state's water quality laws that implemented the federal Clean Water Act, damaging federal resources (John 1994).

Since 1990, a major focus of management in the Everglades has been ecosystem restoration (Davis and Ogden 1994). This effort began with a modest attempt by scientists to synthesize existing information in ways that would help to solve chronic environmental issues, such as decline in wading bird populations, vegetation changes, and changes in aquatic biota, among others. The Davis and Ogden (1994) volume led to a number of formal planning processes such as the US Army Corps of Engineers' Restudy of the Central and Southern Florida Flood Control Project, a state-federal taskforce, and the Governor's Commission for a Sustainable South

Florida, which culminated in the passage of the Everglades Restoration Act in 2000 by Congress. A key goal of the restoration program was to restructure the timing, distribution, and magnitude of water flow to Everglades National Park while still meeting societal objectives of flood control, water supply, and nutrient abatement. The outcomes and practices of this long-term project will be discussed in the subsequent section on adaptive governance.

5.4 The Role of Law in Adaptation and Transformation

The role of law and litigation is varied in the Everglades and can be linked to the system phase of the panarchy model (Gunderson and Holling 2002, Chapter 1). New laws, such as the Flood Control Act (Flood Control Act of 1948) and the Water Resources Acts of 1972, 1992, or 2000, were formed after environmental crises. The laws were crafted in direct response to the reconciling or solving the societal problem such as flood, drought, or water quality crisis. Such laws provided resources from larger scale systems, in the form of money and expertise, and contributed to moving the system from crisis through the reorganization phase and into a new trajectory. As such, they provided stability for collective action (Craig et al. 2017). The role of law in providing cross-scale, legal integration is further developed in Chapter 12 by Gosnell et al. However, litigation and lawsuits can also be viewed as creating instabilities, as depicted by the phase of creative destruction in the panarchy model.

The modern era of Everglades' restoration is characterized by recurring lawsuits (Fumero and Rizzardi 2001). The lawsuit filed in 1988, in which the United States sued the South Florida Water Management District, cited the adverse water quality effects of water management upon Everglades National Park and the Loxahatchee National Wildlife Refuge. One irony of this lawsuit was that the state governmental entity charged with responsibility to operate the regional water system was sued by the federal government for the unforeseen and undesirable consequences of operating the system that the federal government had designed, built, and approved.

The instability caused by the lawsuit served as a critical turning point for the Everglades. It triggered years of multiparty litigation, including affirmative defenses against the US Army Corps of Engineers and disputes over intervention. The settlement agreement was eventually codified as a court-approved federal consent decree. Part of the agreement was that in return for Flo-Sun Land Corporation's agreement to reduce phosphorus flows in the Everglades by complying with the emerging regulatory program, the United States agreed not to sue the sugar company for a period of 10 years, and the court upheld the agreement. Ultimately, the principles of these settlement agreements were codified by state law in the Everglades Forever Act (EFA) of 1994 (EFA 1994). But the consent decree and EFA, rather than sculpting a new vision for the Everglades, simply introduced a new tool that prompted more litigation. Waves of regulation and litigation have relentlessly modified, slowed, or even stopped the restoration progress.

When the South Florida Water Management District, Florida Department of Environmental Protection, and US Department of Justice finally agreed to those historic settlement terms, interested stakeholders quickly sought administrative review of the agency decisions. At the time, the court concluded that administrative review of the settlement was premature. Instead, the court concluded that additional scrutiny would be afforded when the agency took actions to implement the agreement. The scrutiny has never stopped. To begin with, despite the state's investment of $1.8 billion, construction of 57,000 acres of treatment marshes, treatment of more than 1700 tons of phosphorus, and regulation of 640,000 acres of agricultural lands, departures from water quality requirements continue to occur. Indeed, 20 years after the historic settlement, the US Department of Justice continues to pursue enforcement actions against the South Florida Water Management District for consent decree violations.

A corollary state statute, the Everglades Forever Act as previously mentioned, was supposed to be the blueprint for Everglades' restoration. Instead, it too became an independent source of decades of litigation. Permits issued pursuant to the EFA, which recognized the near-term and long-term "schedules and strategies" at issue, have been upheld by Florida courts. The establishment of ten parts per billion of phosphorus as the numeric interpretation of a narrative "no imbalance of flora and fauna" standard also proved time consuming and controversial, albeit ultimately defensible. Yet actual compliance with the water quality standards in the ecosystem has proven difficult, and delays in construction of the storm water treatment areas and other considerations eventually led the state legislature to modify the deadlines in the EFA. The amendments spawned yet another lawsuit. Concerned that the amended EFA was inconsistent with the original consent decree, the court appointed a special master to supervise the process and, later, to determine appropriate remedies.

Through settlement and legislative negotiations, the executive and legislative branches of the state and federal government labored intensely to plan for a sustainable Everglades. But despite the good intentions of the Everglades Consent Decree and Everglades Forever Act, the ecological system continues to surprise. Floods and droughts influence water quality compliance, and the 1800 miles of canals sprawling throughout the entire Central and South Florida Flood Control system—built more than 50 years ago—were not designed with water quality compliance in mind. In the compartmentalized and channelized Everglades ecosystem, sustainability is a difficult task, even under the best conditions.

Single-purpose regulatory programs, like the Clean Water Act (CWA), Endangered Species Act (ESA) or process requirements of the National Environmental Policy Act (NEPA), and the Federal Advisory Committee Act (FACA), become platforms for stakeholders to intervene in the restoration process to tweak it in their favor or block aspects they perceive as counter to their interests. Instead of promoting ecological and social resilience, these laws and programs fragment the SES into issue pigeonholes, such as water quality, species habitat, flood control, and frustrate interagency coordination.

Unsatisfied with the consent decree and EFA process, tribes and other environmentally minded advocates have frequently used the CWA to require even more stringent requirements to be imposed in the Everglades. For example, they challenged the EFA as a change in water quality standards, one that must be reviewed and approved by the US Environmental Protection Agency. Litigation over this aspect of the CWA has been ongoing for more than a decade, leaving continuous uncertainty over the legality of the blueprint for Everglades restoration. Similarly, stakeholders have argued over whether various water management structures throughout the Everglades, which move water, require National Pollutant Discharge Elimination System (NPDES) permits. Litigation over this point has reached the United States Supreme Court, necessitated controversial new federal rulemaking, and once again, has spread over more than a decade. Complicating matters even more, the state of Florida is also implementing additional "numeric nutrient criteria" and Total Maximum Daily Load (TMDL) programs that create new CWA-based water quality requirements for the canals and water bodies of South Florida (Norgart 2004). Those requirements can only complicate the abundant choices and challenges already facing the water managers who manage the Everglades ecosystem.

In fact, NEPA mandates careful consideration of federal actions, and that process creates yet another way for Everglades stakeholders to demand that their preferred alternatives be considered. For example, the lack of an environmental impact analysis, and a NEPA challenge, was used by agricultural interests to challenge the original Everglades Consent Decree. Similarly, when water managers sought to elevate the Tamiami Trail to allow waters to flow underneath the bridge, a NEPA case brought by the Miccosukee tribe was only stopped when Congress passed an appropriations rider demanding the project be completed, "notwithstanding any other law."

The ESA also provides a constant source of controversy in the Everglades. Arguing over the effects of the Everglades restoration on endangered Cape Sable seaside sparrow and the threatened Everglades snail kite, stakeholders have frequently used ESA to second-guess water management decisions. Fights over the Florida panther, thus far, have led to opinions upholding the federal decisions. Yet even when the plaintiffs succeed in court, the influence of those victories has been subtle, at best. A battle over the supplementation of an administrative record, or even a remanded biological opinion requiring the United States Fish and Wildlife Service to calculate a precise number of sparrows that could acceptably and incidentally be harmed, killed, or otherwise "taken," might create new procedural burdens for the agencies but accomplishes little in clarifying how to manage the entirety of the Everglades ecosystem. As both the United States Fish and Wildlife Service and the federal courts have recognized, the choice to offer beneficial protection for a single species can have detrimental consequences for many others.

The legislative history of the Central and Southern Florida Project refers to the Everglades as "a single watershed." Recognizing the complexity of managing an ecosystem, state and federal bureaucrats undertook a renewed comprehensive planning effort during the 1990s and early 2000s, seeking to build on the goals set forth in the Everglades Forever Act, focusing on a sustainable SES. Governor

Chiles' Commission for a Sustainable South Florida envisioned its restoration effort as directly supporting a "sustainable South Florida economy and quality communities." The planning efforts eventually led to the restoration-oriented Water Resources Development Act of 1996 "for the purpose of restoring, preserving, and protecting the South Florida Ecosystem" (Water Resources Development Act 1996).

Implementation of that Comprehensive Everglades Restoration Plan (CERP), however, just like the waters of the Everglades, has moved slowly. The USACOE 2003 Programmatic Regulations required the agency to develop the CERP adaptive management program. Instead of helping restoration, the regulations helped to stop it. Frustrated with the slow pace of federally funded projects, the South Florida Water Management District elected to construct a reservoir in the Everglades Agricultural Area (EAA) with state funds, unconstrained by the Corps' regulations. The Natural Resources Defense Council disagreed and challenged the project permits. Rather than risking liquidated damages on construction contract claims, the South Florida Water Management District stopped the project, and the court eventually declared the lawsuit moot.

Disputes over land acquisition also helped to delay CERP and similar restoration planning initiatives. For example, expansion of the boundaries of Everglades National Park was disputed, as efforts to acquire the 8.5 square mile area built in the historic Everglades, west of the protective levee system, were challenged. The state's unprecedented effort to buy the land holdings of the United States Sugar Corporation offered a particularly interesting series of ultimately unsuccessful lawsuits, including allegations of Government in the Sunshine violations by decision-makers, tribal demands to resume construction of the EAA reservoir instead, and New Hope Sugar's lawsuit to stop the purchase of lands. While much of this land acquisition proceeded anyway, the litigation increased costs, created delays, and frequently, forced modifications to the comprehensive planning efforts.

Battles over environmental policy have also been fought in a context not specifically related to environmental law. The Miccosukee tribe has repeatedly used leasehold rights, Indian trust doctrine, due process, and equal protection claims to oppose United States Army Corps water management efforts, albeit with limited success. Former Florida Governor Claude Kirk unsuccessfully sued the sugar industry, making allegations of campaign-contribution induced conspiracies to allow a continued public nuisance. The Supreme Court of Florida dodged the dispute, applying the doctrine of primary jurisdiction and deferring to the administrative agencies, empowered by Florida law to manage air and water pollution. And a particularly determined group of riparian landowners living adjacent to the Central and South Florida Flood Control Project sought $50 million in compensation for an alleged physical taking of their riparian rights. The lawsuits backfired on the waterfront plaintiffs, leading the federal court to conclude that Florida law did not establish a riparian right to be free from pollution.

Sometimes, entirely procedural arguments associated with the operation of the bureaucracy, wholly unrelated to environmental considerations, have also been used in efforts to alter plans in the Everglades (Gunderson et al. 2014). Violations of the Federal Advisory Committee Act led to a lawsuit and injunction that altered the cooperation of a group of state and federal representatives on managing water levels in the Southern Everglades, but did not succeed in preventing the Army Corps from obtaining expert feedback on wildlife biology from a non-federal conflict resolution group. The content and timeliness of responses to public records requests pursuant to the federal Freedom of Information Act and to state laws have also been the source of litigation.

Finally, even the cost of the Everglades restoration, and who pays for it, became just as controversial as the restoration itself. Citizens have tried to eliminate taxes paid by allegedly nonpolluting parties, relying upon a citizen-passed amendment to the Florida Constitution to demand that the polluters must pay more to fund the Everglades restoration. The litigation proved, once again, to be a distraction. Courts deferred to the legislature and found the constitutional clause not to be self-executing. Challenges to plans to pay for the Everglades restoration using bonds were also unsuccessful. Yet money remains a central force in Everglades litigation; lawsuits must be financed and paid for, too, so attorney's fees routinely become yet another source of litigation, to which staff and restoration dollars must be diverted.

5.5 Adaptive Governance: Integrating Ecology, Organizations, and Law

Adaptive governance incorporates formal organizations, informal groups, and individuals at multiple scales and requires collaboration, communication, and adaptation in response to social and ecological monitoring (Folke et al. 2005; Chaffin et al. 2014). At the essence of many definitions of adaptive governance is the capacity to anticipate, manage, and adapt to ecological regime shifts (Garmestani and Benson 2013; Olsson et al. 2008). As such, governance structures must incorporate scientific and technical understanding of different types of resilience practices (Cosens et al. 2017). These practices fall into categories of (1) maintaining resilience for desired system configurations, (2) intentionally changing ecosystem regimes, and (3) developing transformational capacity. This definition includes the idea that adaptive governance is a framework that allows for adaptive management of natural resources (Gunderson and Light 2006). Law, policy, and intermediaries (e.g., bridging organizations) are also important aspects of adaptive governance, producing networks that can increase political and financial support critical for fostering adaptive management. Intermediaries foment the development of new ideas, facilitate communication between entities, and create the flexibility necessary for the interplay of ecological and social systems for successful environmental governance. Adaptive governance is inclusive of informal organizations, stakeholders, and other relevant actors who participate with formal organizational structure (Olsson et al. 2006; Bruner et al. 2005). Additionally, adaptive governance must include considerations of structure, scale, adaptive capacity, legitimacy, and power (Chaffin et al. 2014).

Adaptive capacity in social systems is characterized by open and frequent lines of communication between both formal and informal entities at multiple scales. The generation of adaptive capacity in management entities is necessary insurance for sustainability (Gunderson 1999). Adaptive capacity refers to the ability of the Everglades SES to respond to ecological regime shifts (Gunderson et al. 2002; Pahl-Wostl 2009).

Identification of adaptive capacity includes both evidence of social learning and the authority to experiment and adapt. Hence, adaptive capacity of a system involves the ability of the governance system to undertake and execute programs of adaptive management (if appropriate; see Allen et al. 2011). For the Everglades, the application of adaptive management is indeed authorized for the restoration program. This requires the capacity to integrate management actions that are structured as much for learning as for achieving social goals and the capacity to monitor appropriate ecological indicators, to evaluate how systems respond to management actions, and to provide pathways and repositories for knowledge and experience. While there is some debate as to the level of experimentation necessary for management to be adaptive, we suggest that many factors create a barrier to adaptive management in the Everglades (Allen and Gunderson 2011). The large experiments that are necessary to test hypotheses of restoration have yet to be conducted (Gunderson and Light 2006). This is evidence of limited adaptive capacity and little or no adaptive governance. Similar arguments apply to necessary experiments needed for adapting to climate change.

There appears to be sufficient authority in the federal and state laws to manage adaptively with respect to resilience (LoSchiavo et al. 2013). Legislation, such as the Water Resources Development Act of 1999, directly defines the social values of a desired ecological regime. In this case, a restored state would be indicated by stable populations of key endangered species, lack of nutrient transformed vegetation, restoration of ecological processes of water flow, nutrient cycling, and landscape disturbances (Water Resources Development Act 1999). A larger number of nesting wading birds as well as the absence of keystone exotic species would also define a restored condition.

5.6 Obstacles and Opportunities for Adaptive Capacity

At the beginning of the twentieth century, Florida Governor Napoleon Bonaparte Broward promised a grand "Empire of the Everglades" (Blake 1980). Anticipating an abundance of agricultural crops, his engineer declared that "it will be impossible to form or declare an adequate idea of the importance and extent of this enterprise." Congress, too, envisioned the Everglades as a vast, interconnected SES when it enacted the flood control programs in the 1940s. Now, after more than a century of executive and legislative efforts to manage the balance between agriculture and the environment and water supply and water quality, we have learned that Broward's engineer was right but unintentionally so. In the Everglades, it has indeed proven impossible to form an adequate idea; instead, the judiciary has swamped the Everglades restoration, and a proliferation of litigation under narrowly focused statutory schemes and provisions has relentlessly altered plans for comprehensive restoration. Even in a world facing ongoing climate change, this dynamic will continue, for better or worse.

Ultimately, for all these reasons, the goal of a comprehensively and adaptively managed SES in the Everglades, benefitting both agriculture and alligators, has been continuously thwarted by the directional shifts caused by legislators, litigators, and the legal system. Indeed, a summary of the many ways that litigation has altered the Everglades restoration contains references to more than 80 cases (see Gunderson et al. 2014). Any one case can alter decades of planning, and every case—even the meritless ones—can bring publicity that still succeeds in undermining public confidence in the overall policy objectives. But in the future, as the forces of climate change reshape the Everglades, countless decisions will be subjected to stakeholder and judicial second guessing (Light 2008). Sea level rise could transform Everglades National Park, so decisions to spend billions of dollars to clean up phosphorus for a freshwater system seem suspect when the ecosystem is at risk of becoming an estuary or salt marsh, yet preserving the Everglades might actually be essential to protecting South Florida. Rising temperature could accelerate evapotranspiration, affecting the entire strategy of using reservoirs to store and treat water. Patterns of rainfall, floods, and drought will change, and the engineering assumptions for the drainage system could all prove fundamentally flawed in an era where past statistical highs and lows are no longer predictive of future results. And at any moment, the huge effects of tropical storms and hurricanes can cause catastrophe, especially for the Lake Okeechobee dike, part of America's most vulnerable infrastructure.

The ongoing Everglades restoration process has been slow to implement an adaptive management program (LoSchiavo et al. 2013) and bereft of any large-scale experimentation (Gunderson and Light 2006). Perhaps one reason is that experimentation is too risky because failure is unacceptable. Even small-scale experimentation to restore sheet flow has been stopped by litigation because of the fear of unwanted ecosystem changes associated with phosphorus-laden water. Another reason has to do with barriers to adaptive governance. One barrier lies with restoration coordination that involves federal agencies who provide knowledge and resources but still must comply with CWA or ESA laws. In other areas, such as the Grand Canyon Adaptive Management Program (Zellmer and Gunderson 2009), federal groups engaged nongovernmental groups to facilitate coordination by establishing FACA committees for technical and policy guidance. Perhaps as a legacy of lawsuits or lack of leadership across groups involved in governance, such coordinating committees have not had the legitimacy and trust needed for experimentation in the Everglades (Cosens et al. 2017).

Neither the Everglades Consent Decree nor the Everglades Forever Act was written with climate change in mind. Eventually, these problems will become the subject of even more Everglades litigation, using laws like the Clean Water Act, Endangered Species Act, National Environmental Policy Act, and countless others named above and not yet even thought of. But in an era of a rapidly changing climate—a world full of competing stakeholder opinions—the Everglades Forever Act, and its goal of a restored and sustainable Everglades system, may forever be an elusive ideal.

Acknowledgments This work was developed in part under the Adaptive Water Governance Project, funded by the US National Socio-Environmental Synthesis Center (SESYNC) under funding from the US National Science Foundation, NSF DBI-1052875. The views set forth by contributors to this volume represent their own and do not represent the views of any public or private entity the contributor is affiliated with.

References

Ali A, Abtew W, Horn S, Khanal N (2000) Temporal and spatial characterization of rainfall over central and south Florida. J Am Water Res Assoc 36:833–848

Allen C, Gunderson L (2011) Pathology and failure in the design and implementation of adaptive management. J Environ Manag 92:1379–1384

Allen C, Fontaine J, Pope K, Garmestani AS (2011) Adaptive management for a turbulent future. J Environ Manag 92:1339–1345

Aumen N, Wetzel R (1995) Ecological studies on the littoral and pelagic systems of Lake Okeechobee, Florida. Arch Hydrobiol Beih Ergebn Limnol 45:343

Blake N (1980) Land into water-water into land: a history of water management in Florida. University Presses of Florida, Tallahassee

Bruner R, Steelman T, Coe-Juell L, Cromley C, Edwards C, Tucker D (2005) Adaptive governance: integrating science, policy, and decision making. Columbia University Press, New York

Chaffin B, Gosnell H, Cosens B (2014) A decade of adaptive governance scholarship: synthesis and future directions. Ecol Soc 19(3):56. https://doi.org/10.5751/ES-06824-190356

Cosens BA, Craig RK, Hirsch S, Arnold CA, Benson MH, Decaro DA, Garmestani AS, Gosnell H, Ruhl JB, Schlager E (2017) The role of law in adaptive governance. Ecol Soc 22(1):30. https://doi.org/10.5751/es-08731-220130

Craig RK, Garmestani AS, Allen CR, Arnold CA, Birgè H, DeCaro DA, Fremier AK, Gosnell H, Schlager E (2017) Balancing stability and flexibility in adaptive governance: an analysis of tools available in U.S. environmental law. Ecol Soc 22(2):3. https://doi.org/10.5751/es-08983-220203

Davis S, Ogden J (1994) Everglades: the ecosystem and its restoration. St. Lucie Press, Deerfield Beach

Douglas M (1947) The everglades: river of grass. Rinehart & Company, New York

Duever M, Meeder J, Meeder L, McCollom JM (1994) The climate of south Florida and it's role in shaping the everglades ecosystem. In: Davis S, Ogden J (eds) The everglades: the ecosystem and its restoration. St. Lucie Press, Deerfield Beach, pp 225–248

Folke C, Hahn T, Olsson P, Norberg J (2005) Adaptive governance of social-ecological systems. Annu Rev Environ Resour 30:441–473

Fumero, Rizzardi K (2001) Everglades ecosystem: From Engineering to Litigation to Consensus-Based Restoration, The Everglades Symposium Issue, 13 St. Thomas L. Rev. 667

Garmestani A, Benson M (2013) A framework for resilience-based governance of social-ecological systems. Ecol Soc 18(1):9. https://doi.org/10.5751/ES-05180-180109

Gunderson L (1999) Stepping back: assessing for understanding in complex regional systems. In: Johnson K, Herring M, Greene S (eds) Bioregional assessments: science at the crossroads of management and policy. Island Press, Washington, DC, pp 45–61

Gunderson L, Holling C (2002) Panarchy: understanding transformations in systems of humans and nature. Island Press, Washington, DC

Gunderson L, Light S (2006) Adaptive management and adaptive governance in the everglades ecosystem. Policy Sci 39:323–334

Gunderson L, Loftus W (1993) The Everglades. In: Martin W (ed) Biodiversity of the Southeastern United States/Lowland terrestrial communities. John Wiley, & Sons, New York, pp 199–255

Gunderson L, Holling C, Peterson G (2002) Surprises and sustainability: cycles of renewal in the everglades. In: Gunderson L, Holling C (eds) Panarchy: understanding transformations in systems of humans and nature. Island Press, Washington, DC, pp 315–332

Gunderson L, Carpenter S, Folke C, Olsson P, Peterson GD (2006) Water RATs (resilience, adaptability, and transformability) in lake and wetland social-ecological systems. Ecol Soc 11:16

Gunderson L, Garmestani A, Rizzardi K, Ruhl JB, Light A (2014) Escaping a rigidity trap: governance and adaptive capacity to climate change in the everglades social ecological system Idaho Law Rev 51:127

Hamman R (2005) Florida's water management framework. In: Scholz J (ed) Adaptive governance and water conflict: new institutions for collaborative. Earthscan, Washington, DC, pp 15–24

Hela I (1952) Remarks on the climate of South Florida. Bull Mar Sci 2:438–447

Holling C (1973) Resilience and stability of ecological systems. Ann Rev Ecol Syst 4:1–23

John D (1994) Civic environmentalism: alternatives to regulation in states and communities. CQ Press, Washington, DC

Light A (2008) Beyond the myth of Everglades settlement: the need for a sustainability jurisprudence. Tulsa Law Rev 44:253–260

Light A, Dineen J (1994) Water control in the everglades: a historical perspective. In: Davis S, Ogden J (eds) The everglades: the ecosystem and its restoration. St. Lucie Press, Deerfield Beach, pp 47–84

Light S, Gunderson L, Holling C (1995) The everglades: evolution of management in a turbulent ecosystem. In: Gunderson L, Holling C, Light S (eds) Barriers and bridges to the renewal of ecosystems and institutions. Columbia University Press, New York, pp 103–168

Loschiavo A, Best R, Burns R, Gray S, Harwell MC, Hines EB, Mclean AR, St. Clair T, Traxler S, Vearil JW (2013) Lessons learned from the first decade of adaptive management in comprehensive everglades restoration. Ecol Soc 18:70

Maloney F, Ausness R, Morris J (1972) A model water code: text and commentary. University of Florida Press, Gainesville

Norgart C (2004) Florida's impaired waters rule: is there a "method" to the madness? J Land Use Environ Law 19:347

Olsson P, Gunderson L, Carpenter S, Ryan P, Lebel L, Folke C, Holling CS (2006) Shooting the rapids: navigating transitions to adaptive governance of social ecological systems. Ecol Soc 11:18

Olsson P, Folke C, Hughes T (2008) Navigating the transition to ecosystem-based management of the Great Barrier Reef, Australia. Proc Natl Acad Sci 105:9489–9494

Pahl-Wostl C (2009) A conceptual framework for analyzing adaptive capacity and multi-level learning processes in resource governance regimes. Glob Environ Chang 19:354–365

Scholz J, Stifte B (2005) Adaptive governance and water conflict: new institutions for collaborative planning. Resources for the Future, Washington, DC

Snyder G, Davidson J (1994) Everglades agriculture: past, present and future. In: Ogden S et al (eds) Everglades: the ecosystem and it's restoration. St. Lucie Press, Deerfield Beach, pp 85–116

Tarlock A (1987) Protection of water flows for national parks. In: Land and water law review. University of Wyoming College of Law XXII, Laramie, pp 29–49

Zellmer S, Gunderson L (2009) Why resilience may not always be a good thing: lessons in ecosystem restoration from glen canyon and the everglades. Nebraska Law Review 87(4). Lincoln

Legal References

Everglades Forever Act (1994) Fla. Stat. § 373.4592
Flood Control Act of June 30 (1948) Law 858-80th Congress
Florida Water Resources Act (1972) Fla. Stat. § 373.012-373.200
River Basin Monetary Authorization and Miscellaneous Civil Works Amendments Act (1970) 84 Stat. 310
Water Resources Development Act (1996) PL 104–303
Water Resources Development Act (1999) PL 106–153

Chapter 6
The Emergence of Adaptive Governance in the Klamath River Basin

Brian C. Chaffin, Hannah Gosnell, and Robin K. Craig

Abstract The Klamath River Basin of south-central Oregon and northern California has been the locus of historic conflict over the governance of water situated at the spiritual, cultural, and economic confluence of fishing and farming livelihoods. In recent years, a series of crises have impacted communities and stakeholder groups across the basin and jeopardized the continued existence of endangered and threatened fish species as well as the dominant economic and social relations in the basin. From these crises, however, a set of human-driven processes emerged that closely resemble the seeds of adaptive environmental governance. This chapter describes social-ecological system structures and dynamics that led to this potential emergence of adaptive governance in the Klamath River Basin. The major aim of this chapter is to critically evaluate the role of law in the basin as a tool for both creating disturbances and opening windows of opportunity through which adaptive processes could emerge. The major insight from the Klamath case is that the distribution and application of political power cannot be underestimated as either a barrier or facilitator of adaptive governance. Without an explicit recognition and analysis of power dynamics, adaptive governance scholarship lacks a critical lens to interrogate the contexts of governance transitions and evaluate the potential for new arrangements to attain explicit goals such as the sustainability of ecosystem services, the fair allocation of resources, and other principles of good governance that promote social and environmental justice.

B. C. Chaffin (✉)
W. A. Franke College of Forestry and Conservation, University of Montana, Missoula, MT, USA
e-mail: brian.chaffin@mso.umt.edu

H. Gosnell
College of Earth, Ocean, and Atmospheric Sciences, Oregon State University, Corvallis, OR, USA
e-mail: gosnellh@geo.oregonstate.edu

R. K. Craig
S.J. Quinney College of Law, University of Utah, Salt Lake City, UT, USA
e-mail: robin.craig@law.utah.edu

B. Cosens, L. Gunderson (eds.), *Practical Panarchy for Adaptive Water Governance*, https://doi.org/10.1007/978-3-319-72472-0_6

Keywords Resilience · Adaptive governance · Adaptive law · Resilience assessment · Climate change · Water management

6.1 Introduction

At a cursory glance, governance of water and associated natural resources over the past century in the Klamath River Basin resembles a quintessential failure of hierarchical laws and policies to work together at the local or regional scale. A myriad of federal policies have played major roles in shaping the distribution of water on the Klamath landscape, including federal irrigation policy, energy regulatory policy, federal-tribal policy, and endangered species policy. Many of these policy directions and associated legal mandates have come into direct and indirect conflict with each other, specifically over the last two decades in the basin, poignantly playing out on the national stage as a narrative of intractable conflict over competing human values for water, fish, and associated livelihoods.

At a second and much closer glance at the basin, however, water governance in the Klamath bioregion may be beginning a transition toward more adaptive environmental governance. Through a unique set of historical, cultural, geographic, climatic, political, economic, and social circumstances, the Klamath social-ecological system seems to be poised for the negotiation and testing of new governance arrangements that have characteristics of adaptive governance. Although ecological conditions in the Klamath River Basin and the allocation and distribution of water on the landscape have not yet been significantly altered through governance processes that unfurled in the Klamath over the past 15 years, change is still on the horizon. In April of 2016, the US Departments of Interior and Commerce, the states of Oregon and California, several basin tribes, the Klamath Hydroelectric Project owner, PacifiCorp, and other invested organizations such as irrigation groups, conservation NGOs, and fishermen's associations signed an amended Klamath Hydroelectric Settlement Agreement (KHSA) which provides a pathway for removing four major hydropower facilities on the Klamath River by year 2020. The agreement to proceed toward dam removal under the US Federal Energy Regulatory Commission (FERC) authority was heralded "as an important part of the resolution of long-standing, complex, and intractable conflicts over resources in the Klamath Basin" and based on the idea that "potential benefits for fisheries, water and other resources of removing the Facilities outweigh the potential costs, risks, liabilities or other adverse consequences of such removal" (Klamath Hydroelectric Settlement Agreement 2016).

To many stakeholders in the basin, this agreement was a victory—a significant step toward restoring historic runs of anadromous fish to tributaries in the basin where they had long been extirpated. Restoring salmon and steelhead trout populations in the basin also represents an effort to maintain cultures and livelihoods, both for tribes who have relied on these species since time immemorial for subsistence, ceremony, and spiritual renewal, and for a contemporary economy partly dependent on commercial and recreational fishing industries. To others in the basin, however, this agreement to pursue the removal of the Klamath dams is just one piece of a much larger and more complex solution that emerged to address the interrelated problems of over-allocated water, social marginalization, economic stagnation, and environmental degradation throughout the basin in a holistic and integrated manner—a solution that by 2016 had at best been neglected and, at worst, failed. In either case, the narrative of conflict and cooperation over water that has played out in the past two decades in the Klamath is illustrative of the barriers and opportunities for the emergence of varying configurations of adaptive governance as a mode of environmental governance more apt to accommodate the complexity of fierce human competition over scarce resources, the legacies and evolving dynamics of environmental modification and degradation, and the exacerbating impacts of a changing climate. In this chapter, one aim is to highlight some important aspects of barriers and opportunities for the emergence of adaptive environmental governance using the Klamath River Basin narrative as an example. The structure and dynamics of the Klamath social-ecological system and the role of law in ongoing efforts to govern water in the basin are uniquely highlighted in the Klamath case. At the conclusion of this chapter, a brief reflection on adaptive governance research needs is presented, from both practical and theoretical perspectives.

6.2 Social-Ecological System Structures

The Klamath River Basin covers 40,790 km^2 and drains snowmelt from the Cascade mountains and the high desert plains of south-central Oregon into the river that runs over 420 km (over 260 miles) to the Pacific Ocean in northern California. The basin is roughly the size of the state of Maryland, shaped like a titled hourglass, and pinched near the California-Oregon border by the southernmost reaches of the Cascade mountains (Fig. 6.1). The upper basin (mostly in Oregon) is dominated by agricultural land use, producing onions, potatoes, alfalfa, mint, and other irrigated crops. The mountainous and rain-driven lower basin is dominated by large tracts of public forests. Historically, snowmelt pooled in the upper basin in a large system of seasonally expanding lakes and marshes that are today represented by a much-reduced set of static lakes regulated via extensive water extraction for irrigation and a dam located at the narrows of the Klamath River between the upper and lower basins. Downstream, a series of large hydroelectric dams have significantly constrained the variability of river flow, which once fluctuated between roughly 3000 and 300,000 cfs annually. Since 1962, these dams have blocked the return of migrating anadromous fish including salmon and steelhead trout to natal spawning habitats (National Research Council 2004).

Salmon are key species in the basin, ecologically as a nutrient source, culturally as a subsistence and spiritual species for Native American communities, and economically as an indicator of healthy commercial fishing populations off the coasts of California, Oregon, and Washington (Most 2006). Although salmon are perhaps

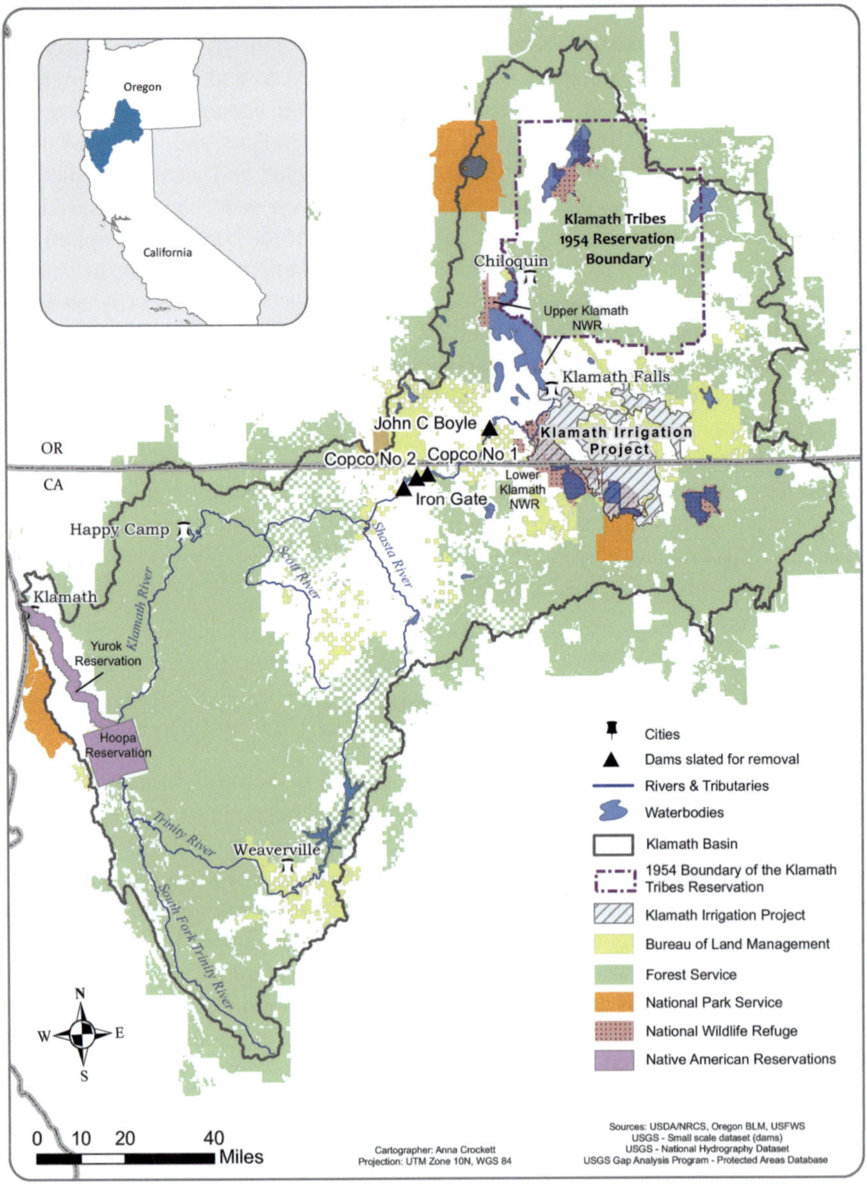

Fig. 6.1 Klamath Basin of Oregon and California. Sources: USDA/NRCS, Oregon BLM, USFWS, Cartographer: Anna Crockett

the most charismatic megafauna in the Klamath Basin, they are not the only species of importance in the contest over water use in the basin; in 1988, two species of suckerfish, the Lost River and shortnose sucker that live in the upper basin lakes and tributaries, were listed as endangered under the US Endangered Species Act (ESA)

(NRC 2004). This action would eventually have consequences for the agricultural industry in the upper basin, particularly irrigators served by water deliveries from the federal Klamath Irrigation Project, a US Bureau of Reclamation (USBR)-operated project initiated in 1905 to expand agricultural production in the upper basin through both irrigation and drainage of seasonal wetlands (Doremus and Tarlock 2008). In addition, the upper basin is home to a large complex of US National Wildlife Refuges, designed to preserve remnant fragments of the Pacific Flyway, pass-through habitat for seasonally migrating bird populations along the migration corridor between Canada and Mexico (Wilson 2010).

Generally, the basin is rural in character, with Klamath Falls, OR (population ~21,000), the largest municipality located in the upper basin, and Weaverville, CA (population ~3600), the largest in the lower basin located at the headwaters of the Trinity River, a major Klamath tributary. Politically, the basin is spread across two states and seven counties and encompasses land historically and contemporarily claimed by four federally-recognized Native American tribes (the Klamath Tribes (Klamath, Modoc, and Yahooskin), the Hoopa Valley Tribe, the Karuk Tribe, and the Yurok Tribe) as well as several other tribes and bands (e.g., the Resighini Rancheria, the Quartz Valley Tribe, and the Shasta Indian Nation).

6.3 Social-Ecological System Dynamics

Similar to many places in the American West, the history of Native and Euro-American relations in the basin is a story of marginalization of Native peoples and one that serves to illustrate the underlying social tensions over allocation and management of water, land, and resources that characterized the basin into the twenty-first century (Chaffin et al. 2014). Euro-American settlers began seeking privatization of land in the basin in the late nineteenth century under provisions of various Congressional land allotment acts. At the same time, Congressional treaties and Executive Orders created two reservations in the basin, the Hoopa Valley reservation in the lower basin and the Klamath Tribes reservation in the upper basin (the Yurok reservation was later split from the Hoopa Valley reservation, so today there are three federally recognized reservations in the basin). In 1887, the Klamath Tribes reservation was allotted under the Dawes Act, beginning the process of fragmenting the reservation by transferring non-titled Indian land to white settlers. In 1954 under the guise of federal termination policy (federal policy aimed at assimilating Native Americans by terminating the US-tribal trustee relationship and dispossessing tribal lands), an act of Congress stripped legal recognition of the Klamath Tribes as a sovereign nation and transferred all communally owned tribal property to the federal government (Most 2006). Throughout the same period, Euro-American irrigators (in large part financed by the federal government) proceeded to modify the landscape by diking and draining seasonal wetlands, channelizing tributaries, and mechanizing the flow and timing of water deliveries on the landscape—mostly at the expense of fish and wildlife habitat.

Although lower basin tribes did not experience as much Euro-American incursion on post-reservation lands, these tribes experienced acute and often violent conflict with the state of California and the federal government over the management of salmon in the Lower Klamath Basin as recently as the 1970s. Tribal fishers refused to stop commercially harvesting fish, despite a state of California mandated closure and confrontations with federal agents. This conflict, known as the Klamath Salmon Wars, led to the arrest of many Hoopa and Yurok tribal members and general hostility and confusion between tribal, non-tribal, commercial, and recreational fishers and among tribal peoples and fisheries managers caught in the middle of the contentious debate (Most 2006).

In 1973, the state of Oregon initiated a general stream adjudication of the Klamath River and tributaries in order to quantify and order (by priority dates) all claims to water rights in the Oregon portion of the upper basin—this was to include all unsettled *Winters v. United States* (1908) water rights claims of the Klamath Tribes based on the Treaty of 1864 and original reservation boundary, which were confirmed in a 1979 federal district court (*United States v. Adair* 1979) case to have survived termination. In 1983, the legal foundation for these water rights was expanded further when the 9th Circuit, a federal court of appeals recognized that tribal water rights can be nonconsumptive in nature and used to support primary purposes of the reservation, reserved by the tribes in treaty, and which included the right to hunt, fish, trap, and gather edible plants on the reservation. Water rights to support these purposes thus entitled the tribe to "prevent other appropriators from depleting the streams waters below a protected level in any area where the non-consumptive right applies" (*United States v. Adair* 1983). The U.S. Supreme Court refused to review the ruling (*Oregon v. U.S.* 1984). At this point, quantifying these rights through the Klamath adjudication process became a pathway for the Klamath Tribes to increase sovereignty over water and fishery resources in the basin, although this wouldn't happen formally until 2013 (Chaffin et al. 2014). At the same time, tribes in both the upper and lower basin, along with the US Fish and Wildlife Service (USFWS) and state wildlife agencies, pursued the listing of both Klamath sucker species as endangered (1988) and the lower basin Klamath coho salmon as threatened (1996) under the ESA (NRC 2004). This created a need for consultation under Section 7 of the ESA for both the operation of the Klamath Irrigation Project and the necessary reauthorization of the federal license for the Klamath Hydroelectric Project dams set to expire in 2006 (Gosnell and Kelly 2010).

The application of the ESA to the operation of the Klamath Irrigation Project reached a head during the spring of 2001 when low water levels in Upper Klamath Lake due in part to drought conditions forced the USBR to shut off water deliveries to over 1,400 farms and roughly 210,000 acres of farmland (Doremus and Tarlock 2008). On May 7, 2001, over 10,000 protestors lined the streets of Klamath Falls in public defiance of the federally-enforced shutoff, passing buckets of water from the headgates of a closed irrigation canal to the nearest farm field over a mile away. Interestingly, Klamath Falls protesters were not limited to Upper Klamath Basin farmers, but also included contemporary proponents of the past wise use movement, the Sagebrush Rebellion, and the County Supremacy Movements from all across the West. The protest represented a public release of tension over the ongoing struggle to manage water and fish in the basin and to balance the competing needs of multiple communities of Klamath River water users dependent on the river for economic vitality, cultural persistence, and spiritual well-being (Gosnell and Kelly 2010; Chaffin et al. 2014).

This agricultural "disaster" of 2001 caused financial hardships for family farms reliant on Klamath Project water deliveries for already planted crops, but was not the only impact of the ongoing water conflict. In 2002, after a speedy and controversial National Research Council (NRC) report, the USBR shifted operations to allow less water than mandated in the previous ESA biological opinion to flow down the Klamath River in early fall, creating conditions that precipitated the death of over 30,000 returning adult, fall-run Chinook salmon (California Department of Fish and Game 2004). The dead fish subsequently washed up on the banks of Yurok and Hoopa Valley reservations and in the communities of the Karuk Tribe. In the late 2000s, marine commercial harvest limits of Klamath salmon were either drastically reduced or closed altogether (Chaffin et al. 2014). Gosnell and Kelly (2010) collectively call these events a "rotating crisis" that affected many, if not all Klamath Basin communities in some way at different times, threatening the security of livelihoods and cultural identities.

What emerged during and after these governance crises in the basin was a set of processes through which the seeds of adaptive governance were sewn. The rotating crises in the Klamath Basin, beginning with historic tribal marginalization and shifting in the water shutoff of 2001, facilitated a social (and to a lesser degree, ecological) release of capital which would be recombined in more innovative ways in the following years. Although the emergence of adaptive governance in the Klamath Basin has not yet been *formally* institutionalized, a shift in governance approach has undeniably occurred, and the potential exists for further transition or possibly complete transformation of the Klamath Basin as a social-ecological system (Chaffin et al. 2014, 2016b; Chaffin and Gunderson 2016). The remainder of this chapter is structured using major themes from this book to highlight the nuances of emergent processes of adaptive governance in the basin.

6.4 The Role of Law in Creating Windows of Opportunity for Adaptive Governance

It is arguable that between Euro-American settlement of the Klamath Basin in the late nineteenth century and the acute, rotating crises and conflicts over water into the early 2000s, environmental governance failed. The various approaches to governance of that time period were unable to balance the sustainability of vital ecosystem functions with competing demands for water and failed to allocate ecosystem services equitably among human communities (Chaffin et al. 2014). Law played a major role in creating this failure as the tool used to entrench irrigation as the dominant value for Klamath River water and by continuing to fragment Native sovereignty when it was used to impede and, in some cases, prevent access to land, water, and resources. As time passed, however, this status quo became fragile, and it was the application of a different set of laws—mainly the Endangered Species Act, but to some degree tribal water law—that served as "the detonator," creating a large enough disturbance to destabilize irrigation 'business as usual' in the basin, subsequently creating a window of opportunity for a potential transition in natural resource governance priorities (Gosnell et al. 2017).

Cosens et al. (2017) define one of the roles of law in adaptive governance as either creating a disturbance or opening a window of opportunity through which adaptive processes can emerge. In the Klamath case, we see evidence of law used to work toward these ends simultaneously. The development of precedent in tribal water law from the *Winters v. United States* (1908) US Supreme Court decision, through quantification processes outlined in *Arizona v. California* (1963), to the *United States v. Adair* (1979, 1983, 2002) federal district and circuit court decisions pertaining specifically to nonconsumptive use of tribal water rights (for species habitat) represents a slow-onset change in the legal control of water flows in the basin. Although still not settled or quantified at the time of the irrigation water shut-off of 2001, the unquantified legal potential of this shift in rights (and corresponding potential political power) was not lost on historically-adversarial irrigators and tribes or on federal natural resource managers as legal battles shifted to negotiation and settlements talks around 2005. At the same time, the powerful ESA (Section 7) worked faster, but in concert, bringing the operation of the Klamath Irrigation Project to a halt and placing the national spotlight on the over-allocation of water in the basin (Gosnell et al. 2017). As Cosens et al. (2017) point out, implementation of laws can create a disturbance when competing regulatory requirements cannot be satisfied by piecemeal compliance. In the Klamath, federal irrigation policy, endangered species law, and the federal-tribal trust responsibility could not be rectified under the governance of the basin at the time, and something had to give.

What followed immediately in the aftermath of disturbance was neither cooperative or collaborative nor did it resemble adaptive governance. The period of environmental governance in the Klamath Basin between 2001 and 2004 was characterized by lawsuits, divisive op-eds in local newspapers, and even acts of physical violence between tribal and non-tribal peoples in the basin motivated by racism, anger, and fear (Chaffin 2014). This period was described by legal scholars Doremus and Tarlock (2008) as one of "macho law, combat biology, and dirty politics." But when the dust settled (or more accurately, when zero-sum options were exhausted by conflicting parties), law was used in another role highlighted by Cosens et al. (2017): to eliminate barriers and facilitate adaptive processes. It was in part the legal requirement to relicense the Klamath Hydroelectric Project that provided a venue of opportunity for exhausted parties (tribes, irrigators, fishermen's associations, environmental and conservation advocacy NGOs, federal and state natural resource management agencies, and community groups) to meet and discuss larger-scale, holistic, basin conflicts over water that precluded any piecemeal governance solutions such as a hydroelectric project reauthorization. During this time, the federal government relied on flexibility in law for both federal hydropower relicensing (alternatives to) and tribal water rights settlements, to provide financial and logistical support and some degree of coordination (facilitation) for emerging discussions around governance changes in the basin—another key role of law in the emergence of adaptive governance (Cosens et al. 2017; Craig et al. 2017).

6.5 The Emergence of Adaptive Governance in the Klamath River Basin

DeCaro et al. (2017a) describe the ways in which legal and institutional contexts provide opportunities for adaptive governance to emerge. Wide variations in social and ecological contexts across a landscape render it extremely difficult to create legal frameworks that do not stifle the inherent self-organization that can lead to the emergence of new, adaptive governance arrangements. Common principles established by traditional centers of environmental governance activity (e.g., governments at different scales, NGOs, corporations, etc.) including "legal, economic, and democratic tools" have been shown to help create contexts more favorable to self-organizing behavior (DeCaro et al. 2017a). In the Klamath transition, however, the initial emergence of adaptive governance was less a product of institutional and legal design, but instead a product of social capital, leadership, and the malleability of social cognition and individual perception (Chaffin 2014).

In 2004, among other, more disconnected collaborative efforts emerging in the upper basin between irrigators and tribal interests, Bob Chadwick, an expert facilitator and conflict mediator living in the basin, initiated and led a series of listening sessions. These sessions were voluntary sessions strategically designed to air emotional and perceptual grievances among conflicting parties (Gosnell and Kelly 2010). During five listening sessions held across the basin from 2004 to 2005, all major parties to the Klamath water conflict were represented (Messier 2012). Although not a product of traditional centers of governance activity, these informal, facilitated sessions highlight some of the design principles suggested by DeCaro et al. (2017a, Chap. 16) such as transparent opportunities for conflict resolution and meaningful participation. Legitimacy of these events was not created by a government authority, but instead by the near-complete participation of stakeholders in the Klamath conflict including representatives from all federal natural resource management agencies, the states of Oregon and California, basin tribes, environmental and conservation NGOs, and irrigation and fishermen's groups (Messier 2012). The individual makeup of each session and the rotation of locations for the sessions from upper to lower basin also represent the incorporation of an institutional design principle highlighted by DeCaro et al. (2017a): well-defined, representative geographic boundaries for facilitating the emergence of adaptive governance. During these "Chadwick Sessions," it became apparent that solutions to water conflict in the basin could best, and perhaps only, be addressed at the basin scale (Chaffin 2014).

The Chadwick Sessions also highlighted the importance of recognizing the role of cognitive bias as either a facilitator or inhibitor of emerging adaptive governance. In an innovative contribution to the literature on adaptive governance, DeCaro et al. (2017b, Chap. 17) point out that cognitive and social cognitive biases can disrupt decision-making and thus affect transitions to adaptive governance; in turn, governance systems must find ways, formally and informally, to cope with personal biases clouding decision-making judgment in natural resource allocation and management. DeCaro et al. (2017b) state that adaptive governance is more likely to emerge when basic cooperative dilemmas are overcome; these dilemmas include barriers to encouraging collaborative problem-solving, garnering social acceptance and commitment, and cultivating a culture of trust and tolerance for change and uncertainty. Klamath stakeholders sitting in the audience at the first Chadwick Session in remote Somes Bar, CA in July of 2004, faced all of these cooperative dilemmas, but by their attendance, they had unknowingly made a commitment to overcome them as a social group. What followed over the next 2 years between Klamath stakeholders—apology, listening, risk taking, and personal transformation of understanding about self, others, and personal roles in the basin conflict (Gosnell et al. 2017; Chap. 12)—can undoubtedly be attributed to the genius of Bob Chadwick as a facilitator but also to the concept that overcoming cognitive and social cognitive barriers to cooperation is a critical element and precursor to the emergence of adaptive governance (DeCaro et al. 2017b).

The Chadwick Sessions encapsulated the necessary social and psychological 'heavy lifting' that needed to be done before a transition in governance could emerge; the process facilitated a level of trust among stakeholders in the basin that allowed more substantive discussion about the management of resources in the basin (Chaffin 2014). As a result of the fortuitous timing of the Chadwick talks, coupled with a significant risk taken by the two most contentious adversaries in the basin (leaders of irrigators and tribes) when they chose to engage in informal, closed door discussions, and the PacifiCorp-initiated alternative to FERC relicensing proceedings, a window of opportunity opened for important and necessary social learning to take place in the basin. Over the next several years (roughly 2006–2008), groups of specialists (e.g., fisheries biologists, hydrologists, and lawyers) met consistently as subgroups of larger negotiations toward settlement of water conflict issues in the basin (Chaffin et al. 2014). These epistemic groups, including hydrologists from tribes and federal and state agencies, synthesized collective knowledge about the state of the basin and together underwent a unique process of collaborative knowledge creation (e.g., creating a better working understanding of basin hydrology, fish habitat requirements, legal options) that directly informed the stakeholders' perception and understanding of the basin and its problems but also directly contributed to politically acceptable solutions crafted through collaborative negotiations (Gosnell et al. 2017; Chap. 12). By 2008, these early, emergent processes of adaptive governance facilitated changes in mental models of basin stakeholders (DeCaro et al. 2017b) and not just among end users of water and advocates for fish, but also among managers from different and even historically competing or adversarial agencies (Chaffin 2014; Gosnell et al. 2017; Chap. 13). Managers and basin stakeholders began to recognize the need to manage the interconnected complexities of the basin as a whole (from headwaters to mouth), not simply as related to single resources or ecosystem services of interest. Evidence of this can be found in the creation of the first joint biological opinion for both endangered suckers (managed by USFWS) and threatened salmon (managed by NOAA Fisheries) in relation to operations of the USBR Klamath Irrigation Project (Gosnell et al. 2017; Chap. 12). Gosnell et al. (2017) highlight the processes that led to this event as it was a precedent setting action to combine an ESA biological opinion for a freshwater aquatic species and a (mostly) marine species, although it aligns with the individual and social transformations that were prevalent in the basin at the time.

6.6 The Klamath Agreements and the Failure to Institutionalize Adaptive Governance?

In 2008 and 2010, respectively, basin stakeholders including government partners at the highest level of federal and state administrations released two agreements: the Klamath Basin Restoration Agreement (KBRA) and the companion FERC settlement, the Klamath Hydroelectric Settlement Agreement (KHSA). Together, these agreements aimed to (1) address over-allocation of water in the basin through water use retirements and habitat restoration; (2) provide predictive certainty of water deliveries for irrigators in the upper basin early enough in the water year for economic planning and sustainability; (3) settle outstanding Indian water rights claims; (4) remove the four main stem Klamath dams; (5) pursue endangered sucker habitat restoration and threatened salmon reintroduction upstream of impounded river sections; (6) address degraded, nutrient-rich water quality in the rivers, lakes, and reservoirs; (7) settle outstanding litigation between parties; and (8) address regional economic depression. The agreements were not agreed upon by all original parties to the negotiation, but heavily supported by those who had worked through the process to find tractable, politically acceptable solutions. There were two main groups of stakeholders who "opted out" of these agreements: a group of environmental NGOs opted out of a preceding 2008 agreement in principle because of the failure to include negotiation of private farming rights on federal wildlife refuges in the upper basin, and the Hoopa Valley Tribe opted out because of failure to include further discussions of the Trinity River, a Klamath tributary along the reservation (Chaffin 2014).

Taken together, the processes that led to the Klamath Agreements and the agreements themselves resembled the emergence of adaptive governance (Chaffin et al. 2016a). The agreements provided mechanisms that align with the legal and institutional design principles adopted by DeCaro et al. (2017a): legal devolution of governance of terms of the agreement to a representative governance body (the Klamath Basin Coordinating Council, members partially elected and partially appointed by stakeholder groups); internal, transparent conflict resolution processes; and transparent and consistent opportunities for general public participation beyond stakeholder representation. A critical component of the agreements included a request for sizeable (originally around $750M) financial support from the US federal government and Congressional authorization for federal agencies to participate in the holistic, basin-scale programs for restoration and settlement of long-standing conflict.

After the public release of the Klamath Agreements, momentum was strong for the negotiating parties to collectively pool political resources and simultaneously work toward introduction and passage of Congressional legislation that would authorize and fund critical aspects of negotiated social-ecological restoration and conflict resolution in the basin (Chaffin et al. 2014). What happened instead was a political backlash at the local and national scales that many of the stakeholders did not expect. As pointed out by DeCaro et al. (2017b), mediating social systems (e.g., news media, popular discourse, group relations) debate and reinterpret environmental governance innovations, transforming their meaning and individuals' perceptions of them. This is precisely what happened in the Klamath after 2010. Although the leaders of key stakeholder groups had spent years sequestered in rooms together building trust, collaboratively creating an intimate understanding of the social-ecological function of the basin, transforming personal perceptions of the basin and other stakeholder groups, and crafting a vision for the future of the basin, the general public had not had the opportunity to engage in this social learning or personal and social transformation. Instead, the public was provided with over 1,000 pages of legal documents detailing the future of water and resource management in the basin, combined with an informal promise from basin leaders that 'we're all on the same page now and this will work' (Chaffin 2014).

Humans are influenced by perceptions of legitimacy and other "social-psychological needs, such as security and procedural justice, i.e., decision-making fairness" (DeCaro et al. 2017b, citing: Sheldon et al. 2001; Tyler 2006), and these elements were not considered in the public release of the Klamath Agreements. Some stakeholders argued that bringing the public along in the process of developing the agreements was not necessary, as the negotiations were mostly over vested legal rights to water, contracts, and other proprietary resources that the public had no right to weigh in on (Chaffin 2014). In 2012, this argument was rebutted in force by the overwhelming passage of a referendum in Siskiyou County, CA, that forced the county to withdraw from its initial approval of the agreements. Just to the north in Klamath County, OR, voters removed all county commissioners from office who had previously supported the agreements. The 2012 election cycle was in part the height of the conservative tea party revolution in America, but it clearly identified procedural and strategic flaws in the attempt to institutionalize the emergence of adaptive governance in the Klamath Basin. In addition, the failure of the US Congress to pass authorizing legislation between 2010 and when the proposed agreements expired (without authorization) at the close of 2015 exposed an underestimation of the slow acceptance of governance change at higher scales and the need for political windows of opportunity to garner support for innovative experiments in governance (Chaffin 2014).

6.7 Conclusion: What the Klamath and Intractable Water Governance Conflicts Need from Adaptive Governance Research

As large-scale resource conflicts resulting from both increased degradation and demand on limited resources and the marginalization of access to resources continue to emerge, groups of actors will continue to seek out new modes of environmental governance to allocate scarce resources, resolve conflict, and collectively pursue desired states for social-ecological systems. The authors of chapters in this volume have synthesized and produced a large, valuable compendium of knowledge that will be useful in analyzing or facilitating transitions toward adaptive governance. The major insights from the Klamath case are that politics and the distribution and application of political power cannot be underestimated as either barriers or facilitators of adaptive governance. Governance, the process of resolving tradeoffs for allocating resources and risks, has even been defined as "the myriad of ways in which power is exercised within society" (Turner 2013). Without an explicit recognition and analysis of power dynamics, adaptive governance scholarship lacks a critical lens to interrogate the contexts of governance transitions and evaluate the potential for new arrangements to attain explicit goals such as the sustainability of ecosystem services, the fair allocation of resources, and other principles of good governance that promote social and environmental justice.

An analysis of power and politics in transitions toward adaptive governance is important to ensure that outcomes reflect the best solutions for governing a system and do not solely reflect the often unavoidable influence of powerful political actors, networks, or organizations, potentially co-opting the processes of emergent adaptive governance for their own benefit. Some have argued that the structure of the Klamath Agreements was not valid because they did not include the demands of some environmental groups and the Hoopa Valley Tribe who refused to sign due mostly to the lack of inclusion of National Wildlife Refuge and Trinity River issues, respectively. An empirical analysis of stakeholder relationships reveals that significant political power existed on behalf of continuing the settlement process without further consideration of these two issues (and by extension, these groups)—thus altering the geographic and participatory scope of the potential solution, but opting for traction toward broad resolution in the foreseeable future (Chaffin 2014). This is not to say that environmental groups or the Hoopa Valley Tribe's claims were not legitimate, but instead suggests that an analysis of power relations is necessary to better understand how adaptive governance emerges, the scope that defines its approaches and outcomes sought, and the potential barriers and opportunities for institutionalization of these approaches. For example, in 2013, after the local political backlash against the agreements, the Klamath Tribes water right was finally quantified in state adjudication court, and they "called the river" in order to keep irrigation water in stream for fish habitat in the upper basin (Chaffin et al. 2014). A political analysis of this action in the context of the continuing transition toward adaptive governance revealed this as a strategic move to entice conservative opponents of the agreements with property and water rights upstream of the tribes' rights to return to the negotiating table (Chaffin 2014).

Fortunately, the emergent changes in social capital, innovations, and reorganizations developed by stakeholders in the Klamath Basin throughout the last decade have not vanished, and instead this capacity lies dormant, waiting for the next window (or venue) of opportunity. By expanding the scope and practice of adaptive governance research—and more importantly, expanding practiced-based governance research—to include analyses of power and politics as potential barriers and facilitators of adaptive governance, there is continued hope for successfully pursuing holistic solutions for social-ecological systems like the Klamath Basin.

Acknowledgments This work was developed in part under the Adaptive Water Governance Project, funded by the US National Socio-Environmental Synthesis Center (SESYNC) under funding from the US National Science Foundation, NSF DBI-1052875.

References

California Department of Fish and Game (2004) Klamath river Fish-Kill: final analysis of contributing factors and impacts. The Resources Agency, California. http://www.pcffa.org/KlamFishKillFactorsDFGReport.pdf. Accessed 16 Apr 2014

Chaffin BC (2014) Reallocating resources, rebuilding community: the Klamath Basin agreements and the emergence of adaptive governance. Dissertation, Oregon State University

Chaffin BC, Gunderson LH (2016) Emergence, institutionalization and renewal: rhythms of adaptive governance in complex social-ecological systems. J Environ Manag 165:81–87

Chaffin B, Craig RK, Gosnell H (2014) Resilience, adaptation, and transformation in the Klamath River Basin social-ecological system. Idaho Law Rev 51:157–193

Chaffin BC, Garmestani AS, Gosnell H, Craig RK (2016a) Institutional networks and adaptive water governance in the Klamath River Basin, USA. Environ Sci Pol 57:112–121

Chaffin BC, Garmestani AS, Gunderson LH, Benson MH, Angeler DG, Arnold CA, Cosens B, Craig RK, Ruhl JB, Allen CR (2016b) Transformative environmental governance. Annu Rev Environ Resour 41:399–423

Cosens BA, Craig RK, Hirsch S, Arnold CA, Benson MH, DeCaro DA, Garmestani AS, Gosnell H, Ruhl JB, Schlager E (2017) The role of law in adaptive governance. Ecol Soc 22(1):30. https://doi.org/10.5751/es-08731-220130

Craig RK, Garmestani AS, Allen CR, Arnold CA, Birgé H, DeCaro DA, Fremier AK, Gosnell H, Schlager E (2017) Balancing stability and flexibility in adaptive governance: an analysis of tools available in U.S. environmental law. Ecol Soc 22(2):3. https://doi.org/10.5751/es-08983-220203

DeCaro DA, Chaffin BC, Schlager E, Garmestani AS, Ruhl JB (2017a) Legal and institutional foundations of adaptive environmental governance. Ecol Soc 22(1):32. https://doi.org/10.5751/es-09036-220132

DeCaro DA, Arnold C.A.T, Boamah EF, Garmestani AS (2017b) Understanding and applying principles of social cognition and decision making in adaptive environmental governance. Ecol Soc 22(1):33. https://doi.org/10.5751/es-09154-220133

Doremus H, Tarlock AD (2008) Water war in the Klamath Basin: macho law, combat biology, and dirty politics. Island Press, Washington, DC

Gosnell H, Kelly EC (2010) Peace on the river? Social-ecological restoration and large dam removal in the Klamath basin, USA. Water Altern 3:361–383

Gosnell H, Chaffin BC, Arnold CA, Craig RK, Benson MH, Devenish A (2017) From detonator to stabilizer: rethinking the role of the Endangered Species Act's Section 7 consultation process in social-ecological system dynamics. Ecol Soc 22(1). https://doi.org/10.5751/es-05116-170428
Klamath Hydroelectric Settlement Agreement (2016) United States Department of Interior. https://www.doi.gov/sites/doi.gov/files/uploads/FINAL%20KHSA%20PDF.pdf. Accessed 2 June 2017
Messier JY (2012) Conflict in the Klamath watershed and a relationship-building framework for conflict transformation. Dissertation, George Mason University
Most S (2006) River of renewal: myth & history in the Klamath Basin. Oregon State University Press, Portland
National Research Council (2004) Endangered and threatened fishes in the Klamath River Basin: causes of decline and strategies for recovery. The National Academies Press, Washington, DC
Sheldon KM, Elliot AJ, Kim Y, Kasser T (2001) What is satisfying about satisfying events? Testing 10 candidate psychological needs. J Pers Soc Psychol 80(2):325–339
Turner MD (2013) Political ecology I: an alliance with resilience? Prog Hum Geogr 38(4):616–623. https://doi.org/10.1177/0309132513502770
Tyler TR (2006) Psychological perspectives on legitimacy and legitimization. Annu Rev Psychol 57:375–400
Wilson RM (2010) Seeking refuge: birds and landscapes of the pacific flyway. University of Washington Press, Seattle

Judicial Opinions

Arizona v. California, 373 U.S. 546 (1963)
United States v. Adair, 478 F. Supp. 336, 345 (D. Or. 1979)
United States v. Adair, 723 F. 2d 1394 (9th Cir. 1983)
Oregon v. United States, 467 U.S. 1252 (1984)
United States v. Adair, 187 F. Supp. 2d 1273 (D. Or. 2002)
Winters v. United States, 207 U.S. 564 (1908)

Chapter 7
Governing the Rio Grande: Challenges and Opportunities for New Mexico's Water Supply

Melinda Harm Benson, Ryan R. Morrison, Dagmar Llewellyn, and Mark Stone

Abstract The Rio Grande in New Mexico provides water to the urban environments of Albuquerque, Santa Fe, as well as surrounding small towns and rural agricultural communities. Long-term climate change projections suggest that New Mexico will experience ongoing drought in the coming decades, placing stress on a system already struggling to meet increasing water demands. Resilience theory provides a lens through which the governance challenges associated with climate change and other factors can be examined. The construction and operation of the many dams, reservoirs, and levees, along with channelization activities, have lowered the overall functional diversity of the river system through channel simplification and bed degradation, with implications for both riverine and riparian habitats. The earlier peak will require more nuanced and intensive water management, including more management flexibility. Changes in water governance strategies will be needed in order to adapt to increased temperatures and other challenges the future will bring. New strategies will include more aggressive management of the upland forest system to decrease the risk of wildfire in the watershed, more operational flexibility for dams and reservoirs, and a new approach to water storage and allocation.

Keywords Resilience · Adaptive governance · Adaptive law · Resilience assessment · Climate change · Water management · Regime change

M. H. Benson (✉)
Haub School of Environment and Natural Resources, University of Wyoming, Laramie, WYO, USA
e-mail: mhbenson@uwyo.edu

R. R. Morrison
Department of Civil and Environmental Engineering, Colorado State University, Fort Collins, CO, USA
e-mail: ryan.morrison@colostate.edu

D. Llewellyn · M. Stone
Department of Civil Engineering, University of New Mexico, Albuquerque, NM, USA
e-mail: ldagmar@unm.edu; stone@unm.edu

B. Cosens, L. Gunderson (eds.), *Practical Panarchy for Adaptive Water Governance*, https://doi.org/10.1007/978-3-319-72472-0_7

7.1 Introduction

The American Southwest is facing unprecedented challenges to traditional forms of water governance. This chapter centers on portions of the Rio Grande watershed struggling to meet the challenges of supplying scarce water to human populations. In central New Mexico, the Middle Rio Grande (MRG) provides water to the urban environments of Albuquerque, Santa Fe, as well as surrounding small towns and rural agricultural communities. The MRG is experiencing the pressures of increasing urbanization demands and water supply constraints from international treaties and interstate compacts, in context of highly variable and unpredictable water availability. Long-term climate change projections suggest that New Mexico will experience ongoing drought in the coming decades, with water shortfalls and extended dry intervals expected to become increasingly common (Llewellyn and Vaddey 2013).

This chapter begins an overview of the MRG as a complex and dynamic social-ecological system (SES) that must adapt to expected hydrologic changes. Employing concepts from resilience theory, the challenges facing water governance in the MRG are described, placing particular emphasis on the current function and capacity of existing governance structures. After a basic overview of resilience as a paradigm understanding the dynamics of an SES, the current governance framework for the MRG is described. Having identified the key social and ecological elements involved and interactions among them, the chapter presents potential tipping points or thresholds that, if crossed, may threaten the long-term viability of the MRG.

7.2 Social Ecological System Structures of the Middle Rio Grande

Viewing the MRG as a dynamic SES, it is important to focus on the amount of *change* the system can undergo while keeping its identity, including the system's adaptive capacity. From a management perspective, promoting resilience involves (1) evaluation of the current trajectory of the system state and (2) fostering the ability of the system to resist perturbations. The abilities to influence both of these factors are determined by a combination of attributes of both the social and the ecological aspects of the system (Walker and Salt 2012).

Resilience theory acknowledges the possibility of "surprise," the unpredictable qualities of SES (Folke 2006). A critical component of a resilience orientation is the recognition that regime shifts can occur. As a result, a resilience-based approach to management is more realistic than traditional approaches because it acknowledges nonlinear change and provides a way of thinking about how to foster the SES components and dynamics we value and want to protect. The emphasis of resilience-based management is on building adaptive capacity rather than maintaining stationarity.

Where regime shifts occur, transformation results, and a fundamentally new system with different characteristics is created. Intentional transformation involves a conscious and deliberate negotiation from one system state to another. A system's transformative capacity is defined by (1) the degree to which managers of the SES are prepared for a change (as opposed to managers being in a state of denial), (2) the identified options for change or possible new trajectories for the system, and (3) the capacity to change or the ability to make choices among possible new trajectories (Walker and Salt 2012). The concept of transformation is particularly helpful in cases, such as the MRG, which are approaching potential ecological thresholds (Gunderson et al. 2017, Chapter 10) and provides managers with a framework for approaching impending change.

Adaptive capacity and transformative capacity are related attributes within a SES. These capacities are crucial both when the management orientation is to maintain the current system state *and* when SES dynamics are such that transformation should or will occur. Transformative capacity highlights an important element of resilience theory that is often overlooked in policy discussions. That is, a resilient system state is not inherently good or bad (Zellmer and Gunderson 2009). There are many examples of relatively stable and resilient SESs that are not desirable. Any notion of building resilience must therefore be followed by the questions—resilience *of* what and *to* what (Carpenter et al. 2001). In other words, it is necessary to first identify whether overarching system states are desired (referred to as general resilience) and/or elements of the system we want to keep (specified resilience), as well as those that we would prefer to lose. Once the desired outcomes are recognized, managers can perform an assessment of the perturbing factors and disturbances and assess whether they constitute potential or existing threats, as well as their capacity to control those threats (Carpenter and Brock 2008). This chapter examines the *specified* resilience of the MRG's existing water supply and allocation strategies to climate change.

Understanding the current water supply and allocation strategies in the MRG requires a basic overview of the current governance structure and key actors. It is multifaceted and involves formal legal mandates and institutions, as well as stakeholder-based and collaborative institutions. The headwaters of the Rio Grande are located in Colorado and northern New Mexico. The river then bisects New Mexico from north to south before exiting the state to form the border between Mexico and Texas. This chapter focuses on the 250 km reach between Cochiti Dam and Elephant Butte Reservoir in central New Mexico because of its vital importance for the region. Figure 7.1 shows the area with its major watershed features, including the interbasin transfer from the San Juan River to the Chama River, which is operated in accordance with the Colorado River Compact.

Human occupation of the MRG has a rich and complex history. Several Native American Pueblo communities live in the MRG and have their own, extensive history of water use. In the mid-1500s, the Spanish brought the first wave of European colonialism (Phillips et al. 2011). Along with them came Spanish settlements in the form of agricultural communities that were granted large tracts of land by the Spanish Crown. These communities brought communal irrigation

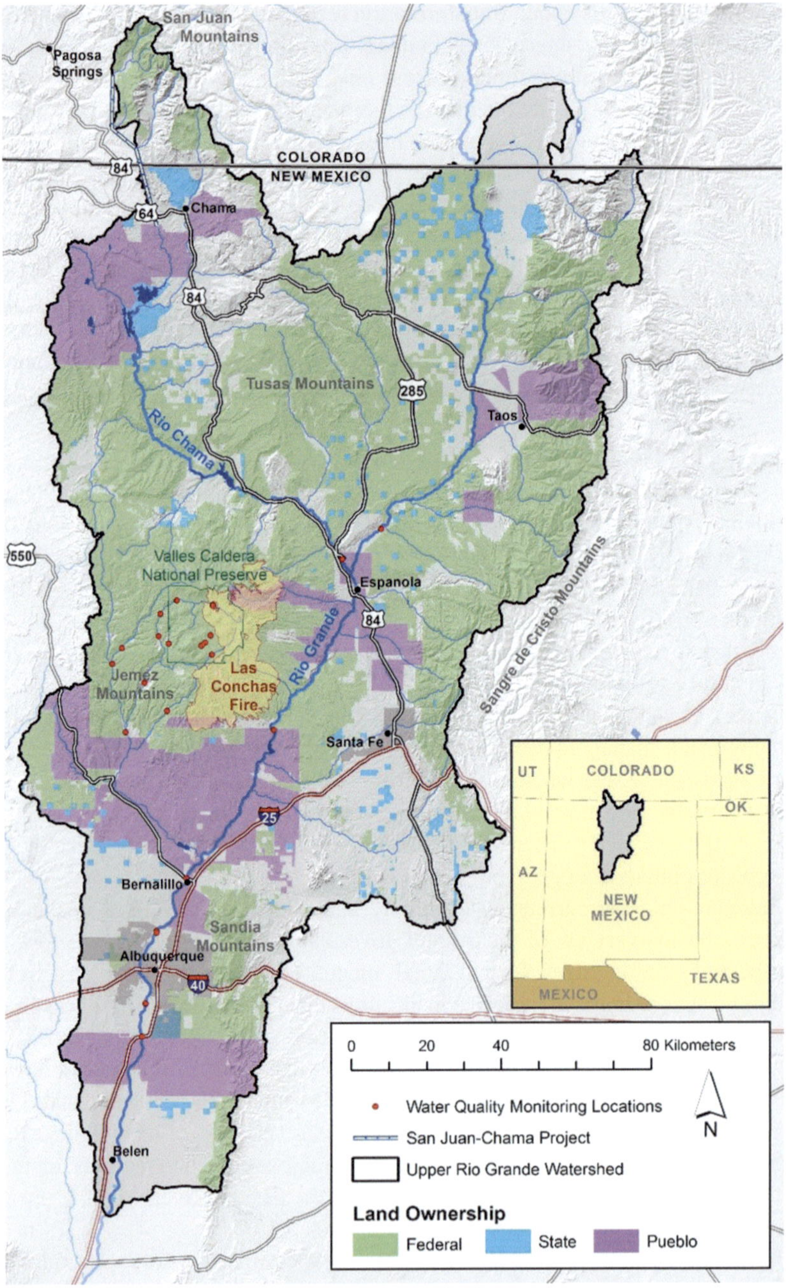

Fig. 7.1 The Middle Rio Grande Basin

systems of governance called acequias, resulting in water diversions from a public waterway and divisions of water among acequia members. For a brief period, the MRG became part of Mexico; the United States gained possession of the landscape with the signing of the Treaty of Guadalupe Hidalgo in 1848.

This treaty led to another wave of Euro-American colonialism—this time spearheaded by Anglo ranchers and farmers arriving on the newly constructed railroad (Phillips et al. 2011). Anglo settlement in New Mexico followed a pattern familiar across the American West, including the encouragement of settlement by various homesteading provisions and federal water projects supporting the development of large-scale irrigated agriculture.

Today, the MRG is stressed due to the effects of continuing human modifications to the landscape and river system along with both natural and human constraints and demands. These stressors include population growth, biodiversity loss, and cyclical droughts, all occurring in an over-allocated system under constraints of interstate water compact obligations. The effects of climate change are being felt in the basin, with water demands increasing due to higher temperatures. Climate change impacts are projected to be particularly acute in this region, exacerbated by the over-allocation of the historic water supply.

The MRG is subject to a number of interstate compact agreements among states. Of primary importance are the Colorado River Compacts and the Rio Grande Compact. The states with rights to the Colorado River were divided into two categories—Upper Basin and Lower Basin. New Mexico is part of the Upper Basin, and its allotment of water was determined by the Upper Colorado River Basin Compact of 1948, which granted New Mexico the right to approximately 11.25% of the Upper Basin's water—estimated at approximately 0.84 million acre-feet per year (Hogan et al. 2014). Major infrastructure, such as the San Juan-Chama (SJC) Diversion Project, has been developed to convey a portion of this water to the MRG. The SJC Project is a trans-basin diversion that moves approximately 96,000 acre-feet of water annually from the Colorado River Basin to the Rio Grande watershed to allow the MRG to take advantage of a portion of New Mexico's allocation under the Upper Colorado River Compact (Benson et al. 2013). The other major interstate compact influencing governance in the MRG is the Rio Grande Compact among Colorado, New Mexico, and Texas.

At the federal level, the US Bureau of Reclamation (Reclamation) and US Army Corps of Engineers (Corps) have numerous responsibilities in the MRG. Reclamation operates two projects for water storage and delivery: the SJC Project, which includes the Heron Reservoir, and the Middle Rio Grande Project, which includes the El Vado Reservoir (Fig. 7.1). The Corps has primary responsibility for key flood control operations for the MRG, including the Abiquiu and Cochiti reservoirs. Each of the federal water operations of these agencies has its own statutory mandates and corresponding constraints on management operations. Table 7.1 provides a summary of the most significant dams and reservoirs in the system.

Table 7.1 Existing dams and reservoirs in the MRG and existing operational authorizations and constraints (Kelly et al. 2007). The Galisteo Dam and Jemez Canyon dams are operated by the Corps for flood and sediment control within the MRG basin, but are not listed here because they are not on the main stem of the river

Dam/ reservoir	Current reservoir manager and water project, construction date, and capacity	Description of current operational authorization—including date and statutory citation	Current legal and regulatory mandates and constraints
Heron Dam and Reservoir	Reclamation (San Juan-Chama Project; completed 1971, 401,320 acre-feet capacity)	1962 PL 87-483. Stores NM allocation under Colorado River Compact. 96,200 acre-feet delivered to 13 contractors plus federal uses on an annual basis	Filling is subject to water availability in San Juan tributaries and to availability under Colorado River Compact. Undelivered contractor allocations revert to federal pool at the end of the year
El Vado Dam and Reservoir	Reclamation and MRGCD (Middle Rio Grande Project; completed 1935, rehabilitated in the 1950s, storage rights assigned 1963; approx. 190,000 acre-feet capacity remaining)	1927 Conservancy Act, NMSA $73-14-1 through 73-14-88; Act of 1928, 45 Stat. 312 (appropriating federal funds for Pueblo share). Store native water for use by six Middle Rio Grande Pueblos and non-Indian irrigators of the MRGCD. Provide power generation for Los Alamos County	Subject to Article VII storage restrictions under Rio Grande Compact. Both MRGCD and Reclamation claim title. Regulates flows through National Wild and Scenic Reach of the Rio Chama. Hydropower generation does not dictate flows
Abiquiu Dam and Reservoir	Corps of Engineers (completed 1962; 551,000 acre-feet capacity for flood and sediment control; 183,099 acre-feet storage of SJC Project water)	Flood Control Act of 1948, PL-81-858; Flood Control Act of 1960, PL 86-645; PL 97-140 (1981) SJC Project storage; PL 100-522 (1988) native storage	Releases limited to 1800 cubic-feet-per second due to channel capacity restrictions. Storage of native water is legal, but environmental clearances and property owner permission not yet assured
Cochiti Dam and Reservoir	Corps of Engineers (completed 1975; 50,000 acre-feet recreation pool refilled with SJC Project water; 590,000 acre-feet flood control space)	Flood Control Act of 1960, PL-86-645; PL 88-293 (50,000 acre-feet pool for recreation, fish, and wildlife)	Release limited to 7000 cubic-feet-per second at Albuquerque gage due to channel capacity restrictions. Must pass all inflow except during flood operations. Floodwaters must be released as soon as practicable
Elephant Butte Reservoir	Reclamation (Rio Grande Project; completed 1916; approx. two million acre-feet storage capacity for native and SJC Project storage plus flood control)	Rio Grande Reclamation Project, enacted 1905, PL No. 58-108	Upper Reservoir pool/delta is a critical habitat for endangered southwestern willow flycatcher

The US Fish and Wildlife Service, the federal agency primarily responsible for implementation and enforcement of the Endangered Species Act (ESA), is also a key player in the MRG, which is home to two listed species under the ESA— Rio Grande silvery minnow (*Hybognathus amarus*) and the southwestern willow flycatcher (*Empidonax traillii extimus*). As discussed later in this chapter, implementation of the ESA under the limited water supply and water management constraints in the MRG has been a key driver of social system dynamics.

As in many Western states, the Office of the State Engineer is responsible for administration of water in accordance with the prior appropriation doctrine in New Mexico. These duties include, where necessary, the adjudication of water rights in order to help the state define its existing water rights, meet its interstate compact obligations, manage shortages, and protect the state's waters. To date, however, there has been no adjudication of water rights in the MRG, and the actual administration of water supply is more complex in practice than it would be under strict adherence to the prior appropriation doctrine.

At the local level, the cities of Albuquerque, Santa Fe, and Rio Rancho are all major water users and are key players in land-use management decisions that impact water use in municipal areas. Albuquerque's water is managed by the Albuquerque Bernalillo County Water Utility Authority (Water Authority). Both the City of Santa Fe and the Water Authority receive water from the San Juan-Chama Project and have drinking water supply projects that divert from the Rio Grande. The Water Authority receives the largest share of San Juan-Chama Project water, with an annual allocation of 48,200 acre feet (Kelly et al. 2007).

Irrigation districts have a quasi-municipal status as political subdivisions of the state in New Mexico, and no description of MRG governance would be complete without the Middle Rio Grande Conservancy District (MRGCD). The MRGCD was created in 1925 to both construct the dams and levees necessary to drain the historic floodplain for agricultural use and to deliver water to district members. It is the largest agricultural water user in the MRG (U.S. Bureau of Reclamation 2013).

The listing of species native to the MRG as endangered created a need for the federal, state, and local entities discussed to cooperate to meet the needs of the listed species. The Middle Rio Grande Endangered Species Collaborative Program (Collaborative Program) coordinates the efforts of the various actors to achieve ESA compliance and fund research related to species recovery efforts. Notably absent from the list of Collaborative Program participants are environmental groups. Environmentalists used litigation as a tool to force species protection efforts under the ESA, bringing a citizen suit enforcement action against Reclamation and the Corps in 1999, which led directly to the creation of the Collaborative Program. But after some initial participation, environmental groups pulled out of the process. Environmental groups are still concerned about species protection, however, and are using litigation to ensure ESA compliance.

7.3 Social and Ecological System Dynamics

Similar to most watersheds in the American West, the SES dynamics of the MRG are driven by the social and ecological elements in the system. It can be difficult to parse the social from the ecological, as they often co-emerge and intertwine. For example, the built system including dams, levees, and other human-engineered changes to the Rio Grande river system are human in the sense that they were created by and serve social processes but also ecological in the sense that they alter the river in physical ways that create new ecological dynamics.

7.3.1 Built Infrastructure: Dams and Levees

Perhaps the most immediate and direct driver of change that has affected the MRG is the construction of physical infrastructure. The dams and levees control the volume, timing, and location of water flow and sediment in the river system. Between the Cochiti Dam and Elephant Butte Reservoir headwaters, 235 miles of levees exist (Fig. 7.1). Systematic levee construction in the MRG began in the late 1920s as a response to flooding and increased in the 1950s with funding support from the Flood Control Act of 1948. Besides the expansion of levees, the Flood Control Act of 1948 funded additional large flood control measures in the MRG, including clearing and straightening of the Rio Grande channel, construction of Cochiti Dam, and the installation of bank stabilization structures in the floodplain. Table 7.1 outlines the most significant dams and reservoirs influencing the MRG and their current management authorizations. This built infrastructure is necessary for both the agricultural and municipal use of the watershed's historic floodplain. Both water delivery and flood control are dependent on this complex network of ditches, levees, and dams.

7.3.2 Water Allocation Regimes and Agricultural Practices

The physical infrastructure is operated by governance structures that move and disperse water. Like in many Western states, water allocation in New Mexico is governed by the prior appropriation doctrine. Prior appropriation is a historically based allocation system that anticipates scarcity. The doctrine of prior appropriation states that when shortages occur, the right to use water is determined by the chronological order in which the water was first put to beneficial use. Senior appropriators are served first; and in a water-short year, junior appropriators may receive a reduced amount or no water, depending on the supply.

Water rights are usufructuary and vest only when applied to a beneficial use. In addition, water rights are subject to forfeiture if not used, which creates a general disincentive for conservation strategies. When the prior appropriation doctrine was formally established in New Mexico in 1891, there was no recognition of the values associated with leaving water in stream for wildlife and other uses; and until recently, leaving water in stream for fish and wildlife was not recognized as a beneficial use (Bardwell and Oglesby 2014). Instream flow rights remain relatively limited and, to date, have only been held in the MRG on a temporary leasing basis by Reclamation. Purchase of water rights for instream flow purposes is now authorized under the Strategic Water Reserve by the Interstate Stream Commission, although the Strategic Water Reserve has yet to be put into use for the Middle Rio Grande (Bardwell and Oglesby 2014). Water rights are considered a property right that can be bought and sold similar to other forms of real estate.

Today, the majority of water use supports agriculture in a traditional system of gravity-fed flood irrigation. Much of this system originated as a network of acequias, which served as local governance structures. Today, over 70 of the individual acequias have been consolidated into the MRGCD, which provides irrigation water for about 53,000 acres of crops, primarily alfalfa, and supports a thriving local dairy industry. Agriculture is mainly small scale and family owned or associated with the six MRG Pueblos that are members of the irrigation district.

7.3.3 Urban Growth and Associated Development

Like much of the American West, the MRG has seen a steady increase in human population and concomitant increase in municipal and industrial water demand. Santa Fe and Albuquerque each rely on ground and surface water supplies. While water conservation programs are an important part of the overall supply strategy, both municipalities are constantly seeking new sources. One source of supply is the purchase of senior water rights from willing sellers, and several municipalities in the MRG are purchasing water rights from farmers to meet their growing needs. As a result, there has been a shift in many water rights in the MRG from their original agricultural use to municipal use. Many of these transactions involve purchases of senior surface water rights to offset the impacts to the river of municipal groundwater pumping. Between 1982 and 2011, 21,000 acre-feet of water rights were transferred, most of which were transfers of agricultural rights to cities such as Albuquerque and Santa Fe (Bushnell and Armstrong 2014). The competing demands for a limited water supply in the MRG were highlighted in Reclamation's 2025 report, which highlights concerned areas in the western USA where existing water supplies are, or will be, inadequate to meet the water demands of people, cities, farms, and the environment even under normal water supply conditions (Llewellyn and Vaddey 2013).

7.3.4 Species Protection and Instream River Flow

The ESA is also a major societal factor in the MRG as it is the major driver of ecological restoration efforts in the watershed. Under Section 7 of the ESA, all federal agencies are required to consult with the appropriate wildlife agency to ensure that its actions are not likely to jeopardize the continued existence of listed species or result in destruction or adverse modification of critical habitat. If a jeopardy determination is made, the Fish and Wildlife Service, in its Biological Opinion, seeks to identify reasonable and prudent alternatives that would allow the action agency to move forward with the proposed activity while avoiding jeopardy for the species.

In the case of the silvery minnow, a jeopardy determination was made regarding the water operations for Reclamation, the Corps, and a number of nonfederal actors (such as the MRGCD) whose activities are closely linked to federal water operations. Activities that took place under the accompanying 2003 Biological Opinion included establishment of ramp-down rates for river flows that facilitate silvery minnow spawning; salvage operations for rescuing silvery minnow from isolated pools when necessary; establishment by Reclamation of a supplemental water program, in which it leases water on a willing seller-willing buyer basis to enhance instream flow; and further development of the Collaborative Program. Consultation is currently underway for a new Biological Opinion, and there is currently litigation brought by environmental groups challenging ESA implementation in the MRG. Whatever the outcome, it is safe to say that the silvery minnow will continue to be a major focus of management efforts, serving as a proxy for overall river health. One of the major challenges moving forward will be finding willing sellers for water transfers in order to increase the amount of water left instream for the minnow. During the past decade, the Water Authority was one of the main entities leasing water to Reclamation's Supplemental Water Program. That was because it was not yet using its allocation of San Juan Chama Project water—now it is.

7.3.5 Upland Forest Systems and Watershed Integrity

Important ecological drivers also are present in the forest upland areas associated with the MRG basin. These high-elevation forests are natural reservoirs that capture snow during the winter and release the moisture as runoff in the spring and summer. Two of the most dominant drivers of the recent changes to this system are increased activity of several species of bark beetle, which has killed many acres of pine trees, and an increased vulnerability to catastrophic wildfire, due to current moisture stress in combination with forest management practices of the past century (Allen 2007). In 2012 more than 172,000 acres of pinyon pine, ponderosa pine, or Douglas fir forest experienced mortality due to one or more species of bark beetle

(U.S. Department of Agriculture 2012). The beetle outbreaks cause large areas of upland forest to be susceptible to wildland fires (Allen 2007) with the greatest wildfire risk occurring shortly after the infestation and dropping off thereafter (Hicke et al. 2012). When combined with drought conditions currently gripping the Southwest, positive feedback loops are created between bark beetle outbreaks, forest dieback, and forest fires, leading to greater areas of forest mortality. This feedback loop and the threat of fires are expected to increase in intensity due to climate shifts in the future (Abatzoglou and Williams 2016; Moritz et al. 2012).

The degraded condition of the MRG's forested headwaters has resulted in a dramatic increase in fire frequency, severity, and size over the past decade (Schoennagel et al. 2004). A period of rapid forest harvesting around the turn of the century, followed by decades of fire suppression, resulted in tree densities and fuel loads that are drastically higher than they were historically (Moore et al. 2009). These high fuel loads combined with stress associated with sustained drought (Allen et al. 2010), increased air temperatures, reduced winter snowpack that is melting off earlier in the year (Westerling et al. 2006), forest dieback (Williams et al. 2010), and increased human activity in the wild land/urban interface (Theobald and Romme 2007) have all contributed to this trend. Additionally, grazing pressure in the lowlands under dry and hot conditions results in increased dust emissions that in turn reduce snowpack albedo, accelerate snowmelt, and reduce water availability (Beschta et al. 2012).

In the MRG headwater watersheds, the wildfire season (May through July) is followed by the monsoon rains (July through September). The timing of these events can lead to extreme flash flooding, debris slides, severely degraded water quality, and associated negative impacts on the natural and human systems that depend on the MRG and its tributaries (Schoennagel et al. 2004). Post-fire peak flows have been shown to be 10- to 100-fold higher than baseline conditions.

7.3.6 Climate Change and Drought

Humans have profoundly altered the water distribution, landscape, and ecosystems of the MRG. Current hydrologic analyses project that the impact of each of these changes will be exacerbated by the impacts of global climate change on this already overstressed system. These changes are expected to decrease overall water supplies, increase water demand from local agriculture and other outdoor uses due to higher temperatures, and intensify both droughts and floods. Longer growing seasons will put additional moisture stress on our forests, while increasing the survival of the bark beetles that have been preying on them (Raffa et al. 2008). Impacts of droughts and floods on human communities have been increasing in recent years, and these increasing trends are expected to continue (Showstack 2014).

7.4 Cross-Scale Interactions, Trajectories, and Tipping Points

The system elements discussed above are obviously only part of the complex and dynamic SES known as the MRG. The establishment of dams and levees, historically based and narrowly construed water supply allocation regimes and water operations, and increasing tensions between agricultural and municipal uses highlight some of the basic characteristics that make water governance challenging in this basin. The ecological elements such as impacts of channelization, stresses on upland forest systems, climate change, and drought provide a conceptual model of the ecology of the MRG. One advantage of using resilience theory to understand the MRG is that it emphasizes the dynamics among various elements of the SES, with particular emphasis on key interactions between the elements that comprise the watershed's system dynamics. Based on the necessarily cursory description of the social-ecological system elements above, two ecological processes will have an important impact on the MRG's identity: changes to the system's hydrograph and regime change in the upland forest system.

7.5 Changes in Hydrograph: Earlier Peak Runoff and Water System Dynamics

Climate change is causing the peak spring runoff in the MRG to occur earlier in the calendar year, and the portion of the year in which snowpack can be maintained is decreasing. Snowmelt runoffs are occurring 1–4 weeks earlier in many portions of the American West, and these trends are projected to continue (Rauscher et al. 2008). Many ecological events, including aquatic species reproduction, are tied to spring runoff events, often in concert with other cues, such as the length of the day or the timing of an insect hatch. For example, the timing of seed dispersal for various riparian plant species, such as cottonwoods and willows, is synchronous with the occurrence of peak river flows during the spring and early-summer seasons (Stella et al. 2006).

The endangered silvery minnow may also be impacted by a decreasing snowpack and earlier peak runoff. The silvery minnow has already suffered due to loss of riverine habitat and inadequate river flows. Reductions in snowpack and earlier spring peak flows will lead to even less available water to support summertime river flows (Cowley 2010). In addition, because silvery minnow spawn during elevated flows, shifts in spring flows have the potential to disrupt spawning patterns and further threaten the species.

Changes in the hydrograph are also affecting the social system. Built water storage operations may need to play many of the roles natural systems once played. Milder winters and hotter summers are likely to result in longer growing seasons for agriculture and urban landscaping, making for more intense and prolonged demands for water. Traditional surface water storage may become less practical, leading to increased experimentation with aquifer storage and recovery. The Water Authority, for example, recently completed the Bear Canyon Storage Project, the first large-scale effort in aquifer storage and recovery in the basin. Releasing small amounts of San Juan-Chama water into the Bear Canyon arroyo is part of this attempt to recharge the aquifer. The water authority intends to move from this more passive form of water infiltration to direct injection to get the water into the aquifer. The goal is to put up to 40,000 acre-feet back into the aquifer for use when surface water is less available. This strategy is an example of functional redundancy or building adaptive capacity by making the human community less reliant on one source of water.

7.6 Regime Changes for Upland Forest Systems and Implications for Downstream Communities

In the upland forest systems of the MRG watershed, many indicators suggest that the forest ecosystems have already crossed a threshold and are undergoing a transformation. The main driver of the current changes is wildfire, and the implications have been particularly noticeable in the Jemez Mountains of central New Mexico. When the Cerro Grande fire occurred in the Jemez Mountains in May 2000, it was the second largest fire in New Mexico's recorded history, with a total burn area of 47,650 acres; today, it ranks as the 20th largest wildfire. The 2011 Las Conchas Fire is now the second largest in New Mexico state history at 156,000 acres burned. It is the largest ever recorded in the Rio Grande watershed. The combination of bark beetle infestation, drought, and fire will have profound hydrological impacts.

7.7 Adaptive Governance

The effective management of resources within the watershed, including wildlife and pests such as the bark beetle, is key to the MRG's health, water supply, and water quality. Some steps are being made toward more effective management and are beginning to address the underlying drivers for increased fire risks in the upland forests of the MRG watershed, as large-scale fuel reduction (thinning and controlled burns) programs have been initiated. For example, the City of Santa Fe initiated a municipal water user fee to collect funds for watershed protection that has resulted in $7 million in forest treatments. However, for implementation of forest treatment programs to take place at meaningful scales, substantially more investment, both financially and institutionally, are required. In general, other stressors on the forest system, such as extended drought, climate change, and forest die-off, are not currently being addressed at the watershed scale.

The City of Santa Fe's efforts to protect their portion of the watershed provide a potential model for the MRG. The Nature Conservancy is leading a recent effort to do just that via a program called the Rio Grande Fund. This is a collaborative project that seeks to generate the funding necessary to conduct a large-scale forest treatment and watershed restoration program over a 10–30-year period with the help from stakeholders in the MRG. The Nature Conservancy estimates that $21 million a year—or close to $420 million over the next 20 years—will be needed. This is precisely the type of transformative adaptive capacity needed because it identifies the ecological threshold (upland forest regime change) and creates the capacity for choices among the different trajectories in order to build resilience.

7.8 Discussion

Resilience theory helps society recognize the need to build transformative capacity, when a system's current state cannot or no longer should be maintained. In the case of the MRG, future water governance must accommodate regime shifts in upland forest and riparian ecologies, increased stress from climate change and drought, and decreasing water availability. The MRG is a complex SES facing major challenges. The river itself represents the lifeblood of the system, upon which all other social and ecological elements of the system depend. Meeting these challenges will require reconsideration and restructuring of both the MRG's built water systems and its institutional structures of governance.

The MRG is a highly managed hydrological system. The construction and operation of its many dams, reservoirs, and levees, along with channelization activities, have lowered the overall functional diversity of the river system through channel simplification and bed degradation, with implications for both riverine and riparian habitats. In light of the system changes that have been made, as well as the developing hydrologic changes resulting from climate change, much of this built infrastructure is now needed to play part of nature's role. For example, the dams can be operated to release high flows that support the life cycle of native aquatic and riparian species, including endangered species, while also providing water for agricultural and municipal users.

The earlier peak runoffs that the system is beginning to experience will require more nuanced and intensive water management, including adaptive capacity in the form of increased management flexibility for Reclamation and the Corps, which operate the major reservoirs in the basin. All of the water projects managed by these two agencies have significant operational constraints (summarized in Table 7.1). Generally, these dams are authorized for flood control or water storage, but not both. They also have limits on the amount of water that can be stored at a given time, regardless of the dam's actual physical capacity.

The social system will need to make radical changes to its water governance strategies and its management of the upper watershed. Because the main regime changing drivers in the MRG—climate change and wildfire—are inherently unpredictable, how quickly these changes need to occur remains uncertain. There are some indications, however, that a historical climate pattern known as the Pacific Decadal Oscillation (PDO) may provide a window of opportunity. The PDO is a shift in the temperature pattern of the North Pacific Ocean and occurs on a 20- to 30-year cycle, providing relatively high amounts of precipitation to the Southwest, including New Mexico. Although the climatic mechanisms of the PDO are still being studied, there are indications that we are at the beginning of a PDO cool phase. If so, this temporary climate reprieve should lower average temperatures and provide higher-than-average precipitation, allowing those groups already at work in the New Mexico forests a bit of time to reduce fuel loads in these forests and restore riparian areas. The governance question becomes whether the social system can adapt and change old governance norms to take advantage of this possible window of opportunity.

Acknowledgments This work was developed in part under the Adaptive Water Governance Project, funded by the US National Socio-Environmental Synthesis Center (SESYNC) under funding from the US National Science Foundation, NSF DBI-1052875. The views set forth by contributors to this volume represent their own and do not represent the views of any public or private entity the contributor is affiliated with.

References

Abatzoglou J, Williams A (2016) Impact of anthropogenic climate change on wildfire across western US forests. PNAS 42:11770–11775

Allen A (2007) Interactions across spatial scales among forest dieback, fire, and erosion in northern New Mexico landscapes. Ecosystems 10(5):797–780

Allen CD, Macalady AK, Chenchouni H, BacheletD MDN, Vennetier M, Kitzberger T, Rigling A, Breshears DD, Hogg EH, Gonzalez P, Fensham R, Zhang IZ, Castro J, Demidova N, Lim JH, Allard G, Running SW, Semerci A, Cobb N (2010) A global overview of drought and heat-induced tree mortality reveals emerging climate change risks for forests. Forest Ecol Manag 259:660–684

Bardwell B, Oglesby A (2014) Water for New Mexico Rivers. In: Oglesby A (ed) Water matters! 17-1-17-8. Utton Transboundary Resources Center, University of New Mexico, Albuquerque, NM

Benson M, Morrison R, Stone M (2013) A classification framework for running adaptive management rapids. Ecol Soc 18(3):30

Beschta RL, Donahue DL, DellaSala DA, Rhodes JJ, Karr JR, O'Brien MH, Fleischner TL, Deacon Williams C (2012) Adapting to climate change on western public lands: addressing the ecological effects of domestic, wild, and feral ungulates. Environ Manag 51(2):474–491

Bushnell D, Armstrong S (2014) Groundwater. In: Oglesby A (ed) Water matters! 6-1-6-14. Utton Transboundary Resources Center, University of New Mexico, Albuquerque, NM

Carpenter S, Brock W (2008) Adaptive capacity and traps. Ecol Soc 13(2):40

Carpenter S, Walker B, JM AN (2001) From metaphor to measurement: resilience of what to what? Ecosystems 4:765–781

Cowley D (2010) Strategies for ecological restoration of the middle Rio Grande in New Mexico and recovery of the endangered Rio Grande silvery minnow. Rev Fish Sci 14:169–186

Folke C (2006) Resilience: the emergence of a perspective for social–ecological systems analyses. Glob Environ Chang 16:253–267

Gunderson L, Cosens B, Chaffin B, Arnold CA, Fremier AD, Garmestani AS, Craig RK, Gosnell H, irge HE, Allen CR, Benson MH, Morrison RR, Stone MC, Hamm JA, Nemec K, Schlager E, Llewellyn D (2017) Regime shifts and panarchies in regional scale social-ecological water systems. Ecol Soc 22(1):31. https://doi.org/10.5751/ES-08879-220131

Hicke JA, Johnson MC, Hayes JL, Preisle HK (2012) Effects of bark beetle-caused tree mortality on wildfire. For Ecol Manag 271:81–84

Hogan J et al (2014) Water quality regulation. In: Oglesby A (ed) Water matters! 18-1-18-8. Utton Transboundary Resources Center, University of New Mexico, Albuquerque, NM

Kelly S, Augusten I, Mann J, Katz L (2007) History of the Rio Grande reservoirs in New Mexico: legislation and litigation. Nat Resour J 47:525–613

Llewellyn D, Vaddey S (2013) West-wide climate risk assessment; Upper Rio Grande impact assessment. http://www.usbr.gov/WaterSMART/wcra/docs/urgia/URGIAMainReport.pdf. Accessed 29 July 2017

Moore MM, Covington WW, Fulé PZ (2009) Reference conditions and ecological restoration: a southwestern Ponderosa Pine perspective. Ecol Appl 9:1266–1277

Moritz MA, Peresien MA, Batllor E, Krawchuk MA, Van Dorn J, Ganz DJ, Hayhoe K (2012) Climate change and disruptions to global fire activity. Ecosphere 3:1–22

Phillips FM, Hall GE, Black ME (2011) Reining in the Rio Grande: people, land, and water. UNM Press, Albuquerque. 264 pp

Raffa KF, Aukema BH, Bentz BJ, Carroll AL, Hicke JA, Turner MG, Romme WH (2008) Cross-scale drivers of natural disturbances prone to anthropogenic amplification: the dynamics of bark beetle eruptions. Bioscience 58:501–557

Rauscher SA, Pal JS, Diffenbaugh NS, Benedetti MM (2008) Future changes in snowmelt-driven runoff timing over the western US. Geophys Res Lett 35:L16703. https://doi.org/10.1029/2008GL034424

Schoennagel T, Veblen TT, Romme WH (2004) The interaction of fire, fuels, and climate across Rocky mountain forests. Bioscience 54:661–676

Showstack R (2014) Concerns about extreme weather focus on the need for better resilience. Eos 95:69–70

Stella JC, Battles JJ, Orr BK, McBride JR (2006) Synchrony of seed dispersal, hydrology and local climate in a semi-arid river reach in California. Ecosystems 9:1200–1214

Theobald D, Romme W (2007) Expansion of the US wildland–urban interface. Landsc Urban Plan 83(4):340–354

U.S. Bureau of Reclamation (2013) Joint biological assessment bureau of reclamation and non-federal water management and maintenance activities on the middle Rio Grande, New Mexico, Part IV – The Middle Rio Grande Endangered species collaborative program recovery implementation program. http://www.usbr.gov/uc/albuq/envdocs/ba/MRG/Part4/BA-Part-IV.pdf. Accessed 29 July 2017

U.S. Department of Agriculture (2012) PR-R3-16-8, forest insect and disease conditions in the southwestern region. http://www.fs.usda.gov/Internet/FSE_DOCUMENTS/stelprdb5406441.pdf. Accessed 29 July 2017

Walker B, Salt D (2012) Resilience practice: building capacity to absorb disturbance and maintain function. Island Press, Washington, DC

Westerling AL, Hidalgo HG, Cayan DR, Swetnam TW (2006) Warming and earlier spring increase western U.S. forest wildfire activity. Science 313(5789):940–943

Williams AP, Allen CD, Millar CI, Swetnam TW, Michaelsen J, Still CJ, Leavitt SW (2010) Forest responses to increasing aridity and warmth in the southwestern United States. Proc Natl Acad Sci U S A 107(50):21289–21294

Zellmer S, Gunderson L (2009) Why resilience may not always be a good thing: lessons in ecosystem restoration from Glen Canyon and the Everglades. Nebraska Law Rev 87:93–849

Chapter 8
Resilience and Law in the Platte River Basin Social-Ecological System: Past, Present, and Future

Hannah E. Birgé, Craig R. Allen, Robin Kundis Craig, and Dirac Twidwell

Abstract A characteristic of the Anthropocene is an acceleration in the rate of change of many global environmental resources, including loss of biodiversity and increased freshwater use. However, societal response to accelerated environmental change often does little to prevent the undesirable and sudden social-ecological system changes that occur in response to relatively incremental resource depletion. Resilience theory provides a framework for evaluating the interactions among social-ecological systems and the policies meant to guide them toward desirable outcomes. This chapter examines the resilience of the Platte River Basin system through time, assessing linkages among environmental change and governmental institutions, policies, and geophysical realities of the region during three distinct social-ecological regimes: pre-European settlement, heavy modification of the river and adjacent land, and the Platte River Recovery Implementation Program (PRRIP). Policy guided by resilience theory accommodates the potential for rapid, nonlinear change characteristic of complex systems such as the Platte River Basin. With increasingly extreme floods and droughts predicted for the Great Plains in coming decades as climate change progresses, a resilience approach to policy and

---

H. E. Birgé (✉)
Nebraska Cooperative Fish and Wildlife Research Unit, School of Natural Resources, University of Nebraska, Lincoln, NE, USA

The Nature Conservancy, Omaha, NE, USA
e-mail: hannah.birge@tnc.org

C. R. Allen
U.S. Geological Survey, Nebraska Cooperative Fish and Wildlife Research Unit, School of Natural Resources, University of Nebraska, Lincoln, NE, USA
e-mail: callen3@unl.edu

R. K. Craig
S.J. Quinney College of Law, University of Utah, Salt Lake City, UT, USA
e-mail: robin.craig@law.utah.edu

D. Twidwell
Department of Agronomy & Horticulture, University of Nebraska, Lincoln, NE, USA
e-mail: dirac.twidwell@unl.edu

B. Cosens, L. Gunderson (eds.), *Practical Panarchy for Adaptive Water Governance*, https://doi.org/10.1007/978-3-319-72472-0_8

decision-making will contribute to desirable outcomes for people and nature in the next iteration of the Platte River Basin.

Keywords Resilience · Adaptive governance · Adaptive law · Resilience assessment · Climate change · Water management · Restoration

8.1 Introduction

A characteristic of the Anthropocene is an acceleration in the rate of change of many global environmental resources, including loss of biodiversity and increased freshwater use (Steffen et al. 2011). However, societal response to accelerated environmental change often does little to prevent the undesirable and sudden social-ecological system changes that occur in response to relatively incremental rates of resource depletion. Resistance to rapid change characterizes the judicial, executive, and legislative branches of the United States (US) government in the beginning decades of the twenty-first century, although the judicial branch may have capacity to be marginally more adaptive (Craig and Ruhl 2014). Science is similarly reluctant to adopt new ideas, requiring overwhelming evidence before adjusting or replacing paradigms and beliefs. As a result, new ecological knowledge is easily eschewed by decision-makers. Fortunately, emerging methods that integrate science and policy may provide decision-makers with tools for better adapting to complex environmental change.

Resilience thinking (Chap. 1) is a framework for evaluating the interactions among natural resource institutions, law, policy, and management. The hope is that resilience thinking-guided policy will introduce new natural resource management that addresses the dynamic and uncertain nature of ecological systems, allowing decisions to proceed while avoiding hidden, pathological outcomes for people and nature.

Resilience is an emergent property of a complex system that describes its capacity to withstand disturbance without losing defining and self-reinforcing structures and processes (i.e., system components). It is not a normative property, as evidenced by the resilience of many non-desirable system states that remain entrenched regardless of significant management intervention. Resilience includes the ability of the system to absorb disturbance through self-reorganization so that disturbances in the future are more efficiently absorbed, i.e., adaptive capacity. If a system's resilience is overcome by disturbance, initial reinforcing processes and functions are replaced with a novel set that are difficult or impossible to replace, and the system enters an alternative stable state or regime (Holling 1973; Scheffer et al. 2001). A social-ecological system's resilience is sometimes compromised during the quest to optimize production of ecosystem services, which are the benefits people obtain from nature. Optimizing for a single or few ecosystem services in the initial system state, for example crop production, often requires management to exaggerate a target process, for example irrigation, that may ultimately erode system resilience, leaving the system vulnerable to some proximate shock. In this way, system reconfigurations result in a sudden and unexpected loss of desired ecosystem services, including the target service of interest. Using resilience theory to understand how complex systems respond to human intervention, specifically that arising from laws, may provide decision-makers with tools for adapting to environmental change while proceeding with management and avoiding catastrophic loss of ecosystem services.

This chapter explores the resilience of the Platte River Basin system, assessing linkages among environmental change and governmental institutions, policies, and geophysical realities of the region during three distinct social-ecological regimes: pre-European settlement, heavy modification of the river, and the Platte River Recovery Implementation Program (PRRIP). In each historical period, decisions influenced, explicitly or otherwise, social-ecological trade-offs.

8.2 Study Area: The Platte River Basin

The Platte River (Fig. 8.1) drains an area of some 90,000 square miles in Colorado, Nebraska, and Wyoming (Palmer 1996). Its headwaters emerge in its westernmost states, where runoff from the eastern side of the continental divide flows into the South and North Forks of the Platte. These tributaries unite in western Nebraska to form the main stem, which is also fed by smaller tributaries and the Ogallala Aquifer (Platte River EIS Team 2000). The Platte River's South and North Forks run 424

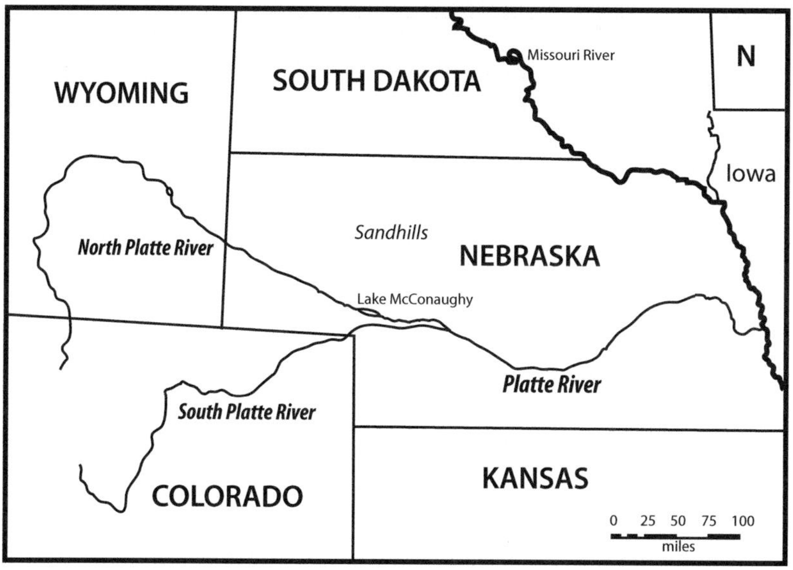

Fig. 8.1 The North Platte, South Platte, and main stem of the Platte river as it flows across the United States of Wyoming, Colorado, and Nebraska

and 618 miles, respectively. The Platte main stem runs 310 miles, ending as it flows into the Missouri on the eastern edge of Nebraska. The main stem currently delivers an average flow of 7052 cubic feet per second (NWIS 2009).

During the spring, snowpack on the east side of the Continental Divide of the Rocky Mountains melts and flows into the Platte. Prior to storage and electrification projects, snowmelt drained the slopes and recharged wetlands along the front range of the mountains into larger tributaries and laterally into the floodplain. During the year, high rain events drove occasional floods (Ring 1999), but the flow regime was dominated by seasonal snowmelt. After water from the North and South Platte stems combine into the main stem of the Platte, it flows down a shallower topographic gradient, averaging about 7 ft of elevation loss per mile as it transects the plains. Prior to major human interference, the Platte was effectively a drainage delta for the Rocky Mountains, with its shallow, braided flow path driving high functional connectivity to the land, like other plains rivers of the world (Poff et al. 1997). The periodic wetting and drying and lateral movement of water across the plains create a shifting spatial and temporal mosaic of ecosystem services, including waste removal, carbon sequestration, and nutrient transformation (Thomas et al. 2001). The dynamic river system provides habitat for native species, including those reliant on high-flow events to scour sandbars of vegetation and maintain floodplains for critical feeding and breeding grounds. The stream and the ecosystem services it generated provided sustenance and well-being to Indigenous People for millennia. However, the relationship between humans and these ecosystems changed with European settlement of the Americas.

Over the past century, temperatures across the Great Plains have been rising and are projected to increase (Karl et al. 2009). Average annual temperatures in the Great Plains were 0.8 °C higher in 2000 than for the 1960–1979 reference period and are expected to be 1.4–7.2 °C higher than this reference period by 2100. Because the majority of surface water in the Platte River Basin originates from winter snowpack, increasing temperatures that affect snowpack can have widespread impacts on water resources in the basin. High springtime temperatures have already resulted in consistently earlier snowmelt in much of the western United States (Shinker et al. 2010). Even earlier spring snowmelt is projected in the future, which may increase the lengths of summer droughts, alter wildfire management, and impact habitat adapted to late springtime floods.

8.3 Pre-European Settlement to the Late Twentieth Century

For more than 10,000 years prior to European colonization of the Platte River Basin, human populations inhabited a system characterized by periodic fires, floods, disease outbreaks, droughts, war, and sparse resources relative to today (Goebel et al. 2008). European settlers introduced novel disease, deforestation, and European agricultural traditions (Starkey 1998), causing a new set of anthropogenic alterations to the system (Krech 1999). Within a century, the millennia-old social-ecological system was transformed to maintain human populations that were orders of magnitude larger than the Indigenous settlements. Pre-European Indigenous societies were highly adapted to the ecological and biophysical realities of the system (Hart and Hart 1997; Wishart 1979), while European settlers altered millennia-old ecological (i.e., land cover) and biophysical (i.e., stream flow) system feedbacks to support a new configuration of human well-being. Whether the replacement of these endemic processes altered social-ecological system resilience to fires, floods, droughts, and disease, especially considering the novel stressors of the Anthropocene, remains uncertain.

Although European pioneer settlers quickly spread westward, establishing small settlements and clashing with native populations, the federal government considered itself the sole proprietor of the western land for roughly the first century of European settlement (Opie 1987). The US Congress of the Confederation adopted the Land Ordinance of 1785 (U.S. Cong. Conf. 1785) both to create the public land survey system for the West and to generate federal income from the sale of public lands (U.S. Cong. Conf. 1785), even though, largely because of the cost to purchase, no land was sold until 1796 (Carstensen 1968). For the next decades, land prices set by the federal government were prohibitively high for poor settlers, who chose instead to stay on the land without ownership. These European settlers scratched out a living in defiance of Indigenous inhabitants, and federal efforts to remove them failed.

Population growth rapidly increased in what is now Colorado, Nebraska, and Wyoming, driven by gold rushes of the mid-nineteenth century (Hobbs 1997) and sustained by agriculture (Smith 1992). The Homestead Act (Pub. L. No. 37–64, 12 Stat. 392) offered settlers of European descent and free African Americans 160-acre parcels of land at a trivial fee on a showing of 5 years of continuous inhabitance and development, further emphasizing agriculture as central to the development of the region. The Kinkaid Act of 1904 (33 Stat. 547) amended the Homestead Act in Nebraska to allow 640-acre parcels of land per homesteader. The Homestead Act also contributed to further displacement of Indigenous Peoples.

As populations expanded, so did cultivation of grasslands for crop production in arid and semiarid regions, requiring increasingly complex irrigation systems (Hobbs 1997). Settlers were familiar with agriculture but not with water scarcity, and they had no experience with irrigation-dependent agriculture (Radosevich 1976). Indeed, most settlers' prior experience with water rights and sharing was based on a system of riparian law developed in water rich areas (Rose 1994). Under riparian law, water rights reside in owners of land adjacent to rivers or lakes. These landowners are expected to make reasonable use of water and share equally in reductions during rare times of scarcity (Adler et al. 2013). Nebraska originally adopted the legal system of riparian rights (Wasserburger v. Coffee 1966), but riparian law was ill-suited to the Platte River Basin and other semiarid systems where the location of minerals, fertile soil, and associated towns were not always distributed in proximity to the riverbed. This required that water be preferentially diverted to settlements far from the stream itself and placed emphasis on a first-come first ability to develop water rather than proximity to water.

The unsuitability of riparian rights in western states led to development of the legal doctrine of prior appropriation (Irwin v. Phillips 1855; Coffin v. Left Hand Ditch Co. 1882). Under the prior appropriation system, water is allocated based on the seniority of water rights rather than proximity to the river, and the burden of scarcity is not proportionally shared (Coffin v. Left Hand Ditch Co. 1882; Rose 1994; Adler et al. 2013). The shift from riparian law was based on an economy and property rights associated with the resource that first attracted settlers to the western United States—gold (Trelease 1971). In the gold rush era of the western United States, the guiding principle was "first in time, first in right" (Irwin v. Phillips 1855). All three of the Platte River Basin states—Colorado, Nebraska, and Wyoming—adopted prior appropriation as their water allocation systems. Colorado adopted prior appropriation as the exclusive method for acquiring water rights long before statehood (Coffin v. Left Hand Ditch Co. 1882), and upon statehood the Colorado Constitution explicitly defined prior appropriation as the means by which water would be governed (Colo. Const. Art. XVI § 6). As in Colorado, the Wyoming Supreme Court traced that state's prior appropriation doctrine to its earliest history of territorial mining and irrigation, concluding that "in this and other jurisdictions, [...] the common law in respect to the use of water and the right thereto is altogether ignored" (Willey v. Decker 1903). Upon statehood, prior appropriation became part of the Wyoming Constitution (Wyo. Const. art 8 §§ 1, 3). Nebraska's history of prior appropriation is more convoluted. Nebraska become a state in 1867 and first adopted prior appropriation as a non-exclusive means to acquire water rights in the Irrigation Act of 1889, effective March 27, 1889 (Neb. Laws 1889, ch. 68, p. 503). However, according to the Nebraska Supreme Court, the state legislature did not fully displace the common law of riparian rights until the Irrigation Act of 1895, effective April 4, 1895 (Neb. Laws 1895, ch. 69, p. 244; Wasserburger v. Coffee 1966). Prior appropriation then became part of the Nebraska Constitution in 1920 (Neb. Const. Art. XV §§ 4–6 (1920); Wasserburger v. Coffee 1966).

For European settlers, beneficial use meant irrigation and industry. Growers in the Platte River Basin continue to produce crops that require high water inputs, including corn, soybean, and wheat. In the western half of the arable zone of the basin, this production is reliant on irrigation. In the eastern half, irrigation supplements rain-fed cropping. As the prior appropriation systems became entrenched in the governance of *within* state water appropriations, conflicts arose between states regarding shared water resources. Such conflicts led to higher-level jurisdiction to determine interstate water allocation (Adler et al. 2013). In the Platte River Basin, compacts and US Supreme Court decrees currently comprise the large share of interstate water allocation agreements. The South Platte River Compact of 1923 between Nebraska and Colorado (South Platte River Compact 1923) was a cooperative approach for handling water disputes. The compact was ratified by Congress in 1926 and committed Colorado to deliver minimum water to Nebraska during the irrigation season. Yet, in the 1930s Nebraska filed a suit against Wyoming (with Colorado joining as a defendant) for equitable apportionment of the Platte River with the US Supreme Court. The Court first denied Wyoming's motion to dismiss (Nebraska v. Wyoming 1935) and then Colorado's (Nebraska v. Wyoming 1945). In

June 1945, the Court issued one of the very few equitable apportionment orders in its history, resolving in 1945 (Nebraska v. Wyoming 1993; Nebraska v. Wyoming 1995) a long-standing dispute over allocation of the North Platte between Nebraska and Wyoming by establishing a relatively clear set of allocation rules wherein Nebraska received 75% and Wyoming received 25% of the river flow (Nebraska v. Wyoming 1945 (enforcing the decree)).

To expedite the process of Western development and settlement, Congress enacted the Reclamation Act of 1902 (Act of June 17, 1902, ch. 1093, 33 Stat. 388, *codified as amended at* 43 U.S.C. §§ 372–498), which founded the Bureau of Reclamation and funded extensive dam and irrigation projects throughout the West, including Nebraska, Colorado, and Wyoming (U.S. Bureau of Reclamation 2014). As industry and human settlements proliferated in the Platte River Basin during the late nineteenth and early twentieth centuries, there was a growing societal demand to reduce variability in water availability. Specifically, mitigating droughts and controlling floods were the social objectives of these water management projects. Currently, 15 major reservoirs or dams and roughly 200 smaller diversion or storage schemes divert and store an average of more than 7.1 million acre-feet along the Platte River (Eisel and Aiken 1997).

Dams and diversions modify overall flow, flow variability, and high-flow events. High-flow events are critical to ecological function of braided streams like the historical Platte. High-flows transport and deposit sediment, scour vegetation from sandbars and banks, mobilize logjams, and transport water and materials across the floodplain and downstream (Poff et al. 1997). With more dams and diversions, the stream also has lower overall flow, less variable flow, and fewer high-flow events. As a result of these modifications, the Platte River is increasingly more channelized, floodplains are reduced, and dynamic riverine wetlands and sandbar island habitat are disappearing (Hynes 1975). Wetlands within or adjacent to the river have declined by an estimated 73% relative to historic cover, and natural sandbars and meander loop wetlands are all but gone from the system (PRRIP 2006). The modern Platte has static, well-defined edges and islands armored with woody vegetation resistant to scouring from high-flow events (Brooker 1985). This altered flow regime drives other less visible ecological features including nutrient cycling, temperature stratification, water velocity, and turbidity (Brooker 1985). Dams and diversions may also serve to isolate aquatic populations, creating population bottlenecks (Benstead et al. 1999) and putting populations of at-risk and endangered species more vulnerable to extinction (Elmqvist et al. 2003).

Due to this profound functional and structural shift, we consider the Platte River to occupy a stable state alternative to the initial state characterized by a braided, shallow, meandering stream with high functional connectivity to the floodplain and sandbar islands. Critically, even with historic peak flows and flooding events, the Platte cannot return to its previous regime due in part to its channelization and armored islands, which both contribute to its stability, i.e., self-reinforcing (negative) feedbacks, in this alternative state. In other words, restoration of the historical flow regime alone will not restore the ecosystem to its initial stable state, because removal of historical flows allowed novel feedbacks to establish which are now highly resilient to management intervention—including the reintroduction of historical high flows.

The Platte River social system also occupies a state that is alternative to that of pre-European settlement, with its own set of self-reinforcing feedbacks. Water provisioning to humans is largely unresponsive to small fluctuations in precipitation and flow, as storage and diversion plans decouple water availability from small-scale (i.e., ten year) droughts. However, the system is not invulnerable to extreme droughts and floods, which may increase in frequency and duration with climate change (Ojima et al. 1999). Beyond the significant social (i.e., cultural, technological, political, and economic) intervention required to possibly revert the system to its pre-European state, that system state supported much smaller-sized human populations with low *immediate* resilience to starvation, flooding, drought, and conflicts over sparse resources—although perhaps with more resilience in the entire population of the Great Plains to inherent ecological and biophysical variability. Here, components of the social and ecological systems have transformed together since European settlement in the system, often through tight and complex coupling.

8.4 Late Twentieth Century–Present: Platte River Recovery Implementation Program (PRRIP)

Beginning in the late twentieth century, the Platte River Recovery Implementation Program (PRRIP) was established partly in recognition that the social-ecological system of the Platte River Basin was so transformed that species reliant on features of the historical stream were at risk of extinction. The federal Endangered Species Act (ESA; 16 U.S.C. §§ 1531–1544), as reformulated in 1973, provided a clear and robust mandate to "halt and reverse the trend toward species extinction, whatever the cost." Resilience theory illustrates that undesirable outcomes in complex systems may not always be reversible simply by removing the offending process due to new reinforcing feedbacks in alternative configurations of the system. By contrast, legislative bodies adopt a command-and-control approach toward natural resources management such as the ESA which, while less effective than a resilience-type approach, provides what appears to be a more efficient decision-making tool. However, the ESA marked a strong departure from earlier conservation policy of the twentieth century that subordinated environmental needs to economic growth and was transformative for environmental policy.

Under the ESA, once a species is listed for protection (16 U.S.C. § 1533(a)(1), (b)(1)), Section 7 requires all federal agencies to, "in consultation with and with the assistance of the [U.S. Fish & Wildlife Service or the National Marine Fisheries Service], insure that any action authorized, funded, or carried out by such agency … is not likely to jeopardize the continued existence of any endangered species or threatened species or result in the destruction or adverse modification of [its critical] habitat" (16 U.S.C § 1536(a)(2)). One of the greatest operational challenges of the ESA is enforcing and defining the meaning of "to jeopardize." Under the US Fish and Wildlife Service and National Marine Fisheries Service's joint regulations, "[j] eopardize the continued existence of means to engage in an action that reasonably would be expected, directly or indirectly, to reduce appreciably the likelihood of both the survival and recovery of a listed species in the wild by reducing the reproduction, numbers, or distribution of that species" (50 C.F.R. § 402.02). Thus, the process of determining "jeopardy" can be open to agency subjectivity or discretion, inviting criticism once a determination is or is not made. In addition, the listing of species triggers a response on the part of the US Fish and Wildlife Service (USFWS) and the National Marine Fisheries Service to then recover the species to a level where the Act's protections are no longer required (16 U.S.C. § 1533(f)). Both the jeopardy consultation requirements and the species recovery obligation can effectively change the operation of rivers with federal dams when listing of aquatic or riparian species occurs. For example, if the USFWS concludes that a proposed project will jeopardize the continued existence of a listed species, Section 7 requires that it identifies in its resulting Biological Opinion "reasonable and prudent alternatives" (RPAs) to offset or reduce the threat to endangered species (16 U.S.C. § 1536(b)(3)(A)). If the RPAs are deemed sufficient to eliminate the jeopardy concern, the project can move forward—but generally only in compliance with the RPAs (Bennett v. Spear 1997). Moreover, the consultation and RPAs set a precedent for similar projects to move more easily through the regulatory system.

In the Platte River Basin in Nebraska, four species have been listed for protection under the ESA: the whooping crane, piping plover, interior least tern, and pallid sturgeon (Platte River ESA listings 1967, 1985, 1990). Efforts by the three Platte River states (Colorado, Nebraska, and Wyoming) to address the degradation of the river to recover endangered and threatened species have been ongoing since 1997 (Freeman 2008). The impetus for this effort was the 1994 relicensing of Lake McConaughy's Kingsley Dam. This was the first relicensing of a major hydroelectric project in the basin since the 1978 declaration of endangered species by the USFWS. Prior to the dam's relicensing, the endangered species listing had stopped, deferred, or substantially modified all new proposed water projects in the basin (Aiken 1999), but this was the first time an existing project was susceptible (Freeman 2008). Nebraska, Colorado, and Wyoming faced serious restrictions in water allocations and, after consultation with USFWS and facing the threat of litigation, agreed in 1994 to negotiate a basin-wide agreement known as the Platte River Recovery Implementation Program (PRRIP).

The primary objectives of the PRRIP are to (1) increase stream flow in the main stem during critical periods, (2) restore habitat for target bird species, and (3) implement adaptive management—a management strategy rooted in resilience theory and designed to reduce system uncertainty while proceeding with management. The agreement satisfied the relicensing requirements for the Kingsley Dam, which continued in its capacity to provide for municipal, industrial, and agricultural water consumption. PRRIP is divided into phases of activities and projects directed at species recovery and founded in adaptive management. The first increment is intended to cover 13 years, from 2007 through 2019, but the phases may be adjusted depending on the performance of earlier increments (Freeman 2008). At that point, the performance of the program will be evaluated, and additional and different activities may be agreed upon to further support species' protection and recovery.

The PRRIP agreement takes key steps in making more water available to the river, restoring and protecting critical habitat, and explicitly incorporating adaptive management into its governance structure. The first increment of the program consists of mitigating pre-1997 water uses by the three states by providing an additional 130,000–150,000 acre-feet of water to the river at strategic points in space and time to best meet the needs of recovery efforts (Freeman 2008). For instance, through the Tamarack Recharge Project located near the Nebraska border, Colorado captures South Platte surface water and pumps it into ponds at varying distances from the river. The water percolates into groundwater which is predicted through hydrological modeling efforts to return to the stream at the times identified by biologists as having the greatest impact on species recovery. Wyoming meets its commitment of 34,000 acre-feet by expanding storage at the Pathfinder Dam, located on the North Platte approximately 111 miles from the Nebraska border and performing strategic releases (Autobee 1996). Finally, Nebraska meets its commitment by devoting 10% of the water captured in Lake McConaughy during the winter storage season (November through April) to the Program's surface water requirements.

In addition, the states agreed to replace all new depletions of water from 1997 forward, with each state tailoring a depletion plan that meets both its own needs and the requirements of the PRRIP agreement. Every year Wyoming measures and compares current water use against the thresholds, and if water use exceeds the thresholds, the state covers the excess depletions by releasing water previously stored in reservoirs back to the stream (Freeman 2008). Nebraska has developed a depletions plan that centers on conjunctive management of surface and groundwater by diverting water into irrigation canals during the non-irrigation season, which seeps into the groundwater and eventually flows to the river, supplementing this with reservoir releases as needed (PRRIP 2012). Nebraska also agreed to cover additional depletions incurred by groundwater wells installed between 1997 and 2005. Since 2005, well owners and/or the state—depending on the mechanism of depletion—have been responsible for replacing depletions (Freeman 2008). Like Nebraska, Colorado relies on conjunctive water management to cover post-1997 depletions (Freeman 2008).

In addition to covering new and existing water diversions, PRRIP acquired more than 10,000 acres of riparian habitat in the central Platte River Basin between Lexington and Chapman in Nebraska for various conservation projects using a willing buyer/willing seller model. The land itself is managed according to a Good Neighbor policy, which includes removing weeds, maintaining fencing, and paying taxes (PRRIP 2009, 2010). Each parcel of land is managed according to a plan designed for species recovery purposes.

Adaptive management is critical to the success of PRRIP. Adaptive management (AM) treats interventions as experiments, using results to learn about system dynamics and revising subsequent activities to more effectively and efficiently reach management goals (Holling 1978). The PRRIP's AM goals center around how best to restore habitat of the four endangered and/or threatened species in the basin. PRRIP's AM plan is meant to provide crucial knowledge for improving states' compliance and management activities (Freeman 2008; PRRIP 2012).

The Platte River Recovery Implementation Program is guided, monitored, and governed by a committee consisting of representatives from Colorado, Nebraska, Wyoming, the US Bureau of Reclamation, the USFWS, relevant end user stakeholders, and environmental interest groups. The committee meets regularly to review the program performance and state compliance (PRRIP 2012). The committee is assisted in its efforts by an executive director who oversees day-to-day operations; an environmental account manager employed by the USFWS who oversees releases of water from reservoirs; advisory committees devoted to the water, land, and adaptive management plans and programs; and an independent scientific advisory committee (PRRIP 2012).

The performance of the PRRIP ultimately relies on the actions of the signatory states. Each state's ability to meet its commitments depends on the state's own water laws and policies and the extent to which their water administrators have the authority to implement the required activities. Water law in Wyoming, Colorado, and, most recently, Nebraska acknowledges the hydrologic connectivity between groundwater and surface water (i.e., allocating one influences the availability of the other) and gives water officials the authority to regulate wells in ways that minimize their impact on surface water flows (Hobbs 1997; Bryner and Purcell 2003). In Nebraska, prior to 2004, ground and surface water were not recognized by the Legislature as interdependent hydrologic resources and were thus managed separately, by natural resources districts (NRDs) and the Nebraska Department of Natural Resources (NDNR), respectively (Aiken 2006). The 1976 Groundwater Management and Protection Act (Neb. Laws 1975) granted NRDs the authority to develop groundwater management plans, subject to the approval of NDNR (Neb. 1975 § 46–707). Most districts developed groundwater quality plans, but only a single district chose to actively regulate groundwater pumping as well (Stephenson 1996); the NDNR could not compel districts to adopt groundwater management plans that strictly regulated pumping or that took into account surface water impacts of pumping (Neb. Laws 1975 §§ 46–707, 46–713).

As a consequence of Nebraska's legal and administrative separation of groundwater and surface water management (Kelly 2010), the NRDs were not required to consider stream flow (i.e., surface water) needs in issuing permits, and managers did not have policy tools to bridge the two water sources and manage them in an integrated fashion (Schlager and Blomquist 2008). Thus, between 1997 and 2004, Nebraska representatives negotiating the PRRIP could not credibly commit to directing the water provided by the states upstream of the central Platte River Basin to meet in stream flow objectives, nor did they have legal and administrative tools in place to implement a realistic depletion plan.

This changed in 2004 when the Nebraska legislature adopted LB 962 (Neb. Laws 2004), which gave the NDNR the authority to declare river basins as over- or fully appropriated (Neb Laws 2004 § 5; Kelly 2010). Under the Act, a designation triggers integrated water management, obliging NRDs and NDNR to develop integrated management plans that recognize the hydrologic connection between surface and ground water. Furthermore, the bill allowed the NDNR to designate both surface and ground water as overappropriated (Neb. Laws 2004 § 53). NDNR has since designated many parts of the Platte River as overappropriated. For example, "[i]n the North Platte NRD, the North Platte River valley and the Pumpkin Creek basin were declared over appropriated, and the rest of the District was designated as fully appropriated" (NPNRD 2017). By the time the PRRIP was adopted, Nebraska's NRDs and the NDNR had an integrated water management plan in place for the Platte River Basin (NDNR 2009; Schlager and Blomquist 2008). Thus, Nebraska could commit to its partners that it can protect program water supplies and implement an effective depletions plan.

The PRRIP is one of the first multistate, collaborative agreements triggered by the ESA. The contemporary, PRRIP-era Platte River does not represent a return to the pre-European, braided stream system state and the specific suite of services it provided to people and nature. It does, however, provide increased land-river connectivity, creation of habitat, and a shift away from unfettered appropriation of surface and ground water.

8.5 Synthesis

Natural resources legislation of the past 50 years are based on an understanding of ecosystems as hierarchical, predictable systems that can be managed effectively through top-down, command-and-control approaches. In this view, an undesirable outcome is reversed by removing the offending process (Holling and Meffe 1996). Current understanding suggests that complex systems of people and nature rarely behave in a predictable manner to top-down control over the long-term. Instead, attempts to optimize or reverse outcomes in systems result in slow changes that erode resilience of the system over time, with an unwanted reconfiguration often proximately triggered by a disturbance or natural disaster. Thus, in order for the next generation of natural resources policy in the Platte River Basin to effectively regulate social interaction with ecological systems, policy must become "cross-scale, interdisciplinary, and dynamic" (Ruhl 2012). This includes rejecting the notion that ecosystems can be controlled from the top-down.

Reframing policy to a resilience perspective will accommodate the inherent complexity and uncertainty of natural systems, improving how we manage natural resources. To do this, experiments that are congruent with law and policy should be designed so that they are safe to fail while fostering learning about complex systems. However, institutional change, which can be slow and difficult to initiate, is required to meaningfully transform systems to accommodate or adopt adaptive governance. Instead, developing an adaptive governance framework (Craig and Ruhl 2014) so that knowledge can be integrated during windows of opportunity could be the most efficient path toward integration into the courts.

8.6 Conclusions

The need for laws that incorporate flexibility and adaptive governance principles is urgent as global scale climate change begins incurring local, unexpected changes in the Platte River Basin, which is already contending with profound transformations. Specifically, climate change-mediated changes in flow regime will likely affect current legal allocations within the basin, since the amount of water allocated to different states is based on river levels from the mid-to-late 1900s, whereas river flows will invariably shift with higher temperatures, longer droughts, changes in snowpack size and melt timing, and heavier precipitation events spaced further apart in time. Like many policy and management plans, PRRIP was formulated on an assumption of centralization-which is an efficient way to formulate transparent policy. Yet this approach may not perform as expected as interactive, complex global change stressors reduce our predictive capacity. Policy guided by resilience theory accommodates the potential for rapid, nonlinear change, establishes organizational and procedural norms, does not dictate outcomes, and transforms as learning emerges (Cumming 2013). Such governance is needed if people and nature are to persist in the next iteration of the Platte River Basin.

Acknowledgments This work was developed in part under the Adaptive Water Governance Project, funded by the US National Socio-Environmental Synthesis Center (SESYNC) under funding from the US National Science Foundation, NSF DBI-1052875. The Nebraska Cooperative Fish and Wildlife Research Unit is jointly supported by a cooperative agreement among the US Geological Survey, the Nebraska Game and Parks Commission, the University of Nebraska, the US Fish and Wildlife Service, and the Wildlife Management Institute.

This work is dedicated to J. Kenney, who led PRRIP as executive director for over a decade with transparency, fairness, and acuity; mending deep rifts and showing the power of authentic collaboration.

References

Adler RW, Craig RK, Hall ND (2013) Modern water law: private property, public rights, and environmental protections. Foundation Press, St Paul

Aiken JD (1999) Balancing endangered species protection and irrigation water rights: the Platte River cooperative agreement. great plains. Nat Resour J 3:119–158

Aiken JD (2006) Hydrologically-connected ground water, section 858, and the Spear T Ranch decision. Nebr Law Rev 84:962

Autobee R (1996) North Platte project. US Bureau of reclamation. https://www.usbr.gov/projects/pdf.php?id=145. Accessed 23 Oct 2017

Benstead JP, March JG, Pringle CM, Scatena FN (1999) Effects of low-head dams and water abstraction on migratory tropical stream biota. Ecol Appl 9:656–668

Brooker MP (1985) The ecological effects of channelization. Geogr J 151:63–69

Bryner G, Purcell E (2003) Groundwater law sourcebook of the Western United States. Natural Resources Law Center/University of Colorado School of Law, Boulder

Carstensen V (1968) The public lands: studies in the history of the public domain. University of Wisconsin Press, Madison

Craig RK, Ruhl JB (2014) Designing administrative law for adaptive management. Vanderbilt Law Rev 67:1–87

Cumming GS (2013) Scale mismatch and reflexive law. Ecol Soc 15. http://www.ecologyandsociety.org/vol18/iss1/art15/. Accessed 23 Oct 2017

Eisel L, Aiken JD (1997) Platte River Basin study: report to the Western Water Policy Review Advisory Commission. U.S. Department of Commerce. http://digitalcommons.unl.edu/cgi/viewcontent.cgi?article=1034&context=ageconfacpub. Accessed 23 Oct 2017

Elmqvist T, Folke C, Nystrom M, Peterson G, Bengtsson J, Walker B, Nor J (2003) Response diversity, ecosystem change and resilience. Front Ecol Environ 1:488–494

Freeman DM (2008) Negotiating for endangered and threatened species habitat in the Platte River Basin. In: Doyle M, Drew CA (eds) Large-scale ecosystem restoration: five case studies from the United States. Island Press, Washington, DC, pp 59–67

Goebel T, Waters MR, O'Rourke DH (2008) The Late Pleistocene dispersal of modern humans in the Americas. Science 319:1497–1502

Hart RH, Hart JA (1997) Rangelands of the Great Plains before European settlement. Rangelands 19:4–11

Hobbs GJ (1997) Colorado water law: an historical overview. Univ Denver Law Rev 1:1

Holling CS (1973) Resilience and stability of ecological systems. Annu Rev Ecol Syst 4:1–23

Holling CS (1978) Adaptive environmental assessment and management. Wiley, London

Holling CS, Meffe GK (1996) Command and control and the pathology of natural resource management. Conserv Biol 2:328–330

Hynes HBN (1975) Edgardo Baldi memorial lecture: the stream and its valley. Verhandlungen der Internationalen Vereinigung für Theoretische und Angewandte Limnologie 19:1–15

Karl TR, Melillo JM, Peterson TC (2009) Global climate change impacts in the United States. Cambridge University Press, New York

Kelly M (2010) Nebraska's evolving water law: overview of challenges & opportunities. Platte Institute for Economic Research. https://www.platteinstitute.org/Library/docLib/20100927_Kelly_Paper_-_FINAL.pdf. Accessed 23 Oct 2017

Krech S III (1999) The ecological indian: myth and history. Norton, New York

National Water Information System [NWIS] (2009) Platte River at Louisville, Nebr. In: Water-data report 2009. U.S. Geological Survey. https://wdr.water.usgs.gov/wy2009/pdfs/06805500.2009.pdf. Accessed 23 Oct 2017

Nebraska Department of Natural Resources [NDNR] (2009) Basin-wide plan for joint integrated water resources management of overappropriated portions of the Platte River Basin, Nebraska. https://dnr.nebraska.gov/sites/dnr.nebraska.gov/files/doc/water-planning/upper-platte/BasinWidePlan_2009.pdf. Accessed 23 Oct 2017

North Platte Natural Resources District [NPNRD] (2017) Integrated water management plan. http://www.npnrd.org/water-management/integrated-management-plan.html. Accessed 30 Sept 2017

Ojima D, Garcia L, Elgaali E, Miller K, Kittel TGF, Lackett J (1999) Potential climate change impacts on water resources in the Great Plains. J Am Water Resour Assoc 35:1443–1454

Opie J (1987) The law of the land: two hundred years of American farmland policy. University of Nebraska Press, Lincoln

Palmer T (1996) America by rivers. Island Press, Washington, DC

Platte River EIS Team (2000) Central Platte River 1998 land cover/use mapping project Nebraska. U.S. Bureau of reclamation. https://www.platteriverprogram.org/PubsAndData/ProgramLibrary/USBR%202000_Central%20Platte%20River%201998%20Land%20Cover%20Mapping%20Project.pdf. Accessed 23 Oct 2017

Platte River Recovery Implementation Program (2011) Bi-annual report, 2009 & 2010. Headwaters Corporation. https://www.platteriverprogram.org/PubsAndData/ProgramLibrary/PRRIP%202011_BiAnnual%20Report%202009-2010.pdf. Accessed 23 Oct 2017

Poff NL, Allan JD, Bain MB, Karr JR, Prestregaard KL, Richter BD, Sparks RE, Stromberg JC (1997) The natural flow regime: a paradigm for river conservation and restoration. Bioscience 47:769–784

PRRIP (Platte River Recovery and Implementation Program) (2006) Available at http://dnr.ne.gov/Media/iwm/PDF/PRRIP_Document_2006.pdf. Discussing the final Platte River implementation program

PRRIP (Platte River Recovery and Implementation Program) (2009) Available at https://www.platteriverprogram.org/PubsAndData/ProgramLibrary/PRRIP%202010_%202009%20WAC%20Update.pdf.

PRRIP (Platte River Recovery and Implementation Program) (2010) Available at https://www.platteriverprogram.org/News/Documents/PRRIP%20BiAnnual%20Report%202009-2010.pdf Biannual Report

PRRIP (Platte River Recovery and Implementation Program) (2012) Platte River Recovery Program, United States Fish & Wildlife Service, http://www.fws.gov/mountain-prairie/wtr/PlatteRiver.htm. Last visited 3 Dec 2014. [hereinafter USFW PRRP 2012]

Radosevich G (1976) Evolution and administration of Colorado water law: 1876–1976. Water Resources, Fort Collins

Ring R (1999) Saving the Platte. High Country News. http://www.hcn.org/issues/147/4744. Accessed 24 Oct 2017

Rose CM (1994) Property and persuasion: essays on the history, theory, and rhetoric of ownership. Westview, Oxford

Ruhl JB (2012) Panarchy and the law. Ecol Soc 17(3):31. https://doi.org/10.5751/ES-05109-170331

Scheffer M, Carpenter S, Foley JA, Folke C, Walker B (2001) Catastrophic shifts in ecosystems. Nature 413:591–596. https://doi.org/10.1038/35098000

Schlager E, Blomquist W (2008) Embracing watershed politics. University of Colorado Press, Boulder

Shinker J, Shuman B, Minckley T et al (2010) Climatic shifts in the availability of contested waters: a long-term perspective from the headwaters of the North Platte River. Ann Assoc Am Geogr 100:866–879

Smith DA (1992) Rocky Mountain West: Colorado, Wyoming and Montana 1859–1915. University of New Mexico Press, Albuquerque

Starkey A (1998) European and Native American warfare 1675–1815. University of Oklahoma Press, Norman

Steffen W, Persson Å, Deutsch L, Zalasiewicz J, Williams M, Richardson K, Crumley C, Crutzen P, Folke C, Gordon L, Molina M, Ramanathan V, Rockström J, Scheffer M, Schellnhuber HJ, Svedin U (2011) The Anthropocene: from global change to planetary stewardship. Ambio 40:739–761

Stephenson K (1996) Groundwater management in Nebraska: governing the commons through local resource districts. Nat Resour J 36:761–778

Thomas SA, Valett HM, Mulholland PJ, Fellows CS, Webster JR, Dahm CN, Peterson CG (2001) Nitrogen retention in headwater streams: the influence of groundwater-surface water exchange. Sci World J 1:623–631. https://doi.org/10.1100/tsw.2001.272

Trelease FJ (1971) Federal-state relations in water law, Legal study. U.S. National Water Commission, Arlington

U.S. Bureau of Reclamation (2014) Projects and facilities database. http://www.usbr.gov/projects/. Accessed 24 Oct 2017

U.S. Congress of the Confederation (1785) Ordinance of May 20, 1785

U.S. Geologic Survey, Water Resource Data, Nebraska, Water Year 2003 267 (2003) Available at http://pubs.usgs.gov/wdr/wdr-ne-03-1/aar2003bookG.pdf; Platte River EIS Team, supra note 5, at 2–6

Wishart D (1979) The dispossession of the Pawnee. Ann Assoc Am Geogr 69:382–401

Legal Documents

Colorado Constitution article XVI, § 6
Nebraska Constitution (1920) article XV, §§ 4–6
South Platte River Compact (1926) 44 Stat. 195
Wyoming Constitution (1889) article 8, §§ 1, 3

Statutes and Regulations

Homestead Act (1862) Public Law No. 37–64, 12 Stat. 392
Interior least tern 50 Fed. Reg. 21,784 (June 27, 1985) (endangered)
Irrigation Act of 1889 (1889) Nebraska Laws 1889, ch 68
Irrigation Act of 1895, effective April 4, 1895, Neb. Laws 1895, ch. 69
Kincaid Act (1904) 33 Stat. 547
Neb. Laws 1975, LB 577, § 24, *codified as amended at* Neb. Stat. Ann. §§ 46–701 to 46–756
Neb. Laws 2004, LB 962 April 15, 2004
Pallid sturgeon (*Scaphirhynchus albus*; 55 Fed. Reg. 36,641 (Sept. 6, 1990) (endangered).
Piping plover (1985) 50 Federal Register 50, 726 (Dec. 11, 1985) (endangered in Great Lakes region, threatened in rest of range)
Platte River ESA listings (1967, 1985, 1990)
US Endangered Species Act (ESA), 16 U.S.C. §§ 1531–1544
US Fish and Wildlife Service, National Marine Fisheries Service (ESA) 50 C.F.R. § 402.02
US Reclamation Act (1902), Act of June 17, 1902, Chapter 1093, 33 Stat. 388, *codified as amended at* 43 U.S.C. §§ 372–498
Whooping crane (1967) 32 Federal Register 48 (threatened)
Whooping crane (1970) 35 Federal Register 199 (endangered)

Cases

Bennett v. Spear, 520 US 154 (1997)
Coffin v. Left Hand Ditch Co., 6 Colo. 443 (1882)
Irwin v. Phillips, 5 Cal. 140 (1855)
Nebraska v. Wyoming, 295 US 40 (1935)
Nebraska v. Wyoming, 325 US 589 (1945)
Nebraska v. Wyoming, 507 US 584 (1993)
Nebraska v. Wyoming, 515 US 1 (1995)
Wasserburger v. Coffee, 141 N.W.2d 738 (Neb. 1966)
Willey v. Decker, 73 P. 210 (Wyo. 1903)

Chapter 9
Assessing Adaptive Water Governance for Lake Eyre Basin and Linked Portions of the Great Artesian Basin in Australia

Barbara Cosens and Angela H. Arthington

Abstract The Lake Eyre Basin in the heart of the outback of Australia is a place in which the social-ecological system is not only highly dependent on but is also defined by the intermittent presence and absence of water. Tributary rivers of this hydrologic system arise in Queensland and the Northern Territory and feed the landlocked Kati Thanda-Lake Eyre in central Australia. Due to decadal cycles of rain, the basin has one of the most variable flows in the world; in any given year, the lake may be a vast expanse of blue against the red soils of the arid lands or a white salt flat. The Great Artesian Basin that underlies portions of the Lake Eyre Basin has been an important source of fresh water for humans, initially serving Aboriginal inhabitants, then the railroad system, and today large pastoral farms of the basin. The setting provides an opportunity to consider the link among resilience, governance, and law outside the North American context yet within a federal system. The efforts of the Australian Commonwealth and state governments to engage in catchment planning and involve both scientists and local stakeholders can be characterized as a governmental effort to achieve adaptive governance across jurisdictions. This much more formal and intentional approach has indeed enhanced the adaptive capacity of Australian water basins and moved them beyond their North American counterparts in which emergence of adaptive governance remains *ad hoc* in response to a social or ecological disturbance. At the same time, the Australian approach is top-down and, without both greater authority and capacity at the local level, may not achieve the level of adaptive governance needed to navigate the changes to come.

Keywords Resilience · Adaptive governance · Adaptive law · Resilience assessment · Climate change · Water management · Groundwater · Dryland rivers

B. Cosens (✉)
College of Law and Waters of the West Program, University of Idaho, Moscow, ID, USA
e-mail: bcosens@uidaho.edu

A. H. Arthington
Australian Rivers Institute, Griffith University, Brisbane, QLD, Australia
e-mail: a.arthington@griffith.edu.au

B. Cosens, L. Gunderson (eds.), *Practical Panarchy for Adaptive Water Governance*, https://doi.org/10.1007/978-3-319-72472-0_9

9.1 Introduction

The Lake Eyre Basin (LEB) and its connection to portions of the Great Artesian Basin (GAB) lie in the heart of Australia's iconic outback where the juxtaposition of vast arid landscapes with artesian waters, mound springs, and extensive episodic wetlands gives the region high conservation value. The assessment of LEB/GAB governance followed the initial assessments of the six North American basins discussed in this volume and was thus the first opportunity to apply a more systematic approach and to test its transferability outside the North American setting yet in a setting with a similar history of colonization and a federal governmental structure. The LEB also provides an opportunity to explore the interaction of governance, law, and resilience in the context of a relatively undeveloped water resource.

This chapter will begin with the history of human interaction with the basin to understand the legacy of that history in its current trajectory. The role of law in adaptive governance of the LEB/GAB will follow, providing an opportunity to explore the interaction of law and governance and its importance in adaptation in a setting outside the US legal system. Finally, this chapter will consider the role of mental models in future adaptation and the importance of both legal and governance mechanisms to build capacity and authority at the local or landscape scale, while maintaining the oversight, provision of knowledge, and resources from higher levels of government.

9.2 The Social-Ecological Context of the Lake Eyre and Great Artesian Basins

The LEB is one of those places in the world uniquely defined by both the presence and absence of water. The LEB covers 1.14 million square kilometers of east-central Australia (Fig. 9.1) and is one of the largest internally draining basins in the world (Thoms et al. 2009). Its terminus, Lake Eyre (known to the Arabana people, the traditional owners of the land, as Kati Thanda), is located in the driest portion of Australia, yet with headwaters in the monsoonal north, the LEB exhibits extreme variability in surface water flow (Kingsford et al. 2014).

The major rivers of the LEB, Cooper Creek, the Diamantina, and Georgina rivers with headwaters in Queensland and the Finke, with headwaters in the Northern Territory, are dryland rivers, meaning that their inflow is generally less than the rate of evaporation (Thoms et al. 2009; Leigh et al. 2010). The rivers that feed the LEB are currently free flowing and thus have not been regulated through the construction of dams and major diversions as is the case in the basins studied in North America. Unlike major internally draining rivers in North America's Great Basin that are fed by spring runoff following snowmelt, the LEB receives water during major flood events in the monsoon season of Northern Australia in which precipitation tends to come in short-duration high-intensity storms.

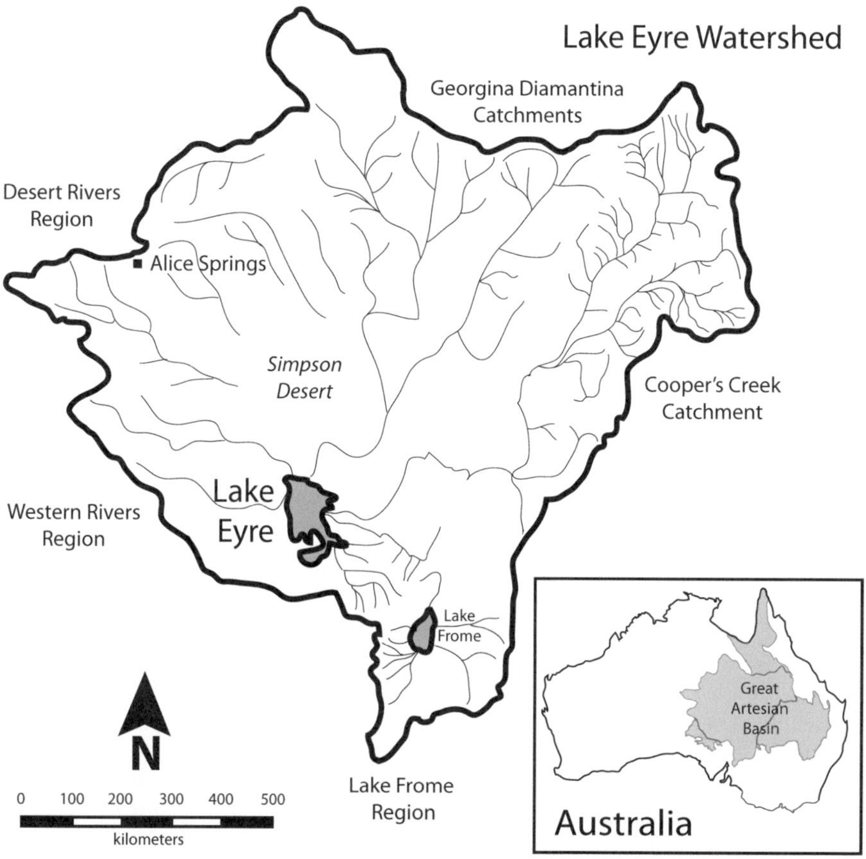

Fig. 9.1 Catchment regions of the Lake Eyre Basin, with inset showing location of the basin within central Australia (Source: The Lake Eyre Basin teacher guide, Australian Government Department of Sustainability, Environment, Water, Population and Communities)

Flow variability and extreme flood events drive ecological cycles in the LEB (Leigh et al. 2010; Kingsford et al. 2014). Large floods that occur on a decadal frequency connect and rewater isolated wetlands, transport sediment and nutrients, and basically flush and reset the aquatic system (Leigh et al. 2010; Arthington and Balcombe 2011). More frequent, moderate floods replenish waterholes which provide habitat for aquatic species (Balcombe and Arthington 2009; Leigh et al. 2010; Kerezsy et al. 2011). Frequent, sequential flood events in upstream areas fill wetlands, allowing subsequent floods of moderate discharge to reach further into the basin, thereby increasing connectivity among aquatic systems (Leigh et al. 2010).

Floods transport nutrients from channel banks and floodplains, which increases biological productivity. This drives increase in populations of most fish species and wetland bird species (Leigh et al. 2010; Arthington and Balcombe 2011). Floods also provide the connectivity necessary for dispersal of aquatic species and trigger breeding in some species (Good et al. 2004; Costelloe et al. 2007; Kerezsy et al. 2011). Certain opportunistic fish species in the basin disperse up to 300 km from permanent to intermittent waterholes during floods, while other species remain close to permanent waterholes (Kerezsy et al. 2013). As floods subside, the larger persistent waterholes that generally form at points of flow confluence serve as refugia for fish and other aquatic species. Species ending up in a waterhole that lasts until the next flood serve as pioneer species recolonizing other connected waterholes when the water returns (Leigh et al. 2010; Arthington and Balcombe 2011; Kerezsy et al. 2013). Isolation during dry periods results in a variety of species assemblages in different waterholes and contributes to basin-wide biodiversity patterns (Good et al. 2004; Costelloe et al. 2007).

The Great Artesian Basin (GAB), one of the largest groundwater basins in the world, underlies most of the LEB basin and extends outside the basin to the east ultimately underlying 22% of Australia (Sinclair 2005; Keppel et al. 2013) (Fig. 9.1). Despite limited natural connection to surface water (Keppel et al. 2013), areas where groundwater does discharge at the surface are of substantial ecological and historical importance and are referred to as mound springs (Keppel et al. 2013; Love et al. 2013a; Kingsford et al. 2014). The mounds are formed by carbonate precipitation and sediment deposition from the springs and are one of the unique hydrogeologic features of the Lake Eyre Basin (Love et al. 2013b). Conically shaped and diffuse springs occur in clusters in the landscape, generally along fault lines in the LEB (Sinclair 2005; Keppel et al. 2013; Love et al. 2013a).

Recharge of the aquifers is limited, estimated at around 1000 GL/year or 800,000 acre-feet/year. The recharge occurs mostly outside the LEB along the Great Dividing Range and does not provide significant recharge to GAB aquifers within the LEB portion of the basin (Sinclair 2005; Keppel et al. 2013; Love et al. 2013b). Discharge rates along the western margin of the basin within the LEB are a function of paleo-recharge – basically reflecting the lag time between a time of greater recharge and the discharge that continues today – and exceed current recharge, a phenomenon that is likely to result in decline over time as recharge and discharge reach a new equilibrium (Keppel et al. 2013; Love et al. 2013a, b).

The LEB/GAB presents an opportunity to consider a basin in which the ecosystem is highly adapted to extreme conditions. Species found in the LEB exhibit adaptive traits such as habitat flexibility, dietary generalization, opportunistic breeding, and tolerance of extreme variability, and yet, droughts are periods of local extinction and play a major role in shaping ecological community structure (Arthington et al. 2010; Leigh et al. 2010). While this adaptive capacity suggests high resilience, the highly specific adaptation to extreme conditions that still provide sufficient water for refugia and periodic dispersal, leaves the system vulnerable should dam development on northern tributaries or climate change affect any of the six characteristics considered of ecological relevance to dryland rivers – flow magnitude, frequency, timing, duration, variability, and rate of change (Leigh et al. 2010; Arthington and Balcombe 2011; Kingsford et al. 2014). In addition, due to lag times in response of the biotic system, impacts from slow change in hydrology may not be detected immediately, even with careful monitoring (Good et al. 2004; Costelloe et al. 2007).

Human alteration of the LEB surface water hydrology is currently minimal, and dams have yet to be built within the basin. The LEB contains two Ramsar wetlands (the Coongie Lakes in South Australia and Lake Pinaroo in NSW), and the basin has been considered eligible for World Heritage listing. The free-flowing river system supports a unique biota; 97% of the basin aquatic species are thought to be indigenous. While alteration of water resource by humans is thought unlikely, climate change may nevertheless increase variability and vulnerability.

The patterns of precipitation and temperature affecting the portion of the LEB basin in South Australia (the location of Lake Eyre) are forecast to change over the remainder of the twenty-first century (Charles and Fu 2015). Climate models for the South Australia arid lands indicate drier patterns over the latter half of the century. Precipitation is forecast to decrease in all seasons but particularly in spring (Charles and Fu 2015). Maximum temperature is predicted to increase in the same region by 2–4 °C (Charles and Fu 2015). Annual potential evapotranspiration is expected to increase between 5 and 12%, particularly in spring corresponding with an increase in temperature (Charles and Fu 2015). Forecasts for the northern portion of the basin suggest that precipitation may increase and exhibit greater extremes (Kingsford et al. 2014).

9.3 The Social-Ecological Dynamics of the Lake Eyre and Great Artesian Basins

Although considerable debate surrounds the timing of the first presence of humans in Australia, charcoal remains indicate their presence at GAB springs as much as 5000 years ago, and there are indications of the use of fire to manage habitat (Gotch 2013). As the main dependable water source in the region, the springs played an important role in Aboriginal history in the area (Keppel et al. 2013; Green et al. 2013; Nursey-Bray et al. 2013). Not only were the springs necessary for survival, but they were central to both spiritual and cultural activities (Keppel et al. 2013) of a number of Aboriginal peoples, including Arabana, Lower Southern Arrernte, Wokangurru, Kuyani, Deri, Pirlatapa, Malijangapa, and Yandruwandha groups (Gotch 2013).

The springs were first described by European settlers in 1865. Because of the need for water during travel through the arid outback, the springs shaped the trajectory of early exploration and settlement, including routes for telegraph lines and the railroad (Keppel et al. 2013; Gotch 2013; Nursey-Bray et al. 2013). In the late 1800s, settlers intent on developing large pastoral stations (the equivalent term to a ranch in North America) and began drilling bores (the equivalent term to well) into the artesian portions of the aquifer for livestock watering (Keppel et al. 2013). Thousands of these wells were left to flow freely, lowering the artesian water table. Flow from these bores represents 50% of the total natural and anthropogenic discharge from the GAB (Love et al. 2013a).

Falling groundwater pressure over much of the GAB was addressed with the adoption of the Great Artesian Basin Strategic Management Plan by the Australian, Queensland, New South Wales, South Australian, and Northern Territory governments in 2000 and again in 2008 (Great Artesian Basin Consultative Council 2000, Great Artesian Basin Coordinating Committee [GABCC] 2008). From the plan, a 15-year, 230 million AU$ program, called the Great Artesian Basin Sustainability Initiative, led to capping hundreds of bores, and piping eliminated thousands of kilometers of open bore drains. This program has saved approximately 200,000 ML/y and in 2014 was extended for another 3 years (GABCC 2015). In May 2017, the government of Australia dedicated an additional eight million AU$ to continue this program.

The basin's social system is highly adapted to its arid setting in which thin lines exist among ability to sustain a livelihood, adaptation, and failure. As a result, similar to the ecosystem, the social system may be vulnerable to any change that reduces overall water supply or increases the intensity, timing, and duration of extreme drought. Despite the fact that it covers roughly one sixth the area of Australia (Thoms et al. 2009), the basin is home to less than 60,000 people or less than 1% of the population of Australia (Lake Eyre Basin Secretariat 2011). Pastoral development of arid lands for livestock, small homesteads, and towns uses limited amounts of water. Tourism in the basin is popular and increasing with over two million people visiting in 2001 (Hadwen et al. 2012; Kingsford et al. 2014). As climate change unfolds, the basin is likely to require resources and knowledge production from external sources to adapt or recover from perturbations. Assistance could produce development consistent with the basin's hydrological and ecological cycles and thus enhance the self-sufficiency and resilience of the basin or could seek to optimize for a stable water supply with an end result of loss of biodiversity, ecosystem services, and vulnerability. The discussion of mental models below is particularly relevant to the challenges that may be involved with choosing an appropriate path.

9.4 The Role of Law in Adaptive Governance

Similar to the United States, Australia has multiple levels of government involved in water management. Water allocation and sanitation are matters of state law in Australia (Stoeckel et al. 2012). Nevertheless, the federal government plays a major role in setting water quality standards (Stoeckel et al. 2012) and in environmental law through its participation in numerous international conventions and their incorporation into the Australian Water Act of 2007 (which has only been applied to the Murray-Darling Basin) and the Commonwealth Environment Protection and Biodiversity Conservation Act of 1999 (Commonwealth of Australia 2014). The federal government has played an increasing role in water allocation in recent years through agreements with states and the territory such as The National Water Initiative (Commonwealth of Australia 2004), and a combination of referral of authority from states to the Commonwealth (Commonwealth of Australia 2012), and federal spending authority.

The LEB includes portions of four states; thus the placement of water allocation law at the state/territorial level poses problems for coordination at the basin scale, just as it does in the United States. Australia has two legal mechanisms to address this: (1) referral of authority from the state to the Commonwealth and (2) the passage of mirror legislation regarding the basin by the relevant states/territory. One example of a referral of authority is the creation of an interstate authority for basin-wide management – this was done in the Murray-Darling Basin (Commonwealth of Australia 2007). In the LEB, mirror legislation has been adopted. Such legislation, while keeping control at the state level, provides less certainty by leaving any state/territory with the unilateral authority to dissolve the agreement by altering the mirror legislation.

An intergovernmental agreement among the four state/territory governments addresses the prevention of cross-border harm and provides a process and forum for raising, but not necessarily resolving, issues related to cross-border impacts on water quantity, quality, and flow patterns in the LEB (Commonwealth of Australia 2000). The LEB Intergovernmental Agreement establishes a high-level coordinating body called the Ministerial Forum. The Forum is comprised of one Minister from each state and territory (with the addition of the Northern Territory under the First Amending Agreement of June 10, 2004) and one from the Commonwealth (Commonwealth of Australia 2000). The Commonwealth member serves as chair, and the Ministerial Forum operates on a consensus basis (Commonwealth of Australia 2000).

The LEB Intergovernmental Agreement provides a process and forum for raising, but not resolving, issues related to cross-border impacts on water quantity, quality, and flow (Commonwealth of Australia 2000) and leaves internal management of the basin to each state (Commonwealth of Australia 2000). A group of high-level governmental officials from state-level agencies forms a Senior Officers Group for the LEB and, according to the assessment of LEB governance completed in 2009, has provided a leadership role beyond that contemplated in the LEB Intergovernmental Agreement (Price et al. 2009). Coordination between intrastate and interstate management is also facilitated at the terminus of the basin in South Australia by the participation since 2007 of representatives of the relevant state Natural Resources Management Boards on an advisory committee for the LEB (Price et al. 2009).

The Ministerial Forum is also set up to interact with local interests. It is advised by both community and scientific advisory committees, is served by an LEB Secretariat, and has held Aboriginal forums throughout the basin (Lake Eyre Basin Ministerial Forum n.d.; Lake Eyre Basin Secretariat n.d.; Price et al. 2009). The establishment of community and Aboriginal forums provides a means for the inclusion of local and traditional knowledge in decision-making, and this combined with the retention of strong oversight at the state and interstate levels comes closer to a form of nested governance than the approaches found in the case studies of North American basins.

Nevertheless, one of the purposes of nested governance in achieving adaptive governance is to move the locus of knowledge, response capacity, and decision-making closer to the resource and thereby tighten the feedback between changes in the ecological system and response (Marshall 2008). In the LEB/GAB, decision-making remains at the state and federal level. In addition, any effort to devolve decision-making may require increased attention to consistency in those chosen to play a role at the local level. The dependence of adaptive response time on relationships and networks requires that participants have longevity in their roles in the basin. In the process of water governance reform in Australia, top-down changes to local governance structures, while designed to provide greater integration by eliminating, for example, separate rangeland and catchment community advisory bodies, have also had the result of eliminating long-standing community relations with board representatives from the prior governance structure (Mitchell 2014). This in turn has led to a loss of legitimacy in the decision-making process as well as a reduction in the flow of information to and from the local level (Mitchell 2014). Of even greater importance in moving decision-making closer to the resource in question is the need for local capacity; a shortcoming throughout the basins studied in this volume.

Participatory capacity addresses the ability of those who rely on the water-based social, economic, and ecological aspects of the basin to participate in decision-making concerning the basin future, to have access to and share in the benefits of the resource, and to have sufficient knowledge not only to be able to understand and communicate changes in local feedback resulting from changes in the water-based system but to have the ability to respond (Cosens et al. 2017; Chap. 10). With the sidenote that volunteer members of local boards are typically oversubscribed and, thus, will always need resources, participation by Aboriginal communities is a continuing challenge for participatory capacity apparent in the LEB.

Scholars have argued for greater engagement of Aboriginal communities in the process of stakeholder input to the allocation of environmental flows to capture cultural and subsistence values that may be associated with species that are either not endangered or not identified as of interest to non-Aboriginal stakeholders (Finn and Jackson 2011; Jackson et al. 2012). While this is a good first step, it is necessary to go beyond the concept of Aboriginal communities as a stakeholder group and assert that an Indigenous governmental voice as co-managers is necessary to appropriately represent their interests and to achieve self-determination.

Indigenous peoples including the Arabana continue to have strong ties to the basin as well as claims to native title to lands and waters (Nursey-Bray et al. 2013). The legacy of their exclusion from resource management until recent decades cannot be ignored. Meeting the needs of formerly marginalized and therefore vulnerable populations is critical to the adaptability and resilience of water-dependent communities. Furthermore, as illustrated by both the Columbia and Klamath basin assessments (Chaps. 4 and 6), increased diversity in the voices, knowledge, and values playing a role in water management may be a factor in enhancing our collective ability to respond to change. This level of engagement will require (1) development of governance capacity within Aboriginal communities and (2) a federal/state process to engage with Aboriginal communities at a governmental level.

The North American case studies illustrate a rights-based approach to development of Indigenous governance capacity. Following recognition and definition of treaty rights in court, Native American Tribes have developed the capacity to participate as co-managers of shared resources (Cosens 2012). A different approach is necessary in Australia (Cosens and Chaffin 2016). Although both Australia and the United States began with a legal framework imported via colonization by England, they diverged in their definition of the rights of the Indigenous people inhabiting the land they came to colonize. Australia did not reject the concept of *terra nullis* (the land belongs to no one) until 1992 (*Mabo v. Queensland* 1992). In doing so Australia recognized the concept of ownership of native title, but not sovereignty over the land and water to which native title extends (*Mabo v. Queensland* 1992; Commonwealth of Australia 2017).

Nevertheless, strides are being made in both the Murray-Darling Basin and in Northern Australia to increase consultation with Aboriginal communities (O'Neill et al. 2016). Despite the lack of progress in the recognition of legal rights to water beyond those tied to cultural purposes (Finn and Jackson 2011; Jackson et al. 2012; Hemming and Rigney 2014; O'Neill et al. 2016), through consultation it is possible that greater strides will be made in recognizing Aboriginal governments as co-managers of water resources rather than mere owners of use rights among many.

One novel approach led by the Ngarrindjeri Nation on the lower Murray River and Coorong wetlands in South Australia uses private intergovernmental contracts as the legal means for rendering agreements binding. In the absence of recognition of either native title to water or a sovereign interest, the use of a legally binding agreement fills the current void in state or federal definition of Indigenous rights (Hemming et al. 2011). This approach was made possible through formation of a governance mechanism to serve as the interface between the Ngarrindjeri Nation and local and state government – the Ngarrindjeri Regional Authority (Hemming and Rigney 2014). A window of opportunity for assertion of self-determination by the Ngarrindjeri Nation was triggered by a combination of biophysical and legal drivers of change. Representatives of the Nation leveraged the threat of the Nation's legal claim to native title in the area of interest (Hemming et al. 2007), and the assertion that the Australian State and Federal government's past assertion of *terra nullis* was illegal under English law in the early 1800s (Hemming et al. 2011). Coincidentally the crisis presented by the impact of the Millennium Drought on the Murray River and the Coorong made the solutions brought forward by the Nation attractive. Formation of the Authority and its assertion of sovereignty in reaching out to local, state, and interstate government, including the Murray-Darling Basin Commission, to negotiate agreements mark a major step in capacity building (Hemming et al. 2007).

What the Australian examples have in common with those in the North American case studies involving Indigenous peoples is the development of capacity in governance as the necessary precursor to self-determination (Cosens and Chaffin 2016). Thus, assistance to Aboriginal communities to build capacity should be the first priority in working toward a greater Indigenous voice in water management. A discussion of how to build capacity is beyond the scope of this chapter, but there are key features in both Native American and Ngarrindjeri Nation examples worth noting. First, higher education can play a role in facilitating the development of capacity. The outreach and engagement by a university First Nations Centre have been instrumental in providing research and assistance to the Ngarrindjeri Nation (Hemming et al. 2007, 2011; Hemming and Rigney 2014) and serve as an excellent example of the neutral role higher education might play in capacity building. Second, federal and interstate restoration funding has been a source of capacity building for both Tribes in the United States (Columbia River Inter-Tribal Fish Commission 2014) and the Ngarrindjeri Nation in Australia (Hemming et al. 2011). Assuring that Indigenous organizations are eligible for public funding is thus essential to capacity building.

The final area of significant importance to the relation among law, governance, and resilience in the LEB/GAB is the integration of management of groundwater and surface water. Integration of resource management is intended to ensure that activities in one sector do not have unexpected consequences in another. The LEB Ministerial Forum articulates a policy focus on whole-of-basin management, and the LEB Intergovernmental Agreement guiding principles call for integrated management of surface and groundwater (Commonwealth of Australia 2000). However, intergovernmental cooperation on the GAB groundwater system is currently separate from that of the LEB.

Joint meetings of the GAB Coordinating Committee with the various LEB entities are currently the primary means of coordination between management of the surface and groundwater resource. This combined with the fact that the management of the GAB is scientifically driven, with no regional entity formed to make policy decisions on matters such as trade-offs between mining and oil and gas development and groundwater-dependent ecosystems, or to establish a policy and rules for development of aquifers with limited recharge, suggests that a more formal structure is needed. At the same time, because of the very slow movement and limited yield of groundwater in much of the GAB, issues of impact are basically local. Thus development of a formal basin-wide authority may not be warranted. Instead, coordination of development and assessment with the decision-making on trade-offs and development of limited recharge aquifers could remain at the state/territory level provided the mechanism to resolve intrastate disputes is in place.

The development of coal seam gas (CSG) in the GAB has become an issue in recent years. Regulation of petroleum and natural gas is entirely separate from allocation of water in Australia as it is in the United States and is at the federal level (Gray 2016). Unlike the United States, all petroleum resources are held by the Crown. Concerns over the impact of fracking on water quality and efforts to assure coordination are reflected in policy instruments such as the New South Wales Gas Plan (Gray 2016). But the mere recognition that natural gas development raises concerns for water quality may be insufficient if strict standards and enforcement are not employed (Gray 2016). Thus, it is important to note that while there is a tendency to believe that integration of management of various resource sectors is good, transitions to greater integration must include standards and enforcement if the goal is to ensure actions in one sector (natural gas extraction) do not harm another sector (water and ecosystems).

9.5 The Emergence of Government-Facilitated Adaptive Governance in the Lake Eyre and Great Artesian Basins

The US basins have much to learn from the process of catchment scale to federal scale coordination in the LEB. Even so, there has been limited success in implementation of adaptive approaches to management and governance in complex jurisdictional settings such as river basins (Lee 1993; Pahl-Wostl 2007). A major barrier is the gap between the scientific approach of management experiments, monitoring and continual adjustment, and the need for stability in social and economic systems (Cosens 2013; Craig et al. 2017). Legitimacy in resource management requires a balance between the stability needed for peaceful social engagement and confident economic investment and the flexibility needed to adapt to and manage change complicated by uncertainty (Craig et al. 2017; Chap. 13). It is through adaptive governance that the balance must be achieved (Pahl-Wostl 2007) and where challenges remain in the LEB.

Although the LEB Intergovernmental Agreement was established to prevent cross-border impacts to the LEB water resource, it is primarily a diplomatic agreement for coordination, dialogue, and exchange of information. To provide a "hard law" (Thompson 2016), mechanism for dispute resolution, while retaining the flexibility to adapt, requires addressing the need in the LEB for a decision-making process to address cross-border disputes that is deeply integrated with the local forums. Thus, the LEB appears to require both stricter mechanisms to address cross-border harm and, at the same time, devolution of some degree of authority to the local level. The following lessons from the review of the Columbia River Treaty may provide an approach to creating a common vision for the LEB/GAB.

In the initial stages of the review of the Columbia River Treaty, entities in the basin including the Northwest Power and Conservation Council (an interstate body created for power planning and a coordinated fish and wildlife program (Northwest Power and Conservation Council n.d.)) and Columbia Basin Trust (a provincial entity that brings water education, economic development, and climate adaptation planning to the rural portion of the basin in British Columbia (Columbia Basin Trust n.d.)) recognized that residents of the basin have more in common across the international border than the existence of the border would suggest and that they might be better situated to find solutions to the problems the basin faces than their respective federal governments. Although each country did not have the authority to create a cross-border forum for a dialogue, various entities in the basin including the Council, the Trust, Native American and First Nation organizations, and environmental NGOs began to take on that responsibility. The Universities Consortium on Columbia River Governance made up of representatives of public universities on both sides of the border facilitated annual meetings to bring basin residents together (Universities Consortium on Columbia River Governance 2015). The Treaty review process is not complete; thus, success cannot be claimed, but the lesson for the LEB/GAB is to recognize that whereas governmental entities may have absolutely no reason to compromise, neighbors do.

Similar to the people of the Columbia River Basin, it is likely that the people of the LEB have more in common than their respective state/territory governments share. A cross-jurisdictional dialogue among the people in the basin to establish a vision for true basin-wide coordination might provide a way forward within the existing government structure and even lay the groundwork for a whole-of-basin approach to governance should the inevitable crisis occur. If the respective state and territory governments could be convinced to accept the outcome of a facilitated in-basin process and memorialize it in mirror legislation in each state/territory, a workable outcome might be achieved.

The absence of a mechanism to resolve disputes among states/territory means that in most circumstances, the downstream state bears the risk; thus the question often becomes: under what conditions would an upstream state relinquish some control to a downstream state? This dilemma is as much an issue between states in a federal system, as it is between nation states. In the LEB, the upstream state is Queensland, and the answer to its participation lies in the need for certainty concerning the level of freedom from interference with development of their water resources. Currently the Commonwealth does have the ability to intervene in an interstate dispute under the EBPC Act. Under the Act and its subsequent amendments, actions that might have a significant impact on matters of national environmental significance may be referred to the Australia Department of the Environment for environmental review (Commonwealth of Australia 2014). Of relevance to the LEB, the list of matters of national environmental significance includes Ramsar wetlands, endemic species, migratory birds, and water resources affected by coal seam gas or large coal mine development (Commonwealth of Australia 2014). The vague notion of significant harm embodied in the EPBC Act gives Queensland very little guidance on when its level of development might trigger state action (e.g., Arthington 2009). In contrast, a negotiated agreement setting forth a level of development that may occur without interstate and Commonwealth interference would alleviate that uncertainty.

9.6 Discussion and Conclusions

The location of the LEB/GAB in the iconic outback of Australia conjures up a mental model (DeCaro et al. 2017; Chap. 17) that views northern tributaries as wild rivers and protects the basin's wetland ecosystems and the mound springs. At the same time, the mental model of the outback has a tendency to arrest Aboriginal people in a pre-colonial existence by recognizing only cultural rights to water without consideration of modern economic needs and the possibility of a role to play in co-management. It also carries with it the frontier mentality of the outback as a source of extractive resources. In addition to conflict with the conservation focus, the extractive view contrasts with the large pastoral industry in which a long-standing partnership between European settlement and Aboriginal labor has evolved with both exhibiting deep ties to the land and significant local knowledge. Can these views coexist in modern Australia? Notions of adaptive governance cannot answer that question. Adaptive governance can, however, provide a mechanism to give the people of the basin the authority to decide among competing values and the capacity to evaluate the trade-offs and respond to change as it unfolds. It is this concept that informs the concluding section.

Possibly because of its unique and relatively untouched and iconic setting, the Lake Eyre Basin brings clarity to the contextual and yet universal nature of aspects of adaptive governance. Although not unique to this setting, it is brought into focus by the juxtaposition of the desire to maintain the stark natural beauty and ecosystems of the LEB with the compelling notion that justice demands the same opportunities for its basin's residents to pursue personal, spiritual, cultural, and economic livelihoods that are enjoyed in more highly developed regions. While this illustrates that adaptive governance is not an off-the-shelve solution that can merely be applied in any setting, it also highlights what is likely to be the most important area of research and investment in governance in any natural resource setting seeking to transition to a form of governance capable of navigating change: (1) how to move capacity and governance closer to the resource in question and (2) how to do so while still operating within a larger government framework that facilitates management implementation and achievement of long-term goals including the equitable distribution of access to and benefits from the resource, through oversight, transfer of knowledge and resources for capacity building. It will require a reversal in the flow of resources, talents, and high-paying science and policy jobs currently directed to the national level. It will require a more meaningful and flexible partnership between national, state, and local government. In this context, the progress made in Australia to involve local citizen forums in assessing the status of its catchments and considering their future is a lesson for those who value and rely on the watersheds of North America. Building capacity at the local level tightens feedback between change in social-ecological systems and those with the ability to respond. It increases the resilience of social systems in the face of natural disasters (Bousquet et al. 2016). In fact, the recognition, restoration, and enhancement of local capacity may be the major challenge of adaptive governance and, indeed, of environmental governance in the twenty-first century.

Acknowledgments This work was developed in part under the Adaptive Water Governance Project, funded by the US National Socio-Environmental Synthesis Center (SESYNC) under funding from the US National Science Foundation, NSF DBI-1052875, and in part through a visiting professorship held by lead author Cosens with the ANZOG-Goyder Institute Visiting Professors Program in Public Sector Policy and Management, in the Faculty of Social and Behavioural Sciences, Flinders University, South Australia in 2015.

References

Arthington AH (2009) Australian lungfish, *Neoceratodus forsteri*, threatened by a new dam. Environmental Biology of Fishes 84:211–221

Arthington AH, Balcombe SR (2011) Extreme flow variability and the 'boom and bust' ecology of fish in arid-zone floodplain rivers: a case history with implications for environmental flows, conservation and management. Ecohydrology 4:708–720

Arthington AH, Olden JD, Balcombe SR, Thoms MC (2010) Multi-scale environmental factors explain fish losses and refuge quality in drying waterholes of Cooper Creek, an Australian arid-zone river. Mar Freshw Res 61(8):842–856

Balcombe SR, Arthington AH (2009) Temporal changes in fish abundance in response to hydrological variability in a dryland floodplain river. Mar Freshwater Res 60(2):146–159

Bousquet F, Botta A, Alinovi L, Barreteau O, Bossio D, Brown K, Caron P, D'Errico M, DeClerck F, Dessard H, Kautsky EE, Fabricius C, Folke C, Fortmann L, Hubert B, Magda D, Mathevet R, Norgaard RB, Quinlan A, Staver C (2016) Resilience and development: mobilizing for transformation. Ecol Soc 21(3):40. https://doi.org/10.5751/es-08754-210340

Charles SP, Fu G (2015) Statistically downscaled projections for South Australia. In: Goyder Institute for water research technical report series, No. 15/1. http://www.goyderinstitute.org/_r91/media/system/attrib/file/82/CC%20Task%203%20CSIRO%20Final%20Report_web.pdf. Accessed 26 Aug 2017

Columbia Basin Trust (n.d.) Columbia Basin trust. https://ourtrust.org/. Accessed 26 Aug 2017

Columbia River Inter-Tribal Fish Commission (2014) Habitat and water protection and restoration, 2014 year in review. http://www.critfc.org/about-us/annual-reports/. Accessed 25 July 2017

Cosens B (2012) Changes in empowerment: rising voices in Columbia Basin resource management. In: Cosens B (ed) The Columbia River Treaty revisited, transboundary river governance in the face of uncertainty. Oregon State University Press, Corvallis, pp 61–68

Cosens BA (2013) Legitimacy, adaptation, and resilience in ecosystem management. Ecol Soc 18(1):3. https://doi.org/10.5751/es-05093-180103

Cosens B, Chaffin BC (2016) Adaptive governance of water resources shared with indigenous peoples: the role of law. Water 8(3):97. https://doi.org/10.3390/w8030097

Cosens BA, Craig RK, Hirsch S, Arnold CA, Benson MH, DeCaro DA, Garmestani AS, Gosnell H, Ruhl JB, Schlager E (2017) The role of law in adaptive governance. Ecol Soc 22(1):30. https://doi.org/10.5751/es-08731-220130

Costelloe JF, Hudson PJ, Pritchard JC, Puckridge JT, Reid JRW (2007) ARIDFLO scientific report: environmental flow requirements of arid zone rivers with particular reference to the Lake Eyre drainage basin. University of Adelaide, Adelaide. http://www.lakeeyrebasin.gov.au/sitecollectionimages/resources/4e05add2-56b2-4c09-8de1-d52a05bb1e6b/files/aridflo-exec-summ.pdf. Accessed 26 July 2017

Craig RK, Garmestani AS, Allen CR, Arnold CA, Birgé H, DeCaro DA, Fremier AK, Gosnell H, Schlager E (2017) Balancing stability and flexibility in adaptive governance: an analysis of tools available in U.S. environmental law. Ecol Soc 22(2):3. https://doi.org/10.5751/es-08983-220203

DeCaro DA, Arnold CA(T), Boamah EF, Garmestani AS (2017) Understanding and applying principles of social cognition and decision making in adaptive environmental governance. Ecol Soc 22(1):33. https://doi.org/10.5751/ES-09154-220133

Finn M, Jackson S (2011) Protecting indigenous values in water management: a challenge to conventional environmental flow analysis. Ecosystems 14:1232–1248

Good M, Bailey V, Pritchard J, Wedderburn S (2004) ARIDFLO: building our knowledge of Lake Eyre Basin rivers: a report to the community. SA Dept of Land, Water and Biodiversity Conservation, Adelaide

Gotch T (ed) (2013) Groundwater-dependent ecosystems of the Western Great Artesian Basin. In: Allocating water and maintaining springs in the Great Artesian Basin, volume V. National Water Commission, Canberra. http://webarchive.nla.gov.au/gov/20160615074424/http://archive.nwc.gov.au/library/topic/groundwater/allocating-water-and-maintaining-springs-in-the-great-artesian-basin. Accessed 25 July 2017

Gray J (2016) Trans-jurisdictional water law and governance in the context of unconventional gas mining: the Australian experience. In: Gray J, Holley C, Rayfuse R (eds) Trans-jurisdictional water law and governance. Routledge, London

Great Artesian Basin Consultative Council (2000) Great Artesian Basin strategic management plan. http://www.gabcc.gov.au/sitecollectionimages/resources/17686c7e-ebe2-4701-b41d-1397b6c85972/files/strategic-mgt-plan-2000.pdf. Accessed 25 July 2017

Great Artesian Basin Coordinating Committee (2008) Great Artesian Basin strategic management plan: progress and achievements to 2008. http://www.gabcc.gov.au/sitecollectionimages/resources/68b0bd6f-7c73-47f2-8791-a39a89ea317f/files/gab-strategic-mgt-plan-progress-achievements-2008.pdf. Accessed 25 July 2017

Great Artesian Basin Coordinating Committee (2015) Great Artesian Basin coordinating committee annual report 2014–2015. http://www.gabcc.gov.au/sitecollectionimages/resources/97cb60a8-a7fd-44e2-9637-543ee61ef23b/files/gabcc-annual-report-2014-15.pdf. Accessed 25 July 2017

Green G, White M, Gotch T, Scholz G (2013) Risk assessment process for evaluating water use impacts on the Great Artesian Basin springs. In: Allocating water and maintaining springs in the Great Artesian Basin, vol VI. National Water Commission, Canberra. http://webarchive.nla.gov.au/gov/20160302151814/http://archive.nwc.gov.au/library/topic/groundwater/allocating-water-and-maintainingspringsin-the-great-artesian-basin. Accessed 16 Jan 2018

Hadwen WL, Boon PI, Arthington AH (2012) Aquatic ecosystems in inland Australia: tourism and recreational significance, ecological impacts and imperatives for management. Mar Freshwater Res 63(4):325–340

Hemming S, Rigney D (2014) Indigenous engagement in environmental water planning, research and management: innovations in South Australia's Murray-Darling Basin Region. In: Goyder Institute for water research technical report series, No. 14/21. http://www.goyderinstitute.org/_r123/media/system/attrib/file/114/Goyder%20report%20Hemming%20and%20Rigney%20 2015_web.pdf. Accessed 21 July 2017

Hemming S, Rigney D, Pearce M (2007) Justice, culture and economy for the Ngarrindjeri nation. In: Potter E, Mackinnon A, McKenzie S, McKay J (eds) Fresh water: new perspectives on water in Australia. University Press, Melbourne

Hemming S, Rigney D, Berg S (2011) Ngarrindjeri futures: negotiation, governance and environmental management. In: Maddison S, Brigg M (eds) Unsettling the settler state: creativity and resistance in indigenous settler-state governance. Federation Press, Annandale

Jackson S, Tan P, Mooney C, Hoverman S, White I (2012) Principles and guidelines for good practice in Indigenous engagement in water planning. J Hydrol 474:57–65

Keppel M, Karlstrom KE, Love AJ, Priestley S, Wohling D, De Ritter S (eds) (2013) Hydrogeological framework of the Western Great Artesian Basin. In: Allocating water and maintaining springs in the Great Artesian Basin, volume I. National Water Commission, Canberra

Kerezsy A, Balcombe SR, Arthington AH, Bunn SE (2011) Continuous recruitment underpins fish persistence in the arid rivers of far western Queensland, Australia. Mar Freshw Res 62(10):1178–1190

Kerezsy A, Balcombe SR, Tischler M, Arthington AH (2013) Fish movement strategies in an ephemeral river in the Simpson Desert, Australia. Austral Ecol 38(7):798–808

Kingsford RT, Costelloe JF, Sheldon F (2014) Lake Eyre Basin: challenges for managing the world's most variable river system. In: Squires VR, Milner HM, Daniell KA (eds) River basin management in the twenty-first century: understanding people and place. CRC Press, Boca Raton

Lee KN (1993) Compass and gyroscope: integrating science and politics for the environment. Island Press, Washington, DC

Leigh C, Sheldon F, Kingsford RT, Arthington AH (2010) Sequential floods drive 'booms' and wetland persistence in dryland rivers: a synthesis. Mar Freshw Res 61(8):896–908

Love A, Shand P, Crossey L, Harrington GA, Rousseau-Gueutin P (eds) (2013a) Groundwater discharge of the Western Great Artesian Basin. In: Allocating water and maintaining springs in the Great Artesian Basin, Volume III. National Water Commission, Canberra. http://webarchive.nla.gov.au/gov/20160615074424/http://archive.nwc.gov.au/library/topic/groundwater/allocating-water-and-maintaining-springsin-the-great-artesian-basin. Accessed 21 July 2017

Love A, Wohling D, Fulton S, Rousseau-Gueutin P, DeRitter S (eds) (2013b) Groundwater recharge, hydrodynamics and hydrochemistry of the western Great Artesian Basin. In: Allocating water and maintaining springs in the Great Artesian Basin, volume II. National Water Commission, Canberra. http://webarchive.nla.gov.au/gov/20160615074424/http://archive.nwc.gov.au/library/topic/groundwater/allocating-water-andmaintaining-springs-in-the-great-artesian-basin. Accessed 21 July 2017

Marshall G (2008) Nesting, subsidiarity, and community-based environmental governance beyond the local scale. Int J Commons 2(1):75–97

Mitchell B (2014) Evolving regional, integrated and engagement approaches for natural resources management in South Australia. Flinders University, Adelaide. http://www.goyderinstitute.org/_r88/media/system/attrib/file/79/Bruce%20Mitchell-WEB.pdf. Accessed 21 July 2017

North West Power and Conservation Council (n.d.) https://www.nwcouncil.org/. Accessed 21 July 2017

Nursey-Bray M, Fergie D, Arbon V, Rigney L-I, Palmer R, Tibby J, Harvey N, Hackworth L (2013) Community based adaptation to climate change: the Arabana, South Australia. National Climate Change Adaptation Research Facility, Gold Coast. https://www.nccarf.edu.au/sites/default/files/attached_files_publications/Nursey-Bray_2013_Community_based_adaptation_Arabana.pdf. Accessed 21 July 2017

O'Neill L, Godden L, Macpherson E, O'Donnell E (2016) Australia, wet or dry, north or south: addressing environmental impacts and the exclusion of Aboriginal peoples in northern water development. Environ Plann Law J 33:402–417

Pahl-Wostl C (2007) Transitions towards adaptive management of water facing climate and global change. Water Resour Manag 21:49–62

Price R, Thoms M, Capon S, Watkins D (2009) Lake Eyre Basin Rivers assessment implementation plan project: MILESTONE 3 Report: governance arrangements for the LEBRA. Kiri-ganai, Canberra. http://www.lakeeyrebasin.gov.au/sitecollectionimages/71d27602-9826-4d4f-9004-fbc30cde225b/files/lebra-governance-milestone3_1.pdf. Accessed 21 July 2017

Sinclair Knight Merz (2005) Hydrology of the Lake Eyre Basin: summary brochure. Department of Natural Resources and Mines, Queensland. http://www.lakeeyrebasin.gov.au/publications/hydrology-lake-eyre-basin. Accessed 21 July 2017

Stoeckel K, Webb R, Woodward L, Hankinson A (2012) Australian water law. Thomson Reuters, Pyrmont

Thompson BH (2016) Defragging: overcoming fragmentation in United States water governance. In: Gray J, Holley C, Rayfuse R (eds) Trans-jurisdictional water law and governance. Routledge, London

Thoms M, Capon S, Price R, Watkins D (2009) Lake Eyre Basin Rivers assessment implementation plan project: milestone 2 report: proposed LEB rivers assessment methodology. Kiri-ganai, Canberra. http://www.lakeeyrebasin.gov.au/sitecollectionimages/71d27602-9826-4d4f-9004-fbc30cde225b/files/lebra-methodology-milestone2_1.pdf. Accessed 21 July 2017

Universities Consortium on Columbia River Governance (2015) Universities Consortium on Columbia River Governance. http://www.columbiarivergovernance.org. Accessed 21 July 2017

Legal References

Commonwealth of Australia (2000) Lake Eyre Basin intergovernmental agreement. Available at http://www.lakeeyrebasin.gov.au/collaborativemanagement/intergovernmental-agreement. Accessed 21 July 2017

Commonwealth of Australia (2004) Australian national water initiative 2004. http://webarchive.nla.gov.au/gov/20160615061800/http://www.nwc.gov.au/__data/assets/pdf_file/0008/24749/Intergovernmental-Agreement-on-a-national-water-initiative.pdf. Accessed 26 July 2017

Commonwealth of Australia (2007) Australian Water Act 2007. https://www.legislation.gov.au/Details/C2007A00137. Accessed 21 July 2017

Commonwealth of Australia (2012) Legislative powers of the Parliament section 51(xxxvii). In: Australian constitution, Chapter I, Part V. https://www.legislation.gov.au/Details/C2005Q00193. Accessed 28 July 2017

Commonwealth of Australia (2014) Australian Environment Protection and Biodiversity Conservation Act 1999. http://www.comlaw.gov.au/Details/C2014C00506. Accessed 21 July 2017

Commonwealth of Australia (2017) Australian Native Title Act 1993 as amended. https://www.legislation.gov.au/Details/C2017C00178. Accessed 26 July 2017

Lake Eyre Basin Ministerial Forum (n.d.) Lake Eyre Basin rivers assessment: implementation plan & business governance model 2010–2018. http://www.lakeeyrebasin.gov.au/sitecollectionimages/71d27602-9826-4d4f-9004-fbc30cde225b/files/lebra-implementation-plan-2010-18.pdf

Lake Eyre Basin Secretariat (2011) About the basin. http://www.lakeeyrebasin.gov.au/about-basin. Accessed 25 July 2017

Lake Eyre Basin Secretariat (n.d.) Lake Eyre Basin aboriginal forums. http://www.lakeeyrebasin.gov.au/community-forums/lake-eyre-basin-aboriginal-forums. Accessed 28 July 2017

Judicial Opinions

Mabo v Queensland [No 2] (1992) 175 CLR 1 (Austl.)

Part II
The Role of Law in Adaptive Governance and Resilience

Chapter 10
Legal Pathways to Adaptive Governance in Water Basins in North America and Australia

Barbara Cosens, Robin Kundis Craig, Shana Hirsch, Craig Anthony (Tony) Arnold, Melinda Harm Benson, Daniel DeCaro, Ahjond S. Garmestani, Hannah Gosnell, J. B. Ruhl, and Edella Schlager

Abstract Law dictates the structure, boundaries, rules, and processes within which governmental action takes place and in doing so becomes one of the focal points for analysis of governmental barriers to adaptation as the effects of climate change are felt. Governance encompasses both governmental and nongovernmental participation in collective choice and action. Adaptive governance contemplates a level of flexibility and evolution in governmental action beyond that currently found in the heavily administrative governments in the United States and Australia. Nevertheless, over time, law itself has proven highly adaptive in democracies, evolving to address and even facilitate the emergence of new social norms (such as the rights of women and minorities) or to provide remedies for emerging problems (such as pollution). Thus, law can adapt, evolve, and be reformed to facilitate adaptive governance. In doing so, not only may barriers be removed, but law may be adjusted to facilitate

B. Cosens (✉)
College of Law and Waters of the West Program, University of Idaho, Moscow, ID, USA
e-mail: bcosens@uidaho.edu

R. K. Craig
S.J. Quinney College of Law, University of Utah, Salt Lake City, UT, USA
e-mail: robin.craig@law.utah.edu

S. Hirsch
University of Idaho, Moscow, ID, USA
e-mail: hirs3810@vandals.uidaho.edu

C. A. Arnold
Louis D. Brandeis School of Law, University of Louisville, Louisville, KY, USA
e-mail: tony.arnold@louisville.edu

M. H. Benson
Haub School of Environment and Natural Resources, University of Wyoming, Laramie, WYO, USA
e-mail: mhbenson@uwyo.edu

B. Cosens, L. Gunderson (eds.), *Practical Panarchy for Adaptive Water Governance*, https://doi.org/10.1007/978-3-319-72472-0_10

adaptive governance and to aid in institutionalizing new and emerging approaches to governance. The key is to do so in a way that also enhances legitimacy, accountability, and justice (i.e., good governance), or such reforms will never be adopted by democratic societies or, if adopted, will destabilize those very societies. By identifying those aspects of adaptive governance relevant to the legal system, this chapter presents guidelines for evaluating the role of law in environmental governance and demonstrates their use by applying them to the basin studies presented in Part I of this volume.

Keywords Adaptive governance · Adaptive law · Environmental governance · Resilience · Water law · Nesting · Polycentricity · Legitimacy

10.1 Introduction

Democratic legal systems are inherently adaptive. They respond to new challenges in all three areas of government. Judicial issue-by-issue interpretation of the law is responsive to changing circumstances and fosters societal resilience by providing final resolution to otherwise intractable disputes. Legislative authority to pass law and amend prior law allows response to new problems and fosters societal resilience by bringing resources to bear on new challenges. Executive implementation of policy is responsive to public discourse and fosters societal resilience by providing legitimacy and accountability through an election cycle. Counteracting this inherent

D. DeCaro
Department of Urban and Public Affairs, Department of Psychological and Brain Sciences, University of Louisville, Louisville, KY, USA
e-mail: daniel.decaro@louisville.edu

A. S. Garmestani
School of Natural Resources, University of Nebraska, Lincoln, NE, USA
e-mail: garmestani@hotmail.com

H. Gosnell
College of Earth, Ocean, and Atmospheric Sciences, Oregon State University, Corvallis, OR, USA
e-mail: gosnellh@geo.oregonstate.edu

J. B. Ruhl
Vanderbilt University Law School, Nashville, TN, USA
e-mail: jb.ruhl@vanderbilt.edu

E. Schlager
School of Government and Public Policy, The University of Arizona, Tucson, AZ, USA
e-mail: Schlager@email.arizona.edu

adaptability is the fact that legal systems and the governments they empower are purposely structured to foster stability. Furthermore, a measure of stability is critical to the resilience of social systems and the ability of private actors to innovate. Governance of water systems has moved beyond purely state-centric control to complex public/private action which have been described as collaborative governance (Bingham 2009), adaptive governance (Chaffin et al. 2014), or collaborative co-management (Olsson et al. 2004). This changes the role that law and government must play in governance. Law in its role of establishing the structure, capacity, and process of government is a key component in striking the balance between stability in government and adaptability of governance. Importantly, in trading stable governmental control for innovative adaptive governance, the engagement of private actors in governance should not come at the expense of legitimacy, accountability, equity, and justice. Hence, good governance becomes an integral part of any framework to reform law to facilitate adaptation.

10.2 Analytical Framework and Its Application to Basin Assessment

Change in environmental governance to give rise to adaptive governance has three stages: (1) the stage in which change is triggered or a window of opportunity for change opens, (2) the stage in which adaptive governance emerges, and (3) the stage in which change is institutionalized. Law plays a different role in each stage. These stages form the bases of an analytical framework for identifying the role of law and designing the appropriate legal reform to facilitate adaptive governance. The remainder of this chapter is organized accordingly. The first section explores the regulation of private behavior which is the primary means through which law serves as a trigger. The second section explores the role of law in determining the structure, capacity, and process of government and how changes in these areas may allow government to facilitate and participate in adaptive governance. The third section explores institutionalization of change which involves changes to both the regulation of private behavior and the structure, capacity, and processes of government.

10.2.1 Assessing Law's Role in the Emergence of Adaptive Governance

A window of opportunity for transformation in governance is thought to occur when the appropriate combination of problem, solution, and politics intersect to make change possible (Kingdon 1995; Olsson et al. 2006). A disturbance sufficient to trigger the emergence of new approaches to governance may be related to biophysical change, such as the long-term droughts recently experienced by Australia and

California (Cosens 2016a), or social change, including health, economic, legal, or political disturbance, such as an election. From that broad set of triggers, this section narrows the discussion to water basins and how the imposition of regulatory law on private actors, alone or in combination with other factors, triggers efforts to develop new solutions including new approaches to governance.

Regulatory law serves as a trigger when its application alone or in combination with other pressures on a water system (such as drought) leads to outcomes unacceptable to a significant number of water interests in a basin. This may be the result of conflicting regulatory measures, harsh zero-sum outcomes in which some win and some loose (Ruhl 2016), or outcomes that will severely limit economic pursuits. Law may also create a disturbance by shifting the allocation of power.

The basin assessments in Part I illustrate how the harsh imposition of multiple regulatory laws in the United States led to governance adaptations. Implementation of the Endangered Species Act (ESA) triggered shifts in governance in the Columbia, Everglades, Klamath, Rio Grande, and Platte basins. Regulation under the Clean Water Act (CWA) in the Klamath and Anacostia basins, hydropower relicensing under the Federal Power Act (FPA) in the Klamath basin, and assertion of Native American Water Rights and other treaty rights in the Klamath and Columbia watersheds also resulted in governance innovations. A reallocation of power followed litigation over treaty fishing rights in the Columbia basin (Cosens and Fremier 2014; Chap. 4), and water rights in the Klamath (Chaffin et al. 2014; Chap. 6) ultimately led formerly marginalized indigenous communities to new roles as co-managers of basin fisheries and as senior water rights holders. The Klamath and Columbia basin assessments illustrate response to multiple regulatory issues. Collaborative problem solving emerged in the Klamath in response to the intersection of the ESA, CWA, FPA, and treaty rights. Efforts to change governance in the Columbia River have only begun in the face of review of the Columbia River Treaty between the United States and Canada but have been fueled by the intersection of ESA regulation, treaty rights litigation, and the policy window created by expiration of certain terms of the international treaty. In these examples, top-down regulation triggers innovation in part by presenting narrow solutions that are socially and/or economically unacceptable. Collaborative processes emerged or are emerging to explore more comprehensive solutions as well as innovative solutions that expanded the possibilities beyond that of a zero-sum game.

The importance of this step in the process of basin assessment is to determine if the situation even warrants exploration of legal reform to remove barriers to adaptive governance and to facilitate its emergence. The basin studies have revealed that while theoretical generalizations may be made about adaptive governance, its emergence is contextual. Thus, efforts to move from command-and-control regulation to adaptive governance by wholesale legal reform, even if politically possible, would likely fail. Implementation is contextual and must proceed through the hard work of basin-by-basin assessment and reform.

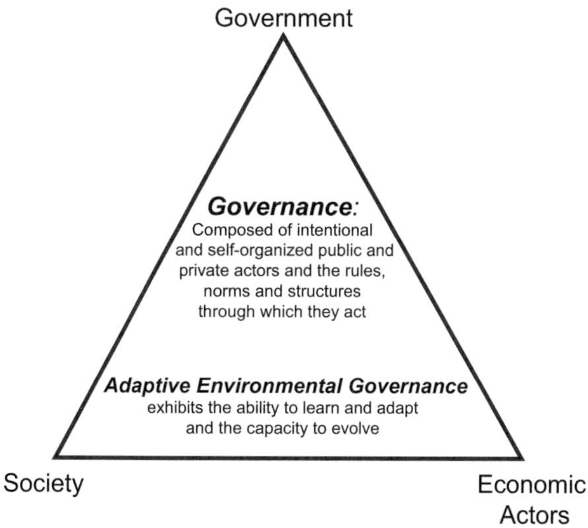

Fig. 10.1 Governance includes laws, policies, regulations, institutions, and institutional structures that reside within governmental, societal, and economic constraints. Governance also includes the informal norms and interactions that influence decisions including those of private and nongovernmental actors. Definitions rely heavily on the work by Rogers and Hall (2003), UN System Task Team (2012) in defining governance, and on a literature review of adaptive governance by Chaffin et al. (2014)

10.2.2 Assessing the Role of Law in Facilitating Adaptive Governance and Assuring Its Legitimacy

Law establishes the structure, authority, and processes for the governmental aspect of governance. Governance results from a complex interplay among a triad of government, society, and economic actors (Fig. 10.1). Consistent with Chaffin et al. (2014), adaptive governance is framed as an emergent or self-organizing response to a problem that may involve a range of governmental, social, and economic actors. We separate economic actors from society because, despite US Supreme Court rulings to the contrary (Citizens United v. Federal Election Commission 2010), their multinational character and primary allegiance to the profit motive distinguish their motivation, capacity, power, and scope of impact from that of persons (with the caveat that some NGOs may mimic this corporate scope of action). With globalization, we risk the inability of state-centric law to effectively regulate the action of economic entities. This in particular is a critical area of concern as we move from state-centered control to new or adaptive governance (Godden 2016) and emphasizes the critical importance of attention to good governance in framing the role of law.

Guidelines for the role of law in governance were developed through integration of conceptual models and definitions of adaptive governance (Dietz et al. 2003;

Ostrom et al. 1961; Ostrom 1999; Folke et al. 2005; Marshall 2008; Huitema et al. 2009) with those for good governance (Dublin statement 1992; United Nations 2003; Rogers and Hall 2003; Lee 2003; Folke et al. 2005; Esty 2006; Huitema et al. 2009; Bingham 2009; Lockwood et al. 2010; Cosens 2013; Borrini-Feyerabend et al. 2013). The development of the guidelines is described in Cosens et al. (2017). The guidelines are grouped into three categories that reflect the type of laws related to governmental action in environmental governance: (1) structure, (2) capacity, and (3) process.

Structure

Structure encompasses both the organizational design of regulatory and management systems and the legal basis of the authority for the managing entities. Structure includes the manner in which law allocates authority among various levels and sectors of government. Assessment of the barriers and opportunities for facilitation of adaptive governance posed by governmental structure requires consideration of (1) polycentricity, (2) integration, and (3) persistence.

Polycentricity is a governance structure consisting of multiple, redundant centers of authority and decision-making at different political scales (Ostrom 1999). The multiple centers are complimentary in that similar authority resides at each scale, such that if one decision body fails to act or acts inappropriately, another body can step up (Ostrom 1999). These structures are nested, increasing the possibility of formation of ad hoc cross-scale networks in response to surprise and at the scale of the problem as well as the potential for local innovation within stable governance at a larger scale (Ostrom et al. 1961; Ostrom 1999; Marshall 2008). Polycentricity allows for subsidiarity or the movement of decision-making authority to the level closest to the resource as possible increasing the likelihood that local knowledge will be used, decisions will be tailored to specific problems, and innovation may occur at the local level supported by stable governance at larger scales (McGinnis 1999; Marshall 2008; Clarvis et al. 2014). It is also critical that any devolution of authority occurs within the context of local governance that is nested within larger scales of governance to foster conditions for implementation of management decisions and assure that local decision-making is not captured by special interests, but meets broader legal standards where applicable. Collaborative processes led to innovative solutions in the Klamath, Platte, and Anacostia basins, and the nesting of those processes within a stable federal framework led to innovation without destabilization (Chaps. 3, 6 and 8). In contrast, communication gaps resulting from the absence of overlap among individuals involved in intra- and interstate advisory bodies (both citizen and scientific) in the Lake Eyre Basin resulted in reduction in trust and a perception that local input had no value (Chap. 9).

Each of the study basins illustrates barriers to adaptive governance created by the mid-twentieth-century development of federal level command-and-control environmental regulation. In fact, this is the flip side of the role of this type of regulation in triggering collaborative response discussed above. The answer is not to eliminate these laws that play an important role in both triggering change and mandating standards but to provide the authority and facilitation of response at the scale of the issue (e.g., the bioregional scale) and through the involvement of private and economic actors. This is not an argument for devolution of the role of state and federal government in mandating certain outcomes but to move the solution closer to the problem while maintaining the nesting within the oversight of higher levels of government.

Integration refers to the elimination of legal fragmentation across both geographic and sectoral boundaries (Cosens and Stow 2014; Arnold 2014; Thompson 2016). In the fragmented approach to water management in the United States, in which the law places water allocation at the state level (and states are only beginning to wrestle with the connections between ground and surface water), land use at the local level, aquatic and riparian endangered species regulation at the federal level, and water quality regulation at both the federal and state levels through cooperative federalism, integration by fiat is highly unlikely. However, our assessments illustrate that it is possible to address fragmentation in specific contexts through emergent adaptive processes. Although, in general, fragmentation itself was not the initial driving force, once collaborative processes begin to emerge, the possibility of addressing multiple regulatory issues through an integrated solution was apparent in both the Klamath (Chaffin et al. 2014; Chap. 6) and Platte River (Birgé et al. 2014; Chap. 8) assessments. Thus, as discussed below, it is the authority to consider and the process to institutionalize new approaches to integration within specific contexts that the law must provide.

Persistence in the actors involved in water management fosters legitimacy and trust, potentially reducing the time needed to respond to surprise. Persistence, in combination with nesting, takes advantage of the power of social networks. Continuity in the actors involved allows networks to build trust and knowledge and facilitate the flow of information and consistency of implementation (Krebs and Holley 2004; Bodin and Crona 2009). All of our basins are located within legal systems in which rules are stable and evolve in a predictable manner, and testing the response to surprise is an experiment yet to come. Looking outside of our assessments, the experience of Australia in its efforts to respond to extended drought through water law reform illustrates the unintended consequences of failure to consider the need for persistence in actors in the process of reform and the resultant loss of trust (Mitchell 2014).

Persistence encompasses the need for stability in rules and players, yet it can also lead to a bias for the status quo which stands as a barrier to adaptation when transformation is warranted (Garmestani and Benson 2013). This is evident in the rigidity trap presented by the Everglades assessment. The legacy effects of both change in ecosystem function and human dependence on local economies are apparent in the failure to transform the status quo in the Everglades. Resources to navigate the transition through restoration of ecosystem function have been made available in both the Everglades and the Columbia basin, but have not been accompanied by the financial and technical assistance to transform economies. This type of resource is generally absent from major changes in federal policy on development of rivers (as well as changes in use of public lands) in the United States, making it difficult to overcome the local bias for the status quo. This in turn fuels litigation. This form of rigidity trap, or engineered legacy, is apparent in all of our basin assessments with engineered transformation of rivers accomplished in the mid-twentieth century to optimize for purposes that excluded values now considered important and will require substantial investment in aging infrastructure and local economies to overcome.

Capacity

Included under the category of capacity is the ability to navigate ecological regime shifts that has been defined as adaptive capacity (Gunderson 2000) and, of equal importance, the right and resources of interested persons and groups to have a role in decision-making, described here as participatory capacity (Huitema et al. 2009; Raadgever et al. 2008; Pahl-Wostl 2009). In considering the role of law, governmental managers must have the legal authority to experiment and the legal authority to facilitate capacity building among local communities.

Adaptive capacity requires the resources (generally appropriated through a legislative act) and legal authority to respond to change and allows a system of governance to adjust during uncertainty regarding the manner in which change will play out. It is reflected in the statutes, regulations, and practices of agencies (Clarvis et al. 2014). Legal sources effectively define, shape, and limit this capacity in water governance, including through definition of the authority of government to (1) adjust water allocations and water quality requirements in response to change, (2) implement adaptive management in appropriate circumstances (Craig and Ruhl 2014), (3) facilitate collaborative adaptive planning (Arnold 2010, 2014), (4) provide resources and flexibility for local innovation, and (5) facilitate the ability of individuals and private entities to adapt through water markets (Cosens 2016a).

Participatory capacity addresses the right and ability of those affected by water management to participate in the development of its goals. Participatory capacity reduces the likelihood of marginalization of portions of society and increases the likelihood that all aspects of a system will be considered in decision-making. It requires that those affected by water management have the legal right of access to decision-making and the knowledge, time, and resources to engage in the substance of decision-making. The role of law in promoting participatory capacity is to provide the requirement for public participation, judicial forums that can recognize and enforce the rights of those without power, and where appropriate, provide an avenue for capacity building through the legislative allocation of resources and authority to facilitate local response (Olsson et al. 2004; Bingham 2009; National Civic League 2013).

Capacity of both types appears to be the weakest link in the North American water basins. It may be surprising that the wealthiest nation in the world scores so poorly in capacity which focuses on "scientific, technological, organizational, institutional and resource capabilities" (United Nations 1992). However, two legal factors apparent in the basin assessments are relevant here. First, in the context of adaptive capacity, there is currently a lack of legal mechanisms at any level that allow for alternatives to traditional environmental enforcement—alternatives that could enhance local innovation while maintaining stability and accountability toward achieving the goals that traditional enforcement mechanisms seek to achieve. Second, in the context of participatory capacity, federal government has greater resources and participatory capacity than local governments and other governance institutions (with the exception of some economic actors). As a result, when federal legal mechanisms—such as the Endangered Species Act and Clean Water Act—apply in a given basin, the absence of state and local legal avenues to develop solutions, the lack of federal level discretion to allow local coalitions authority to create new and more comprehensive solutions, and the lack of capacity of local actors to participate in efforts to tailor solutions may result in unsatisfactory or incomplete outcomes (Bingham 2009, 2010). Nevertheless, there is also evidence that when the law facilitates increased capacity, adaptive governance has a better chance of emerging. The substantial increase in participatory capacity of formerly marginalized communities as a result of recognition of rights through judicial processes coincides with the emergence of local collaborative processes in the Anacostia, Rio Grande, Platte, Klamath, and Columbia basin assessments (Chaps. 3, 4, 6, 7 and 8).

Process

The process of agency implementation of the law is governed by an area of law referred to as administrative law (Stewart 2003), and it is this law that has developed to assure good governance in regulatory bodies that are one step removed from the election process (Esty 2006). In the United States, agencies comply with federal and state administrative procedure acts to guide process, but in the field of environmental regulation, governmental process requirements may also be found in specific statutes and regulations directing agency action, such as the US National Environmental Policy Act (42 USC §§ 4321–4370h) or the consultation requirements in the Endangered Species Act (16 USC §§ 1531–1544). Local governments in the United States also exercise management and regulatory authority. Because this level of government is structured differently than federal and state governments, they are subject to an adapted set of administrative law principles in fields like land use regulation (Juergensmeyer and Roberts 2013) and public health (Richards 2007).

These laws provide a framework for uniform agency action by increasing consistency, transparency, and accountability in implementation of the law by those not elected (Stewart 2003; Cosens 2013; Craig and Ruhl 2014). However, despite the lofty goal of good governance, administrative law in its current form may contribute to the rigid, hierarchical nature of implementation of natural resources law. By drilling down to the purpose underlying different aspect of administrative law, then adjusting it to allow room for more flexible procedures, the guidelines are designed to transform the way in which governmental entities interact with society while maintaining legitimacy (Cosens 2013). Thus, the purpose of the rules governing the process of governmental regulation forms the basis for the guidelines in the process category. The process aspects necessary to assure good governance in the facilitation of adaptive governance can be parsed into six categories: (1) legitimacy, (2) procedural justice, (3) problem-solving approach, (4) opportunity for reflection and learning, (5) balancing stability and flexibility, and (6) dispute resolution. The following paragraphs address each of the six areas.

Legitimacy captures the need for both the perception and the reality of appropriate exercise of governmental authority (Tyler 2006) and underpins the basic level of confidence and trust society holds in those who govern. Legitimacy is necessary for public support of environmental management and generally includes requirements for science-based decision-making, deliberation, accountability, transparency, consistency, stability, and review and recourse for those aggrieved by a governmental action. Our basin assessments illustrate substantial loss of legitimacy when promised management reform fails to materialize often due to the legacy effects of either existing power structures (Everglades and Columbia) or existing infrastructure (all US basin assessments). The legacy of our twentieth century infrastructure is a problem on all basins and will require substantial investment to transform. Experiments on the Klamath and the Platte Rivers have the potential for showing a way forward. As adaptive capacity is increased, legitimacy will be a concern when authorizing greater flexibility, and thus discretion, in agency action (Cosens 2013; Craig and Ruhl 2014), and attention must be given to legitimacy to assure that discretion does not become a pathway to failed experiments in the implementation of flexible approaches such as adaptive management. Legitimacy may also be a challenge in the devolution of authority to the scale of the problem (subsidiarity). As federal and state governments experiment with the devolution of certain authority to local governance, it may be wise to precede on an *ad hoc* basis through legislative approval of place-based measures limited to those situations where participatory capacity is strong (Marshall 2008). Depending on the trajectory taken by both the current settlement process on the Klamath and the Treaty review process on the Columbia, we may see experimentation in implementing local adaptive governance solutions while maintaining the umbrella of international, federal, and state standards and the provision of financial and scientific resources.

Procedural justice provides the right to seek review and engagement at the appropriate level and serves multiple purposes including teasing out unintended consequences of agency action, checking corruption, and avoiding uneven application of the burden of adaptation. The basin assessments reveal procedural justice as critical in giving a voice to formerly marginalized communities. While substantive law on equal protection and due process played a role in redistributing power in the civil rights era (Anacostia), and litigation concerning treaty rights has given voice to Native American communities (Columbia and Klamath), as a matter of process, we also focus on the laws pertaining to governmental engagement with these communities. The empowerment of Native American tribes on the Columbia and Klamath to seek engagement on a government-to-government basis is a critical step in this area but may fall short of the ultimate goal of self-determination expressed in the UN Declaration on the Rights of Indigenous Peoples (Robison et al. 2017). It is only in combination with the building of participatory capacity that we see procedural justice moving in the direction of self-determination in our basins, which in turn increases the likelihood of adaptive co-management (Cosens and Chaffin 2016).

A problem-solving approach and the opportunity for reflection and learning are closely related in application and are thus discussed together. A problem-solving approach requires scientific expertise, openness to traditional and local ecological knowledge, and use of interest-based collaborative processes. Problem solving contrasts with political and ideological approaches to resolving issues, because these frames are generally not subject to compromise. It also contrasts with litigation, which resolves specific legal issues on a piecemeal basis and may not even address the underlying problem. The opportunity for reflection and learning through monitoring and review increases the potential that response to change will be informed, that management will evolve with the changing environment, and that solving the problem, rather than compliance with agreed upon measures, is the measure of success.

The problem-solving approach is reflected in the process of community-based collaborative dialogue to address multiple issues of water allocation, quality and management undertaken in the Klamath settlement process (Chap. 6), and the incorporation of traditional knowledge in fisheries co-management on the Columbia River (Chap. 4). However, our basin assessments also reveal that the opportunity to learn in the process is often missed through failure to devote resources to monitoring, review, and adjustment of agency action (i.e., adaptive management).

Balancing stability and flexibility requires that adaptation timeframes consider both the need for adjustment and the economic need for stability. Balance of stability and flexibility recognizes that while adjustments must occur in the face of change, social systems and particularly economic systems require stability; thus, both must be taken into account. As noted above, striking the appropriate balance of stability and flexibility is a fundamental tension in the application of law. This tension is presented in detail in Chap. 11.

Dispute resolution requires appropriate legal forums, processes, and rules for resolution and is essential as water scarcity and ecological regime shift becomes more common in many parts of the world. Martin and Kennedy (2016) argue that collaborative processes may prove insufficient when major trade-offs are necessary to resolve conflict. The authors identify the need to maintain a case-by-case evolution of judicial dispute resolution which has historically demonstrated adaptive capacity in excess of science-based administration or economic and politically driven collaboration when the unexpected occurs and the conflict is intractable. All of our basins provide examples of resort to judicial resolution of issues when diplomacy failed. The use of judicial process by Native American tribes in both the Columbia and the Klamath basins was necessary to challenge the status quo regarding control of fisheries (Columbia, Chap. 4) and priority over water rights (Klamath, Chap. 6). In circumstances in which conflict is imminent, unless a system for resolving issues is designed and agreed to beforehand, conflict is likely. Providing a forum and process for resolving conflict and making final, binding decisions on trade-offs regarding scarce resources is an essential role for higher levels of government (Ostrom 1990; Dietz et al. 2003), and it is through the law that they do so.

10.2.3 Assessing the Role of Law in Institutionalizing the Results of Adaptive Governance

Finally, institutionalization through legislation (or judicial approval in the case of settlement) of the results of collaborative process is needed to (1) provide the authority for any governmental action needed to implement the result, (2) provide the authority for any new regulation or increased enforcement to achieve the process goals, (3) provide resources to accomplish implementation of the results, and (4) prevent backsliding as those committed to the collaborative effort are replaced by those lacking institutional memory or superseded by a change in power. The provision of authority and resources is evident in the development of the 1964 Columbia River Treaty, and its need is evident for the implementation of the Klamath River settlement and restoration efforts in the Platte River basin and the Everglades. In contrast, watershed movements spurred by the Clean Water Act, such as that described for the Anacostia, rarely are followed by the provision of authority and adequate resources for local implementation (also a factor in the perceived legitimacy of these processes). The current review process of the Columbia River Treaty, while developing considerable momentum during the public process stage, is now experiencing backsliding and, with political change in both the United States and Canada, may see substantial change (Cosens 2016b). Any effort to increase the facilitation of collaborative governance at the federal or state level must include a process for this step if it is to succeed.

10.3 Conclusion

Law is often viewed as a constraint on adaptation but has proven to be highly adaptive through time. Law itself, however, is insufficient to assure the emergence of a form of governance that engages both civil society and economic entities. Thus, an understanding of basin dynamics must first occur through a process of basin assessment (including the evolution of its governance), followed by analysis of the role of law through application of the legal guidelines. This process is context specific, and in any application, only some of the elements may be relevant. The effort to develop guidelines for the role of law in adaptive governance illustrates that while many promising ideas on new and collaborative governance are emerging, their implementation is contextual and will fail if attention is not given to the process elements of good governance. It is also essential that, in the process of application, attention be paid to the front-end need for a trigger before any public or private resources will be devoted to a collaborative effort and the back-end need for institutionalization to assure the results of collaborative processes are realized. By integrating the legal reform needed to facilitate the emergence of adaptive governance with parallel reform to ensure legitimacy and implementation, the likelihood of both acceptance and success in navigating system changes will increase.

Acknowledgments This work was developed in part under the Adaptive Water Governance Project, funded by the US National Socio-Environmental Synthesis Center (SESYNC) under funding from the US National Science Foundation, NSF DBI-1052875. The views set forth by contributors to this volume represent their own and do not represent the views of any public or private entity the contributor is affiliated with.

References

Arnold CA (2010) Adaptive watershed planning and climate change. Environ Energy Law Policy J 5:417–488

Arnold CA (2014) Adaptive water law. Kansas Law Rev 62:101–149

Bingham LB (2009) Collaborative governance: emerging practices and the incomplete legal framework for public and stakeholder voice. J Dispute Resolut 2(2):1–58

Bingham LB (2010) The next generation of administrative law: building the legal infrastructure for collaborative governance. Wisconsin Law Rev 2010(2):297–356

Birgé HE, Allen CR, Craig RK, Garmestani AS, Hamm JA, Babbitt C, Nemec K, Schlager E (2014) Social-ecological resilience and law in the Platte River basin. Idaho Law Rev 51:229–256

Bodin Ö, Crona B (2009) The role of social networks in natural resource governance: what relational patterns make a difference? Glob Environ Chang 19(3):366–374

Borrini-Feyerabend G, Dudley N, Jaeger T et al (2013) Governance of protected areas: from understanding to action. In: IUCN best practice protected area guidelines series no.20. http://cmsdata.iucn.org/downloads/governance_web_1.pdf

Chaffin BC, Gosnell H, Cosens B (2014) A decade of adaptive governance scholarship: synthesis and future directions. Ecol Soc 19(3):56. https://doi.org/10.5751/es-06824-190356

Clarvis MH, Allan A, Hannah DM (2014) Water, resilience and the law: from general concepts and governance design principles to actionable mechanisms. Environ Sci Policy 43:98–110

Cosens B (2013) Legitimacy, adaptation, and resilience in ecosystem management. Ecol Soc 18(1):3. https://doi.org/10.5751/ES-05093-180103

Cosens B (2016a) Water law reform in the face of climate change: learning from drought in Australia and the Western United States. Environ Plan Law J 33:372

Cosens B (2016b) The Columbia river treaty: an opportunity for modernization of basin governance. Colorado Nat Res Energy Environ Law Rev 27(1):1–19. http://www.colorado.edu/law/sites/default/files/CNREELR-V27-I1-Cosens%20Final_0.pdf

Cosens B, Chaffin BC (2016) Adaptive governance of water resources shared with indigenous peoples: the role of law. Water 8(3):97. https://doi.org/10.3390/w8030097

Cosens B, Fremier A (2014) Assessing system resilience and ecosystem services in large river basins: a case study of the Columbia river basin. Idaho Law Rev 51:91–125

Cosens B, Stow C (2014) Resilience and water governance: addressing fragmentation and uncertainty in water allocation and water quality law. In: Garmestani A, Allen C (eds) Social-ecological resilience and law. Columbia University Press, New York

Cosens B, Craig RK, Hirsch S, Arnold CA, Benson MH, Chaffin B, DeCaro D, Garmestani AS, Gosnell H, Ruhl JB, Schlager E (2017) The role of law in adaptive governance. Ecol Soc 22(1):30. https://doi.org/10.5751/es-08731-220130

Craig RK, Ruhl JB (2014) Designing administrative law for adaptive management. Vanderbilt Law Rev 67:1

Dietz T, Ostrom E, Stern PC (2003) The struggle to govern the commons. Science 302:1907–1912. https://doi.org/10.1126/science.1091015

Dublin statement on water and sustainable development (1992) International conference on water and environment, Dublin. http://www.wmo.int/pages/prog/hwrp/documents/english/icwedece.html#principles. Accessed 27 July 2017

Esty DC (2006) Good governance at the supranational scale: globalizing administrative law. Yale Law J 115:1490–1562. https://doi.org/10.2307/20455663
Folke C, Hahn T, Olsson P, Norberg J (2005) Adaptive governance of social-ecological systems. Annu Rev Environ Resour 30:441–473. https://doi.org/10.1146/annurev.energy.30.050504.144511
Garmestani AS, Benson MH (2013) A framework for resilience-based governance of social-ecological systems. Ecol Soc 18(1):9. https://doi.org/10.5751/es-05180-180109
Godden L (2016) Trans-jurisdictional water governance: law's role at multiple levels. In: Gray J, Holley C, Rayfuse R (eds) Trans-jurisdictional water law and governance, Earthscan studies in water resource management. Routledge, London
Gunderson L (2000) Ecological resilience—in theory and application. Annu Rev Ecol Syst 31:425–439. https://doi.org/10.1146/annurev.ecolsys.31.1.425
Huitema D, Mostert E, Egas W, Moellenkamp S, Pahl-Wostl C, Yalcin R (2009) Adaptive water governance: assessing the institutional prescriptions of adaptive (co-)management from a governance perspective and defining a research agenda. Ecol Soc 14(1):26. http://www.ecologyandsociety.org/vol14/iss1/art26/. Accessed 27 July 2017
Juergensmeyer JC, Roberts TE (2013) Land use planning and development regulation law, 3rd edn. West, St. Paul
Kingdon J (1995) Agendas, alternatives, and public policies, 2nd edn. Harper Collins, New York
Krebs VJ, Holley J (2004) Building sustainable communities through network development. The Nonprofit Q 11:46–53. https://nonprofitquarterly.org/2004/03/21/building-sustainable-communities-through-social-network-development/. Accessed 27 July 2017
Lee M (2003) Conceptualizing the new governance: a new institution of social coordination. Paper presented at the institutional analysis development mini-conference, workshop in political theory and policy analysis. Indiana University, Bloomington, 3–5 May 2003. https://pdfs.semanticscholar.org/71ec/0b861a6dad2d93b56ab5f8c6b77bfa415a44.pdf. Accessed 27 July 2017
Lockwood M, Davidson J, Curtis A, Stratford E, Griffith R (2010) Governance principles for natural resource management. Soc Nat Resour 23:986–1001. https://doi.org/10.1080/08941920802178214
Marshall G (2008) Nesting, subsidiarity, and community-based environmental governance beyond the local scale. Int J Commons 2(1):75–97
Martin P, Kennedy A (2016) Intra-national rivalries: a submerged aspect of trans-boundary water governance. In: Gray J, Holley C, Rayfuse R (eds) Trans-jurisdictional water law and governance, Earthscan studies in water resource management. Routledge, London
McGinnis MD (ed) (1999) Polycentric governance and development: readings from the workshop in political theory and policy analysis. University of Michigan Press, Ann Arbor
Mitchell B (2014) Evolving regional, integrated and engagement approaches for natural resources management in South Australia. Flinders University, Adelaide. http://www.goyderinstitute.org/uploads/documents/publications/2014/Bruce%20Mitchell-WEB.pdf. Accessed 27 July 2017
National Civic League (2013) Making public participation legal. http://ncdd.org/rc/wp-content/uploads/MakingP2Legal.pdf. Accessed 27 July 2017
Olsson P, Folke C, Berkes F (2004) Adaptive comanagement for building resilience in social-ecological systems. Environ Manag 34:75–90. https://doi.org/10.1007/s00267-003-0101-7
Olsson P, Gunderson LH, Carpenter SR, Ryan P, Lebel L, Folke C, Holling CS (2006) Shooting the rapids: navigating transitions to adaptive governance of social-ecological systems. Ecol Soc 11(1):18. http://www.ecologyandsociety.org/vol11/iss1/art18/. Accessed 27 July 2017
Ostrom E (1990) Governing the commons: the evolution of institutions for collective action. Cambridge University Press, Cambridge
Ostrom V (1999) Polycentricity. (Parts 1 and 2). In: McGinnis MD (ed) Polycentricity and local public economies. University of Michigan Press, Ann Arbor
Ostrom V, Tiebout CM, Warren R (1961) The organization of government in metropolitan areas: a theoretical inquiry. Am Polit Sci Rev 55:831–842
Pahl-Wostl C (2009) A conceptual framework for analysing adaptive capacity and multi-level learning processes in resource governance regimes. Glob Environ Chang 19:354–365

Raadgever GT, Mostert E, Kranz N, Interwies E, Timmerman JG (2008) Assessing management regimes in transboundary river basins: do they support adaptive management? Ecol Soc 13(1):14. http://www.ecologyandsociety.org/vol13/iss1/art14/. Accessed 27 July 2017

Richards EP (2007) Public health law as administrative law: example lessons. J Health Care Law Policy 10:61–88

Robison J, Cosens B, Jackson S, Leonard K, McCool D (2018) Indigenous water justice. Lewis and Clark Law Review in press.

Rogers P, Hall A (2003) Effective water governance. In: Global water partnership technical committee (TEC) Background Paper 7. Global Water Partnership, Stockholm. http://www.gwp.org/Global/ToolBox/Publications/Background%20papers/07%20Effective%20Water%20Governance%20%282003%29%20English.pdf. Accessed 27 July 2017

Ruhl JB (2016) Seeing past the zero sum game in environmental policy—harder than it looks. In: Environmental law prof blog, ELC essay no. 2. http://lawprofessors.typepad.com/environmental_law/2016/12/elc-essay-2-the-zen-of-nonzerosumness-in-environmental-policy-easier-to-say-than-to-achieve.html. Accessed 27 July 2017

Stewart R (2003) Administrative law in the twenty-first century. New York University Law Rev 78:437–460

Thompson BH (2016) Defragging: overcoming fragmentation in United States water governance. In: Gray J, Holley C, Rayfuse R (eds) Trans-jurisdictional water law and governance, Earthscan studies in water resource management. Routledge, London

Tyler TR (2006) Psychological perspectives on legitimacy and legitimization. Ann Rev Psychol 57:375–400

United Nations (1992) Agenda 21. United Nations conference on environment & development, Rio de Janerio, 3–14 June 1992. https://sustainabledevelopment.un.org/content/documents/Agenda21.pdf. Accessed 27 July 2017

United Nations World Water Assessment Program (UN WWAP) (2003) Water for people, water for life: UN world water development report 1. UNESCO, Paris. http://www.unesco.org/new/en/natural-sciences/environment/water/wwap/wwdr/wwdr1-2003/. Accessed 27 July 2017

UN System Task Team on the Post-2015 UN Development Agenda: Governance and Development (2012) Url: http://www.un.org/millenniumgoals/pdf/Think%20Pieces/7_governance.pdf

Legal References

Citizens United v. Federal Election Commission, 558 U.S. 310 (2010)

16 U.S.C. §§ 1531–1544 US Endangered Species Act

42 U.S.C. §§ 4321–4370h US National Environmental Policy Act

Chapter 11
Stability and Flexibility in the Emergence of Adaptive Water Governance

Robin Kundis Craig, Ahjond S. Garmestani, Craig R. Allen, Craig Anthony (Tony) Arnold, Hannah Birgé, Daniel DeCaro, and Hannah Gosnell

Abstract One of the goals of adaptive governance is to increase management flexibility in the face of a changing social-ecological system. In contrast, one of the key functions of governance systems is to provide stability, predictability, and security for the people subject to that system. This chapter explores this adaptive governance paradox, focusing on the Klamath and Everglades case studies presented earlier in this volume—although the paradox arises in all of the case study river basins and indeed in most adaptive governance projects. It concludes that while the Everglades system has detrimentally privileged stability at the expense of flexibility and

---

R. K. Craig (✉)
S.J. Quinney College of Law, University of Utah, Salt Lake City, UT, USA
e-mail: robin.craig@law.utah.edu

A. S. Garmestani
School of Natural Resources, University of Nebraska, Lincoln, NE, USA

C. R. Allen
U.S. Geological Survey, Nebraska Cooperative Fish and Wildlife Research Unit, School of Natural Resources, University of Nebraska, Lincoln, NE, USA
e-mail: callen3@unl.edu

C. A. Arnold
Brandeis School of Law and Department of Urban and Public Affairs, University of Louisville, Louisville, KY, USA
e-mail: tony.arnold@louisville.edu

H. Birgé
Nebraska Cooperative Fish and Wildlife Research Unit, School of Natural Resources, University of Nebraska, Lincoln, NE, USA

The Nature Conservancy, Omaha, NE, USA

D. DeCaro
Department of Urban and Public Affairs, Department of Psychological and Brain Sciences, University of Louisville, Louisville, KY, USA
e-mail: daniel.decaro@louisville.edu

H. Gosnell
College of Earth, Ocean and Atmospheric Sciences, Oregon State University, Corvallis, OR, USA
e-mail: gosnellh@geo.oregonstate.edu

B. Cosens, L. Gunderson (eds.), *Practical Panarchy for Adaptive Water Governance*, https://doi.org/10.1007/978-3-319-72472-0_11

adaptability, the Klamath Basin system is showing signs that it may be able to appropriately balance stability and flexibility in its governance institutions to better address changing climatic, legal, and political realities.

Keywords Adaptive governance · Balance · Due process · Equity · Fairness · Legitimacy · Nonequilibrium · Procedure · Resilience · Rule · Standard

11.1 Introduction

One of the goals of adaptive governance is to increase management flexibility in the face of a changing social-ecological system (SES) (Cosens et al. 2017; Chaffin et al. 2014b, 2017). However, one of the key societal functions of governance systems is to promote stability and security for the citizens living with that system (Craig et al. 2017). Thus, adaptive governance embraces a key paradox: governance of changing systems must disavow certain traditional forms of governance stability, such as unchanging rules or management regimes, in order to be effective—i.e., to be able to navigate successfully the reality of SES instability (Craig et al. 2017; Camacho and Glicksman 2016; Ruhl 2011; Doremus 2010; Ruhl and Fischman 2010).

The seven case studies presented in this volume all illustrate this paradox to varying degrees, largely because they share key components in their governance histories. Almost all of the river basins and water SESs discussed have been the subjects of human engineering through dams, reservoirs, levees, and other constructed infrastructure, but this engineering can serve both stability and flexibility goals. In places where flooding or flooding and drought alterations are the primary concerns, engineering stabilizes these natural fluctuations to promote human security, even out water supply, and regularize navigation. In other places, however, human engineering was an attempt to add management flexibility by expanding the human uses that a given river could support beyond immediate riparian uses of water—for example, producing commercial-scale electricity to service both local and distant power needs or carrying water far beyond the riverbanks to distant dry lands to produce farms and pastures. In all cases, however, anthropogenic infrastructure has altered the function of natural systems, almost always undermining overall system resilience (Holling and Meffe 1996) and leading to longer-term local emergent problems such as endangered species (Chap. 6) or water quality degradation (Chap. 5). Moreover, almost all of the basins studied are also now being affected by climate change, the impacts of which both create new problems, such as water shortages or rising sea levels, and exacerbate existing problems such as endangered species.

While all of the basin studies presented in this book could thus be the subjects of a stability-flexibility analysis, this chapter will focus on two of the basins studied to illustrate opposing governance responses in the face of a changing SES. In the Everglades in Florida (Chap. 5), adherence to nineteenth-century management goals and fragmentation of management responsibility have impeded any efforts either to restore ecological functionality or to adapt the system to climate change realities. In contrast, in the Klamath Basin straddling Oregon and California (Chap. 6), the residents' willingness to forge new governance institutions and plans may allow their SES to productively adapt to changing water supplies, climatic realities, and re-prioritization of legal rights and goals. Before analyzing these two basins in greater depth, however, this chapter will begin with an overview of the stability-flexibility paradox for adaptive water governance, drawing from Craig et al. (2017).

11.2 Interdisciplinary Insights into the Stability-Flexibility Paradox in Adaptive Water Governance

Because adaptive governance is a form of governance, it must adequately serve the community or communities in which it operates. As stated by Craig et al. (2017), adaptive governance "must not only effectively manage changing SESs but also operate as a legitimate and effective governance system that meets both the psychological needs of the individuals governed and the socio-economic needs of the communities involved." Adaptive governance is generally pursued in order to cope with changing social-ecological conditions—hence the need for flexibility—but it must also avoid having this drive toward increased governance flexibility be perceived or experienced by those governed as economically, politically, or psychologically destabilizing, disruptive, or unfair (Craig et al. 2017). As such, adaptive governance needs to balance management flexibility with governance tools that provide stability.

Stability in the context of adaptive governance refers to the persistence over time, in the same or similar form, of governance structures and processes (Craig et al. 2017). These structures include branches of government and their implementing bodies and nongovernmental governance institutions that exist at multiple political scales, from local to federal (Craig et al. 2017). Adaptive governance processes include the rule of law, substantive rules, standards, and norms, and procedural requirements and opportunities, including public and interest group participation (Craig et al. 2017). Flexibility of governance, in contrast, refers to the degree of latitude possible within a given governance structure or process without fundamentally breaching or displacing the governance system as a whole (Craig et al. 2017). Examples include shifts from conservative to liberal perspectives and control, discretion in interpretation, implementation and application, or exceptions and variances or amendments within substantive rules, standards, and norms (Craig et al. 2017). Flexibility also applies to procedural requirements, such as through discretion to use more or less formal procedures or abbreviated procedures (Craig et al. 2017).

Stability and flexibility in governance offer different strengths and create different risks. In general, stable governance systems allow SESs to function predictably, allowing for personal and social investment in long-term goals and for the promotion of peace through shared understandings and norms regarding what is allowed

Table 11.1 Interactions among governance actions, potential psychological responses by individuals and potential societal consequences of the action. Adopted from Craig et al. (2017)

Governance action	Potential psychological response (individuals)	Potential societal consequences
Innovation in governance structures or goals	Resistance	Rejections of or refusals to enact/vote for changes
Policies and goals that routinely adjust to changing ecological realities	Loss of trust, feelings of personal and economic insecurity	Loss of societal cooperation, failure to invest in social goods or larger projects
Changes in the rules for accessing and using natural resources	Perceptions of winners and losers and of loss of property rights	Litigation or violence
Changes in community priorities as an adaptation strategy	Perceptions of winners and losers, fear of economic insecurity	Social unrest, violence, population shifts out of the community
Decisions about changes made behind closed doors, even if by elected representatives	Lack of transparency undermines the legitimacy of the decision	Discontent, resistance, social unrest, frequent changes in leadership
Decisions about changes made by under-representative groups or single leaders (e.g., mayors or governors)	Lack of full participation undermines the legitimacy of the decision and creates perceptions of marginalization	Fractionalization of the community, loss of social cooperation, discontent, resistance, and social unrest

and what is not (Craig et al. 2017; Levi-Faur 2012). However, governance systems that are too stable and cannot adapt to new circumstances can create "rigidity traps," where the system no longer reflects or adequately responds to the reality that it is supposedly governing (Carpenter and Brock 2008). In contrast, flexibility in governance systems allows complex societies to adjust governance patterns to reflect new social norms and ecological realities—that is, to adapt to new circumstances (Craig et al. 2017; Flatt 2012; Zellmer 2012; Levi-Faur 2012; Ruhl 2011; Camacho 2009, 2011; Craig 2010; Doremus 2010; Ruhl and Fischman 2010; Holling 2001). However, the primary risk from too much flexibility in a governance regime is arbitrariness, which erodes public values such as equal treatment, due process, and procedural fairness (Craig et al. 2017; Cosens et al. 2017; Cosens 2013; Cosens and Williams 2012; Levi-Faur 2012).

In addition, the rebalancing of stability and flexibility in adaptive governance to favor flexibility has psychological and sociological consequences, as is summarized in Table 11.1. Innovations in governance or frequent changes in applicable rules, especially when pursued bureaucratically without public input and buy-in, can lead to resistance and mistrust and occasionally even violence, as was true in the American civil rights movement (Craig et al. 2017; DeCaro et al. 2017; Castro 2012; Swim et al. 2009). Changing social norms and political and legal priorities can also create popular perceptions of "winners and losers" in a variety of contexts, from property rights to economic prosperity. All of these individual and societal reactions, in turn, can undermine and even destroy the legitimacy of the attempted move to adaptive governance, often making attention to process and public participation critical (Craig et al. 2017; DeCaro et al. 2017; Cosens 2013; Tyler 2006; Frey et al. 2004).

Law and policy can make use of a variety of tools and mandate a variety of procedures to enhance the legitimacy of—and hence the individual and societal acceptability of—a more flexible adaptive governance regime, providing social stability (Craig et al. 2017; Biber and Eagle 2015), as summarized in Table 11.2. For example, a classic division in substantive law is between rules and standards. Rules, for example, driving speed limits, establish clear requirements that apply to everyone and are relatively simple to enforce. However, rules cannot adjust to changing conditions, such as when a road turns icy in winter. Standards, such as "don't be negligent," allow the law to flexibly impose duties across changing conditions and circumstances, but they can be subject to differing interpretations, potentially leading to more costly and less fair enforcement (Camacho 2015; Morse 2010). In turn, procedural requirements and burdens of proof can do much to adjust the potential pace of legal change, aiding adaptive governance by matching the governing entity's *ability* to change its management rules and priorities to the projected pace at which such changes are likely to be necessary (Craig et al. 2017; Cosens 2013).

In general, adaptive governance regimes will need to pay increased attention to procedural legitimacy, transparency, and opportunities for public participation to provide the requisite stability, rather than relying on fixed rules and assignments of rights and responsibilities (Craig et al. 2017). In addition, another paradox of adaptive governance comes from the fact that the interaction of stability and flexibility at different scales can also be important. For example, flexibility at the level of the individual (e.g., with respect to liability under unusual circumstances) can enhance the legitimacy and fairness of the governance system as a whole, promoting its overall long-term stability (Craig et al. 2017).

Nevertheless, there is no adaptive governance formula or panacea that can work in all SESs (Craig et al. 2017; Ostrom 2007). Instead, the exact balance between stability and flexibility and the manner in which those two governance drives are balanced—for example, as a matter of substantive law, procedural and public participation requirements, or both—must reflect the exact governance problem being addressed and the specific social and ecological risks and realities surrounding that problem (Craig et al. 2017). As such, the examination of how different SESs have successfully or unsuccessfully rebalanced stability and flexibility in the face of changing conditions can provide useful analyses of different approaches that managers or communities can take to this problem, perhaps revealing combinations of techniques and governance strategies that other SESs may be able to adapt to their own circumstances.

Table 11.2 Legal and policy tools available to accommodate increased substantive flexibility. Adopted from Craig et al. (2017)

Legal or policy tool	Example	Affect on the stability-flexibility balance
Rules	Speed limits	Favors stability and transparency by creating clear rules applicable to everyone but may sacrifice fairness through lack of flexibility for unusual or changing circumstances
Standards	Negligence, which imposes an objective requirement to act reasonably under the relevant circumstances	Favors flexibility and adaptability to a wide range of circumstances but can undermine transparency and promote arbitrariness
Inherently evolutionary law	Common law (judge-made), tort law, and property law	Allows rules and principles of law to evolve over time to fit changing social circumstances, providing the legal flexibility necessary for long-term social stability
Scheduled opportunities for review and amendment of laws	"Sunset" provisions that require statutes and ordinances to be reviewed at a definite future date	Promotes legitimacy through shared decision-making, promotes psychological stability by ensuring that new policies can be revisited, and promotes popular accountability, all of which enhance the stability of the governance system. Amendment and review provisions also provide necessary flexibility to legislative law, allowing it to evolve to suit changing social needs
Regulatory "floors"	States must comply with the minimum requirements of the federal Clean Air Act and Clean Water Act but are free to regulate pollution more stringently if they choose	Promotes stability and fairness by guaranteeing a level of minimum protection/regulation for everyone, while allowing for local flexibility if citizens and their leaders want to go beyond that floor
Regulatory "ceilings"	States cannot protect endangered and threatened species in international trade beyond what the federal government has already provided	Promotes stability and fairness by limiting "excessive" regulation and creating a uniform set of rules that applies to everyone
Evidence-based decision-making	National Ambient Air Quality Standards must be set at levels "requisite to protect the public health" as evidence by medical and epidemiological studies; decisions to list species for protection under the Endangered Species Act must be based on "the best scientific and commercial data available"	Limits arbitrariness and capriciousness in executive or administrative decision-making and provides a basis for judicial review, enhancing the legitimacy and stability of the governance system through public accountability and transparency and by invoking more neutral/ authoritative bases for decision-making than pure politics; however, such requirements also allow flexibility by allowing decisions to change as the evidence evolves

(continued)

Table 11.2. (continued)

Legal or policy tool	Example	Affect on the stability-flexibility balance
Varying burdens of proof	Drug manufacturers must prove the safety and efficacy of new drugs before the FDA will approve them, but the EPA must prove danger before it can prevent a new chemical product from entering the marketplace	The assignment and stringency of a burden of proof can either promote stability by slowing the pace of change (FDA) or increase flexibility and change by speeding the introduction of new products and techniques into society (EPA).
Public participation requirements	Requirements that government agencies provide notice and allow public comment before promulgating new rules; requirements that town council meetings be open to the community and allow for public comment on proposals	Promotes the legitimacy of the resulting decisions and policies/laws by allowing for shared decision-making, popular accountability, enhanced transparency, and an administrative forum for conflict resolution, enhancing the stability of the governance system
Varying procedural requirements before a decision can be implemented	Congress' exemption of the Federal Reserve's interest rate decisions from the Administrative Procedure Act (APA), the APA's own distinction between formal and informal procedures	More extensive procedural requirements generally increase stability and legitimacy by slowing the pace of new decisions and policies and mandating public participation and transparency. In contrast, abbreviated or nonexistent procedural requirements can increase flexibility, nimbleness, and responsiveness in rapidly changing situations at the potential expense of perceived legitimacy
Regularly scheduled elections of leaders	US presidential elections held every 4 years	Allows for popular accountability of elected leaders and allows discontented members of the community regular opportunities to adapt the governance system, promoting the overall stability of that system by allowing for change when demanded
Supermajority voting requirements	The Endangered Species Committee can allow a federal agency to jeopardize the continued existence of an endangered species only through a 5–2 vote	Adds stability to the governance system by slowing the pace of change and/or requiring greater consensus before more radical actions can occur
Judicial review of government action	Provisions that allow affected persons and other citizens to challenge specific government actions within a limited time frame	Provides a conflict resolution mechanism other than self-help and provides for checks and balances on executive and legislative authority, enhancing popular accountability and transparency in decision-making, both of which promote overall governance stability

11.3 Overdetermined Mandates in the Everglades

Although humans have been modifying the Everglades in southern Florida for millennia, "it was during the past century that humans have increased their control over nature in south Florida" (Gunderson et al. 2014; Blake 2010; Douglas 1947). As such, engineering of the Everglades was primarily an attempt to stabilize this system to allow for human development. The resulting system of levees and reservoirs allows for almost complete human control of Everglades hydrology in order to serve a growing list of societal goals, beginning with flood control and now including development, agriculture, and ecological restoration (Gunderson et al. 2014; Light and Dineen 1994). Moreover, this substantially engineered system has experienced a number of abrupt shifts and changes over the last century, which have occurred as a result of three main pathways: human technology altering biophysical processes, leading to ecological or social regime shifts; large-scale events, such as storms or floods, that have overwhelmed the existing technology and led to significant ecological and social changes; and from an erosion of ecological resilience, again changing ecological and social regimes (Gunderson et al. 2014; Light et al. 1995).

Thus, engineering efforts designed to stabilize nature's variation and flexibility—floods (Hamann 2005; Blake 2010) and drought (Hamann 2005; Tarlock 1987) in particular—led to different kinds of problems in the Everglades, such as water pollution and endangered species (Gunderson et al. 2014; Miccosukee Tribe of Indians of Florida v. United States 2007; Aumen 1995; United States v. S. Fla. Water Mgmt. Dist. 1991). These new problems, in turn, led to numerous changes in the law, at both the federal and state levels, to promote environmental goals for the Everglades, and "[s]ince 1990, a major focus of management in the Everglades has been ecosystem restoration" (Gunderson et al. 2014: 134).

However, these restoration efforts have not been successful, largely because of competing interests in Everglades water management (Gunderson et al. 2014). In the absence of a widely shared and agreed-upon prioritization of society's management goals for the Everglades, different interest groups promote their individual agendas, including water supply, flood management, endangered species restoration, water quality restoration, improvement of Everglades National Park, and protection of the Florida Keys coral reef ecosystem in the Gulf of Mexico below the Everglades. The result has been virtually perpetual litigation that has greatly slowed the pace of restoration efforts (Gunderson et al. 2014). In addition, the Miccosukee Tribe, whose reservation sits in the middle of the Everglades, has initiated much of the water quality and Endangered Species Act litigation to protect its homeland in the midst of restoration experimentation and adaptive management, seeking stability for its homeland, which restoration and adaptive management experiments repeatedly flood in pursuit of better ecological management (Miccosukee Tribe of Indians of Florida v. United States 1997, 2003, 2009, 2010).

In terms of the stability-flexibility paradox for changing systems, governance changes within the Everglades SES are probably best viewed as a mandated attempt to increase legal and governance flexibility to promote a new and far-ranging goal—

ecological restoration—that failed to account for the need for re-stabilization at the personal/psychological and societal levels. For example, Everglades restoration goals create very clear losers among the Everglades stakeholders, especially sugar plantations and other agricultural operations. In addition, restoration efforts come with significant trade-offs. One of these is simple cost and the diversion of federal and state monies from, at least potentially, other public expenditures, such as water supply improvements or climate change adaptation measures. Indeed, the *Miami Herald* reported in March 2016 that the costs of Everglades restoration would probably be double the 2000 expectations, increasing to $16.4 billion (Staletovich 2016). In addition, some of the concurrently pursued management goals for the Everglades are simply at odds. For example, flood control measures divert 1.7 billion gallons of Everglades water per day to the ocean (FDEP 2009), even as southern Florida struggles to secure adequate public water supply (SFWMD 2016). Increased flow of water through the Everglades could improve both water supply and ecological issues, but it increases flood risk, and there is also significant mistrust of experimentation to help endangered species.

Everglades restoration goals, in other words, presented *exactly* the kind of shift in governance regimes and infusion of governance flexibility that should warrant increased attention to legitimacy, particularly in terms of public participation. Nevertheless, anthropologist Laura Ogden has concluded that "in practice, with the exception of water management planning, there has been little consideration of the human dimensions of the Everglades ecosystem in the restoration planning process" (Ogden 2006). Instead, the plan for restoring the Everglades has a water management and engineering focus that actually impedes consideration of the humans in the system, because it conceptualizes human activities as drivers of change in the natural ecosystem rather than as participants in a complex SES (Ogden 2006). Indeed, there has been no government research to support an integrated approach to ecosystem management under the restoration plan, even though the National Academy of Science warned about the potential consequences of ignoring social and economic issues and even though Florida residents are known to view Everglades restoration with a variety of attitudes (Ogden 2006). Everglades restoration efforts have thus been, in many senses, mandates of flexibility imposed without the broad public participation and consensus-building procedures that might have created both a shared sense of social-ecological priorities and increased governance legitimacy.

However, Everglades governance offers yet another window on the stability-flexibility paradox with respect to climate change. Despite all of the controversy that arises *within* the context of Everglades restoration, the commitment to Everglades restoration has become, perversely, a hyper-stable governance framework for conceptualizing Everglades management: Everglades restoration is a largely unquestioned—and unchallengeable, legally and politically—goal. As a result, the legal and social commitment to Everglades restoration has undermined this SES's ability to deal more flexibly with new threats to the entire SES—namely, sea-level rise, saltwater intrusion, and other climate change-related impacts on the system. Because of climate change and its attendant sea-level rise, the ocean is starting to invade the Everglades, which could drastically change the area ecologically, socially, and economically (Joyce 2016; Gunderson et al. 2014). Indeed, the Natural Resources Defense Council has listed the Everglades as the third of "9 Places to See Before They're Gone" (NRDC 2016). Climate change thus begs managers to question whether Everglades "restoration" should remain a governance (and financial) priority in the face of a potential transformation of the Everglades ecosystem to something else entirely—but the current governance system lacks the flexibility to even ask that question, let alone comprehensively evaluate possible answers.

11.4 A Move Toward Productive Flexibility in the Klamath River Basin

Engineering in the Klamath River Basin is largely a human effort to increase the flexibility of the system by adding irrigation systems and hydropower dams (Chaffin et al. 2014a). However, like most water basins in the American West, dams in the Klamath Basin also stabilize water supply by providing storage reservoirs that can be tapped in drought years and trapping flood waters in flood years (Chaffin et al. 2014a).

As is generally true of large hydrological engineering projects, engineering in the Klamath River Basin came with ecological impacts, which manifested most obviously in the form of degraded water quality and species of fish being listed for protection under the Endangered Species Act (Chaffin et al. 2014a; USDOI and CDFG 2011). These impacts were exacerbated in the Klamath Basin by climate fluctuations that reduced water supplies and caused repeated drought (Chaffin et al. 2014a; Thorsteinson et al. 2011) but also by the emergence of the Basin's four tribes as political and ecological forces, particularly with respect to legal recognition of the Klamath Tribes' water rights in 1983 and their quantification in 2014 (Chaffin et al. 2014a; OWRD 2013; United States v. Adair 1983).

The first major shock to the hydro-engineered system came in 2001, a drought year, when a federal judge shut down deliveries of irrigation water to farmers and ranchers to provide instream flows to threatened coho salmon (Chaffin et al. 2014a; Doremus and Tarlock 2008). As in the Everglades, new laws imposed on the system injected new goals into the management regime, and the Endangered Species Act lawsuit destabilized social and political components of this SES, namely, farmers' and ranchers' settled expectations of how the governance system "should" work despite changing climatic, ecological, political, and legal realities (Chaffin et al. 2014a; Doremus and Tarlock 2008). As in the Everglades, moreover, more litigation ensued, further destabilizing both the system as a whole and progress toward any of a number of governance goals—water quality improvement, endangered and threatened species recovery, profitable crop yields, and protection of tribal homelands (Chaffin et al. 2014a; Doremus and Tarlock 2008).

Here, however, the story of the Klamath River Basin diverges from that of the Everglades: in the Klamath River Basin, local leaders and community residents got tired of the chaos. Recognizing that the system changes were real and likely permanent (a fact that the 2013 tribal water rights decree only underscored) and after long discussions and exercises in both communication and trust-building, community members came together to design and propose several plans to increase management flexibility in the Klamath River Basin (Chaffin et al. 2014a; Chaffin 2014; Gosnell and Kelly 2010). Infrastructure was again key to the pursuit of this flexibility, because the economics of relicensing four hydroelectric dams in the system created the possibility of removing those dams (Chaffin et al. 2014a).

Dam removal is expensive, and the federal government's cooperation and financing remain critical to institutionalizing and operationalizing the new, more flexible governance system in the Klamath River Basin proposed in two sets of agreements among the stakeholders. While the original negotiated governance agreements for the Basin expired at the end of 2015, the parties quickly renegotiated an Amended Klamath Hydroelectric Settlement Agreement (KHSA) in early 2016. This new agreement created the Klamath River Renewal Corporation (KRRC 2016a), which on September 23, 2016, filed with the Federal Energy Regulatory Commission both a License Transfer Application (joined by PacifiCorp) to transfer ownership of four dams from PacifiCorp and a License Surrender Application, the initial legal steps in removing the dams (KRRC 2016b). Decommissioning of the four dams is expected to cost about $290 million, and a plan is in place to raise the money (Peterson 2016).

Thus, people in the Klamath River Basin took a different path toward system restoration than in the Everglades, invoking a process of negotiation and mutual agreement that was as much about preserving and improving—increasing the resilience of—the social components of the SES equation as it was about improving and restoring the ecological components while recognizing simultaneously that for some members of the community, notably the tribes, these two components are inextricably interlinked (Gosnell and Kelly 2010). Indeed, Mary Milner has characterized the emergence of the new and more flexible governance proposal in the Klamath River Basin as a social movement and exercise of participatory democracy, concluding that "[t]he Klamath Basin's struggle to incorporate environmental principles into water management decisions qualifies as a social movement because it originated in episodes of collective action outside legal and political channels, and because it challenged the existing idea that irrigation and hydroelectric generation trumped all other interests" (Milner 2015).

Of course, nothing is yet certain in the Klamath River Basin, and there is no guarantee that the process there will achieve a workable balance between management flexibility and personal and societal stability. Good will among the participants may be deteriorating; the dams' removal and hence implementation of the new governance plan remain far from certain; and certain individuals and groups remain opposed to the dams' removal. Nevertheless, the participatory procedures that the Klamath community used in its attempt to inject flexibility into its governance future imbues the resulting agreements with far more coherence of purpose and legitimacy than Everglades restoration efforts have enjoyed. Moreover, because climatic shifts have been part of these negotiations from the beginning, restoration in the Klamath Basin, should it occur, will have already considered and incorporated the flexibility necessary to adapt to continuing climate change—especially if the surrounding community structure of increased trust and communication can itself re-stabilize into the new management regime.

11.5 Conclusion

When a community or governance system is trying to evolve, or being forced to evolve, into a more flexible structure capable of adapting to changing social and/or ecological realities, process matters—but so does clear articulation of shared goals and priorities. There is no doubt that the various forms of Everglades restoration governance systems—the federal Everglades Restoration Act and Comprehensive Everglades Restoration Plan and the State of Florida's Everglades Forever Act—are trying to more flexibly manage the Everglades' highly engineered system to achieve new ecological goals while still pursuing the original goals that led to that engineering in the first place. There is also no doubt that the restoration-oriented governance system incorporates various tools to enhance its flexibility, such as the prominent incorporation of adaptive management into the Comprehensive Everglades Restoration Plan (CERP 2016). However, as the endless litigation about this SES demonstrates, these attempted injections of governance flexibility do not operate with a balancing stability gleaned from shared or prioritized goals for the system; instead, differing interests remain free to promote (and litigate) their individual priorities, competing for resources and governance attention and quite effectively retarding overall progress of any sort. At the same time, however, the Everglades restoration governance framework cannot legally or politically entertain the possibility of a future Florida *without* the (freshwater) Everglades, even though climate change and sea-level rise threaten this SES's core identity. The restoration governance system, in other words, is in some ways a rigidity trap, because it provides no procedural mechanisms for questioning its own purpose and existence or for amending its goals to account for climate change.

In contrast, in the Klamath Basin, greater attention to participatory governance in recent years has resulted in a more stable base for experimenting with flexibility in future water resource management. Moreover, the proposed new governance regime provides flexibility to address changing future conditions. The Amended Klamath Hydroelectric Settlement Agreement, for example, contains provisions for its own amendment and sets out dispute resolution procedures as well as "meet and confer" requirements (KHSA 2016). Similarly, the new Klamath Power and Facilities Agreement allows for its own amendment, proving that "[t]he Parties may amend this Agreement but only in writing and only by Consensus. Parties will in good faith entertain proposals for amendment, and Parties shall in good faith refrain from proposing amendments unless there is good cause" (KPFA 2016). However, it also anticipates future changes in pursuit of system restoration, providing general guidelines for how those improvements should be pursued (KPFA 2016). Thus, these agreements provide what appears to be a more effective balancing of stability and flexibility for the system's future governance, promoting both restoration and resilience to future changes.

Acknowledgments This work was developed in part under the Adaptive Water Governance Project, funded by the US National Socio-Environmental Synthesis Center (SESYNC) under funding from the US National Science Foundation, NSF DBI-1052875. The Nebraska Cooperative Fish and Wildlife Research Unit is jointly supported by a cooperative agreement among the US Geological Survey, the Nebraska Game and Parks Commission, the University of Nebraska, the US Fish and Wildlife Service, and the Wildlife Management Institute.

References

Aumen NG (1995) History of human impacts, lake management, and limnological research on Lake Okeechobee, Florida (USA). Adv Limnol 45:1–16
Biber E, Eagle J (2015) When does legal flexibility work in environmental law? Ecol Law Quart 42(4):787–840. https://doi.org/10.15779/z38rz9z
Blake NM (2010) Land into water—water into land: a history of water management in Florida, 2nd edn. Orange Grove Texts Plus, Gainesville
Camacho AE (2009) Adapting governance to climate change: managing uncertainty through a learning infrastructure. Emory Law J 59(1):1–77
Camacho AE (2011) Transforming the means and ends of natural resources management. North Carolina Law Rev 89:1405–1454. https://doi.org/10.2139/ssrn.1852724
Camacho AE (2015) Going the way of the dodo: de-extinction, dualisms, and reframing conservation. Washington Univ Law Rev 92(4):849–906
Camacho AE, Glicksman RL (2016) Legal adaptive capacity: how program goals and processes shape federal land adaptation to climate change. Univ Colorado Law Rev 87(3):711–826
Carpenter SR, Brock WA (2008) Adaptive capacity and traps. Ecol Soc 13(2):40. https://doi.org/10.5751/es-02716-130240
Castro P (2012) Legal innovation for social change: exploring change and resistance to different types of sustainability laws. Polit Psychol 33(1):105–121. https://doi.org/10.1111/j.1467-9221.2011.00863.x
CERP Adaptive Management Program (2016) Comprehensive Everglades restoration plan. U.S. Department of the Interior. http://141.232.10.32/pm/program_docs/adaptive_mgmt.aspx. Accessed 23 Nov 2016
Chaffin BC (2014) Reallocating resources, rebuilding community: the Klamath Basin agreements and the emergence of adaptive governance. Dissertation, Oregon State University. Available at https://ir.library.oregonstate.edn/xmlui/handle/1957/50604. Accessed 29 July 2017
Chaffin B, Craig RK, Gosnell H (2014a) Resilience, adaptation, and transformation in the Klamath river Basin social-ecological system. Idaho Law Rev 51:157–193
Chaffin BC, Gosnell H, Cosens BA (2014b) A decade of adaptive governance scholarship: synthesis and future directions. Ecol Soc 19(3):56. https://doi.org/10.5751/es-06824-190356
Chaffin BC, Gunderson L, Cosens B (eds) (2018) Practicing panarchy: assessing legal flexibility, ecological resilience, and adaptive governance in U.S. regional water systems experiencing climate change. Ecol Soc Spec Issue. https://www.ecologyandsociety.org/issues/view.php?sf=122 Last Visited 22 February, 2018
Cosens BA (2013) Legitimacy, adaptation, and resilience in ecosystem management. Ecol Soc 18(1):3. https://doi.org/10.5751/ES-05093-180103
Cosens BA, Williams MK (2012) Resilience and water governance: adaptive governance in the Columbia river basin. Ecol Soc 17(4):3. https://doi.org/10.5751/es-04986-170403

Cosens BA, Craig RK, Hirsch S, Arnold CA, Benson MH, DeCaro DA, Garmestani AS, Gosnell H, Ruhl JB Schlager E (2017) The role of law in adaptive governance. Ecol Soc 22(1):30. https://doi.org/10.5751/ES-08731-220130
Craig RK (2010) "Stationarity is dead"—long live transformation: five principles for climate change adaptation law. Harvard Environ Law Rev 34(1):9–73
Craig RK, Garmestani AS, Allen CR, Arnold CA, Birgé H, DeCaro DA, Fremier AK, Gosnell H, Schlager E (2017) Balancing stability and flexibility in adaptive governance: an analysis of tools available in U.S. environmental law. Ecol Soc 22(2):3. https://doi.org/10.5751/es-08983-220203
DeCaro DA, Arnold CA, Boamah EF, Garmestani AS (2017) Understanding and applying principles of social cognition and decision making in adaptive environmental governance. Ecol Soc 22(1):33. https://doi.org/10.5751/es-09154-220133
Doremus H (2010) Adapting to climate change with law that bends without breaking. San Diego J Clim Energy Law 2:45–85
Doremus H, Tarlock AD (2008) Water war in the Klamath Basin: macho law, combat biology, and dirty politics. Island, Washington, DC
Douglas MS (1947) The Everglades: river of grass. Pineapple Press, Sarasota
Florida Department of Environmental Protection [FDEP] (2009) Brief history of the Everglades. http://www.dep.state.fl.us/evergladesforever/about/. Accessed 11 Feb 2009
Flatt VB (2012) Adapting laws for a changing world: a systemic approach to climate change adaptation. Florida Law Rev 64(1):269–293
Frey BS, Benz M, Stutzer A (2004) Introducing procedural utility: not only what, but also how matters. J Inst Theor Econ 160(3):377–401. https://doi.org/10.1628/0932456041960560
Gosnell H, Kelly EC (2010) Peace on the river? Social-ecological restoration and large dam removal in the Klamath Basin, USA. Water Altern 3(2):361–383
Gunderson LH, Garmestani A, Rizzardi KW, Ruhl JB, Light A (2014) Escaping a rigidity trap: governance and adaptive capacity to climate change in the Everglades social ecological system. Idaho Law Rev 51:127–156
Hamann R (2005) Florida's water management framework. In: Scholz JT, Stiftel B (eds) Adaptive governance and water conflict: new institutions for collaborative planning. Resources for the Future, Washington, DC
Holling CS (2001) Understanding the complexity of economic, ecological, and social systems. Ecosystems 4(5):390–405. https://doi.org/10.1007/s10021-001-0101-5
Holling CS, Meffe GK (1996) Command and control and the pathology of natural resource management. Conserv Biol 10(2):328–337. https://doi.org/10.1046/j.1523-1739.1996.10020328.x
Joyce C (2016) Rising seas push too much salt into the Florida Everglades. National Public Radio. http://www.npr.org/2016/05/25/477014085/rising-seas-push-too-much-salt-into-the-florida-everglades. Accessed 29 July 2017
Klamath hydroelectric settlement agreement [KHSA] (2016) https://www.doi.gov/sites/doi.gov/files/uploads/FINAL%20KHSA%20PDF.pdf. Accessed 29 July 2017
Klamath power and facilities agreement [KPFA] (2016) Available at. http://www.oregon.gov/owrd/ADJ/docs/Klamath%20Power%20and%20Facilites%20Agreement.pdf. Accessed 29 July 2017
Klamath River Renewal Corporation [KRRC] (2016a) Klamath River Renewal Corporation. http://www.klamathrenewal.org. Accessed 29 July 2017
Klamath River Renewal Corporation [KRRC] (2016b) Frequently asked questions. http://www.klamathrenewal.org/about-us/faq/. Accessed 29 July 2017
Levi-Faur D (ed) (2012) The Oxford handbook of governance. Oxford University Press, Oxford. https://doi.org/10.1093/oxfordhb/9780199560530.001.0001
Light SS, Dineen JW (1994) Water control in the Everglades: a historical perspective. In: Davis SM, Ogden JC (eds) Everglades: the ecosystem and its restoration. CRC, Delray Beach, pp 47–84
Light SS, Gunderson LH, Holling CS (1995) The Everglades: evolution of management in a turbulent ecosystem. In: Gunderson LH, Holling CS, Light S (eds) Barriers and bridges to the renewal of ecosystems and institutions. Columbia University Press, New York, pp 103–168

Milner M (2015) Water law meets participatory democracy: a Klamath Basin example. J Environ Law Litigation 30:87–141
Morse AH (2010) Rules, standards, and fractured courts. Oklahoma City Univ Law Rev 35:559–605
Natural Resources Defense Council [NRDC] (2016) 9 places to see before they're gone. https://www.nrdc.org/stories/9-places-see-theyre-gone. Accessed 29 July 2017
Ogden L (2006) Public participation in environmental decision-making: a case study in the Florida Everglades. Cahiers d' Economie et Sociologie Rurales 3(80):53–74
Oregon Water Resources Department [OWRD] (2013) The Oregon water resources department completes Klamath river basin adjudication. http://www.oregon.gov/owrd/ADJ/docs/2013_03_07_Klamath_River_Basin_Adjudication_Media_Release_Final.pdf. Accessed 29 July 2017
Ostrom E (2007) A diagnostic approach for going beyond panaceas. Proc Natl Acad Sci U S A 104:15181–15187. https://doi.org/10.1073/pnas.0702288104
Peterson M (2016) Will the Klamath river be renewed? owner applies to remove 4 of 5 dams. KQED news. https://ww2.kqed.org/news/2016/09/27/will-the-klamath-river-be-renewed-owner-applies-to-remove-4-of-5-dams/. Accessed 29 July 2017
Ruhl JB (2011) General design principles for resilience and adaptive capacity in legal systems—with applications to climate change adaptation. North Carolina Law Rev 89(5):1373–1403
Ruhl JB, Fischman RL (2010) Adaptive management in the courts. Minnesota Law Rev 95(2):424–484
South Florida Water Management District [SFWMD] (2016) Water supply planning. https://www.sfwmd.gov/our-work/water-supply. Accessed 23 Nov 2016
Staletovich J (2016) Cost of Everglades restoration could double to $16 billion. Miami Herald. http://www.miamiherald.com/news/local/environment/article69122437.html. Accessed 29 July 2017
Swim J, Clayton S, Doherty T, Gifford R, Howard G, Reser J, Stern P, Weber E (2009) Psychology and global climate change: addressing a multifaceted phenomenon and set of challenges: report of the American Psychological Association Task Force on the interface between psychology and global climate change, American Psychological Association, Washington, DC. http://www.apa.org/science/about/publications/climatechange.aspx. Accessed 29 July 2017
Tarlock AD (1987) Protection of water flows for national parks. Land Water Law Rev 22:29–48
Thorsteinson L, VanderKooi S, Duffy W (eds) (2011) Proceedings of the Klamath Basin science conference. U.S. Geological Survey, Oregon, 2010
Tyler TR (2006) Psychological perspectives on legitimacy and legitimation. Annu Rev Psychol 57:375–400. https://doi.org/10.1146/annurev.psych.57.102904.190038
U.S. Department of the Interior [USDOI], California Department of Fish & Game [CDFG] (2011) Klamath facilities removal draft environmental impact statement/environmental impact report 2011. https://klamathrestoration.gov/Draft-EIS-EIR/download-draft-eis-eir. Accessed 29 July 2017
Zellmer S (2012) Wilderness, water, and climate change. Environ Law 42:313–374

Legal References

Miccosukee Tribe of Indians of Fla. v United States, 980 F. Supp. 448 (S.D. Fla. 1997)
Miccosukee Tribe of Indians of Florida v. United States, 259 F. Supp. 2d 1237(S.D. Fla. 2003)
Miccosukee Tribe of Indians of Florida v. United States, 528 F. Supp. 2d 1317,1319 (S.D. Fla. 2007)
Miccosukee Tribe of Indians of Florida v. United States, 656 F. Supp. 2d 1375 (S.D. Fla. 2009)
Miccosukee Tribe of Indians of Florida v. United States, 680 F. Supp. 2d 1308 (S.D. Fla. 2010)
United States v. Adair, 723 F.2d 1394 (9th Cir. 1983)
United States v. S. Fla. Water Mgmt. Dist., 922 F.2d 704 (11th Cir. 1991)

Chapter 12
Finding Flexibility in Section 7 of the Endangered Species Act Through Adaptive Governance

Hannah Gosnell, Brian C. Chaffin, J. B. Ruhl, Craig A. (Tony) Arnold, Robin K. Craig, Melinda H. Benson, and Alan Devenish

Abstract The US Endangered Species Act (ESA) prohibits federal agency actions likely to jeopardize listed species or adversely modify critical habitat. Scholarship on the application of the ESA characterizes the process as unwaveringly rigid, a legal "hammer." This chapter draws on lessons derived from applying the ESA in the Klamath Basin along the Oregon-California border, where an integrated implementation strategy lessened rigidities and barriers to change. Collaboration among leaders in the US Fish and Wildlife Service, the National Marine Fisheries Service, and the US Bureau of Reclamation supported efforts to replace an ecologically and socially fragmented approach to ESA implementation that was fraught with conflict with a more adaptive, flexible, integrated approach to water sharing among

H. Gosnell (✉) · A. Devenish
College of Earth, Ocean, and Atmospheric Sciences, Oregon State University, Corvallis, OR, USA
e-mail: gosnellh@geo.oregonstate.edu

B. C. Chaffin
W. A. Frank College of Forestry and Conservation, University of Montana, Missoula, MT, USA
e-mail: brian.chaffin@mso.umt.edu

J. B. Ruhl
Vanderbilt University Law School, Nashville, TN, USA
e-mail: jb.ruhl@vanderbilt.edu

C. A. Arnold
Louis D. Brandeis School of Law, University of Louisville, Louisville, KY, USA
e-mail: tony.arnold@louisville.edu

R. K. Craig
S.J. Quinney College of Law, University of Utah, Salt Lake City, UT, USA
e-mail: robin.craig@law.utah.edu

M. H. Benson
Haub School of Environment and Natural Resources, University of Wyoming, Laramie, WYO, USA
e-mail: mhbenson@uwo.edu

B. Cosens, L. Gunderson (eds.), *Practical Panarchy for Adaptive Water Governance*, https://doi.org/10.1007/978-3-319-72472-0_12

competing interests. Keys to success included existing collaborative capacity related to improved tribal-irrigator relations and a shift in local agency culture facilitated by empathic leadership which led to a greater sense of shared responsibility for ESA compliance. This effort exemplifies governmental adaptive capacity for flexibility and evolution within constraints of formal law. A truly bioregional approach to endangered species recovery, however, will necessitate greater integration between federal and nonfederal activities.

Keywords Adaptive capacity · Biodiversity governance · Biological opinion · Collaboration · Empathy · Endangered Species Act · Hydrologic modeling · Integrated natural resource management · Klamath Basin · Participatory capacity · Resilience · Section 7 consultation · Trust

12.1 Introduction

Natural resource management is a useful proving ground for testing theories of social-ecological system (SES) resilience. It is hard to imagine a better example of "release" and "reorganization" (Chap. 14) than the almost predictable turbulence regularly caused by implementation of Section 7 of the US *Endangered Species Act* (ESA) (16 U.S.C. § 1536) which prohibits federal actions that may jeopardize species listed under the Act. No incident better illustrates that potential than the events surrounding implementation of Section 7 in the Klamath Basin in 2001. The US Fish and Wildlife Service (FWS) and the National Marine Fisheries Service (NMFS) issued biological opinions (BiOps) that resulted in the US Bureau of Reclamation (BOR) shutting off water to roughly 1200 family farms during a drought, leaving hundreds of thousands of acres of farmland without irrigation and an estimated $35 million in lost farm income (Jaeger 2004; Doremus and Tarlock 2005). The story of how community members and resource managers navigated subsequent aftershocks and ultimately reorganized over the following decade has been the subject of numerous journal articles, books, and documentaries (Doremus and Tarlock 2005; Gosnell and Kelly 2010; Chaffin et al. 2014a). While there is debate about the degree to which lasting transformation of the Klamath Basin SES took place in the wake of the 2001 crisis (Chaffin et al. 2016), many agree that there was a significant increase in collaborative capacity during negotiations surrounding the proposed Klamath Basin Restoration Agreement (U.S. Department of the Interior 2010; Gosnell and Kelly 2010). The improved collaboration aligned in many ways with the principles of adaptive governance – that is, governance institutions that emerge to more flexibly manage SESs for resilience in the face of changing conditions (Folke et al. 2005; Chaffin et al. 2014b). Contributing to that dynamic was the development of an innovative approach to Section 7 implementation that resulted in the nation's first joint BiOp between FWS and NMFS. While it only addressed BOR's actions in the Basin (admittedly only part of the problem), the 2013 joint BiOp stands as a model of integrated natural resource management and provides evidence that, under the right conditions, collaborative strategies aimed at enhancing resilience and adaptive capacity can overcome many of the constraints associated with the Section 7 consultation process. This chapter seeks to characterize the governance conditions that influenced and led to this novel outcome in the Klamath Basin. Focusing on the important role that federal agencies played in better aligning Section 7 implementation in the Klamath with principles of adaptive governance and resilience-based ecosystem stewardship (Chapin et al. 2009), it identifies general principles and strategies that potentially could be used in other contested bioregional management contexts and considers limitations associated with Section 7's focus on federal actions, which challenge bioregional conservation efforts.

The ESA is comprehensive legislation for the preservation of endangered species and has long been one of the most powerful environmental laws in the United States (Sax and Keiter 2006). It plays an important role in the management of river basins throughout the United States (Benson 2012; Craig 2014), where the needs of ESA-listed (endangered or threatened) fish species frequently clash with agriculture, hydropower, and flood control. Since the ESA mandates that federal agencies consider science, not socioeconomic factors, and reverse the trend toward extinction, implementation of the ESA often triggers social and economic crisis by uncompromisingly limiting how resources may be used (TVA vs. Hill 1978). The Act is criticized by many because it forces underlying tensions between development and the environment into courtrooms, news media, and the greater public discourse (Ruhl 2012). Although the ESA is decidedly species-centric, Congress made clear that the purpose of the ESA is "to provide a means whereby the ecosystems upon which endangered species and threatened species depend may be conserved" (16 U.S.C. § 1531(b)). It recognizes the importance of ecological systems by including "the present or threatened destruction, modification, or curtailment of its habitat or range" in the criteria for listing, as well as the designation of critical habitat once listing occurs (16 U.S.C. § 1533(a)(1)). Critical habitat is defined as areas "on which are found those physical or biological features...essential to the conservation of the species" (16 U.S.C. § 1532(5)). The ESA thus requires consideration of broader ecological conditions.

Section 7 of the ESA requires all federal agencies to consult with FWS and/or NMFS when a proposed or ongoing action may jeopardize listed species. Through the lens of SES resilience (Folke 2006) and adaptive governance (Folke et al. 2005), the Section 7 consultation process can be thought of as a legal intervention aimed at identifying species-centric thresholds (i.e., jeopardy) which, when crossed, signal ecological crisis associated with a high risk of species extinction. The ESA is one of the few major US environmental laws explicitly committed to identifying thresholds associated with biodiversity and mandating efforts to stop short of crossing those thresholds. Serving as a canary in the coal mine, an ESA listing and subsequent implementation can be an important lever of change (Saundry 2009).

In spite of its good intentions, the ESA is generally perceived as rigid, myopic, a barrier to adaptive governance, and incompatible with resilience thinking (Benson 2012; Gunderson 2013; Gunderson et al. 2014). Scholars have also noted, however, that crises caused by the implementation of laws like the ESA can eventually lead to positive reorganization through systemic collapse or threat of collapse of SESs followed by transformative change (Folke et al. 2009; Olsson et al. 2006; Moore et al. 2014; Chaffin et al. 2016). The ESA has been observed to trigger innovation by presenting only narrow solutions that are socially and/or economically unacceptable (Cosens et al. 2017). It can act as a catalyst and alter watershed and basin social dynamics by incentivizing collaboration among resource users and land managers to preempt further regulation by solving underlying problems (Prokopy et al. 2014). Thus, ESA implementation may lead to collaboration, more equitable resource reallocation, and flexible resource management by first destabilizing rigid power relationships (Arnold 2004; Karkkainen 2008). While this function as a trigger or catalyst is important, it is what happens next that requires attention. While there are some success stories related to ESA implementation, there is a growing need to identify tools for strategically integrating the goals of such statutes into broader SES governance, addressing legitimate concerns about community socioeconomic stability related to operation of federal water projects, and minimizing social-ecological conflict through deliberate SES transformation (Moore et al. 2014). This will involve finding flexibility in the law to allow for socioeconomic considerations while ensuring ecological goals are not compromised; in other words, flexibility and adaptability are needed against a backdrop of stability for both water users and listed species (Craig et al. 2017).

This chapter uses the history of ESA implementation in the Klamath Basin as a case study to illustrate the potential for collaborative strategies to overcome legal roadblocks to SES management dynamics. It draws from ongoing qualitative social science research, including document analysis and fieldwork in the Klamath River Basin since 2008, to contrast: (1) the Section 7 consultations on listed Lost River and shortnose sucker fishes and coho salmon species undertaken by the BOR, NMFS, and FWS in the early 2000s that resulted in rotating crises (Gosnell and Kelly 2010) for irrigators, tribes, and fishermen with (2) the more integrated approach, including interagency cooperation, hydrologic modeling, and engagement of stakeholders, adopted in 2010 that culminated in the 2013 joint BiOp. Key factors enabling this shift in approach that may be transferrable beyond the context of the ESA are identified, including new cross-scale dynamics featuring trust and empathy and the emergence of a forum for learning and coproduction of knowledge around water sharing led by a unique coalition of stakeholders that included prominent leaders from irrigation groups and basin tribes. This study furthers dialogues on resilience and adaptive governance by discussing the ways in which these enabling factors contributed to an expanded sense of place, an extended sphere of concern, and a shared ownership of the problem among both stakeholders and federal agency employees, all of which facilitated a more integrated approach to managing SES dynamics.

12.2 Section 7 Consultation and the Challenge of Integrated Adaptive Governance

Section 7 of the ESA consists of both prohibitory and affirmative obligations, but it is best known for Section 7(a)(2), which requires that each federal agency carrying out a major action consults with FWS for terrestrial and freshwater species and

NMFS for marine and anadromous species. The consultation is to "insure that any action authorized, funded, or carried out by such agency … is not likely to jeopardize the continued existence of any endangered species or threatened species or result in the destruction or adverse modification of [critical] habitat" (16 U.S.C. § 1536(2)). Under this arrangement, as implemented through joint FWS/NMFS regulations, the onus is on the action agency to (1) determine whether listed species are present in the area of its action and if so (2) develop a biological assessment which describes the action and its anticipated effects on the listed species; and if the action and consulting agency conclude, the action is likely to adversely affect a listed species and (3) request formal consultation with FWS or NMFS which develops a draft BiOp stating whether the action is likely to cause jeopardy and/or adverse modification of critical habitat. The draft BiOp may include one or more alternatives modifying the action to avoid jeopardy or modification of critical habitat. The development of the alternative is subject to negotiation between the action agency and FWS or NMFS, which ultimately issues a final BiOp determining whether the action can go forward without violating Section 7.

BiOps are just opinions, nearly always based on incomplete information about the species and the larger biophysical system, so there is some flexibility and discretion in how FWS and NMFS (collectively "the Services") seek compliance with Section 7. That flexibility is tempered, however, by the dynamics between action agencies and the Services as well as institutionalized caution. While generally the lines of authority for FWS and NMFS are separate, water basins with listing of both freshwater and anadromous species (i.e., many west coast rivers) implicate both agencies. Yet Section 7 implementation tends to occur in a fragmented way that results in a reactive, inefficient approach to biodiversity governance.

In addition, Section 7 is designed to use science-based decision-making as the means for assuring legitimacy and is relatively silent about the need for and role of stakeholder input during the consultation process. While this approach to achieving legitimacy, adopted generally in US public land laws in the early part of the twentieth century, works in a stationary system with narrow management goals, it is inadequate in management of a nonstationary, complex system (Cosens 2010). As a result, it leads to an adversarial dynamic, prone to litigation, between federal agencies and stakeholders, which results in risk avoidance and lack of innovation in the Services' preparation of BiOps. As Benson (2012) observes, "ESA management efforts are often geared towards avoiding lawsuits rather than building resilience."

The problems of a lack of interagency integration and absence of robust stakeholder collaboration are central to reform in ESA implementation. Benson (2012) argues for "a more integrated approach to governance that includes a willingness to reassess demands placed on ecological systems by our social systems." It may be that many of the problematic aspects of the ESA have more to do with lack of integration among federal agencies implementing the law and poor relations with stakeholders than with the law itself and that the degree to which the ESA promotes or inhibits adaptive governance is largely contingent on interagency dynamics as well as stakeholder-agency dynamics.

Inattention to such principles of integrated natural resource management is perhaps the most common criticism of the ESA (Gunderson 2013). An integrated, problem-oriented approach increases capacity for learning and adaptation in the conservation planning process and is an important aspect of adaptive governance (Benson 2012; Carpenter and Brock 2008; Grantham et al. 2010; Pahl-Wostl 2009; Pahl-Wostl et al. 2013; Cosens et al. 2017). Law and resilience scholars describe integration in the law as "mutual cognizance and legal authority for coordination across all governance institutions that influence environmental management, and regulation of physically connected resources" (Cosens and Stow 2014). One way law can facilitate adaptive governance is through cooperation of agencies in taking an integrated approach to what have formerly been fragmented approaches to solving a complex problem (Arnold 2015; Arnold et al. 2014). Resilience scholars also assert that participatory capacity, defined as the right and resources of interested persons and groups to have a role in decision-making (Cosens et al. 2017; Chap. 10), is a prerequisite to collaboration, particularly if the voice of formerly marginalized groups is to be heard. Chap. 10 discusses participatory capacity in detail and its role in the legitimacy of a governmental action. Importantly, nothing in Section 7 precludes a more participatory and integrated approach.

The Klamath case study illustrates the ways in which tendencies that foster segregation in Section 7 consultation manifest but also how these dynamics can shift while agencies are still operating within the bounds of the law. Given that the process was intended to promote problem-solving, we suggest that differences in agency culture and poor governance, not lack of authority, are mainly to blame for problems with agency coordination. The following sections suggest that the multifaceted, cross-scale approach to implementing Section 7 that culminated in the 2013 BiOp in many ways embodies the shifts called for by law and resilience scholars (Cosens et al. 2017; Benson 2012; Gunderson 2013), as well as demonstrates the ability of the ESA to support rather than undermine SES resilience and adaptive governance.

12.3 Section 7 Implementation in the Klamath Basin, 1988–2008: Fragmentation and Conflict

Since development of the federal Klamath Irrigation Project (Fig. 12.1; see Chap. 6 for an overview of the basin) in the early twentieth century, the BOR has cultivated a strong relationship with the farmers associated with the Klamath Water Users Association who use project water for irrigation. Two species of sucker fish were listed in the Upper Basin in 1988, which forced BOR to begin consulting with FWS on project operations. The federal agencies developed a dynamic like the one described in the previous section, in which BOR worked with irrigators to develop a plan for project operations based on their irrigation needs, and then delivered it along with a biological assessment to the FWS (and later to NMFS when the coho

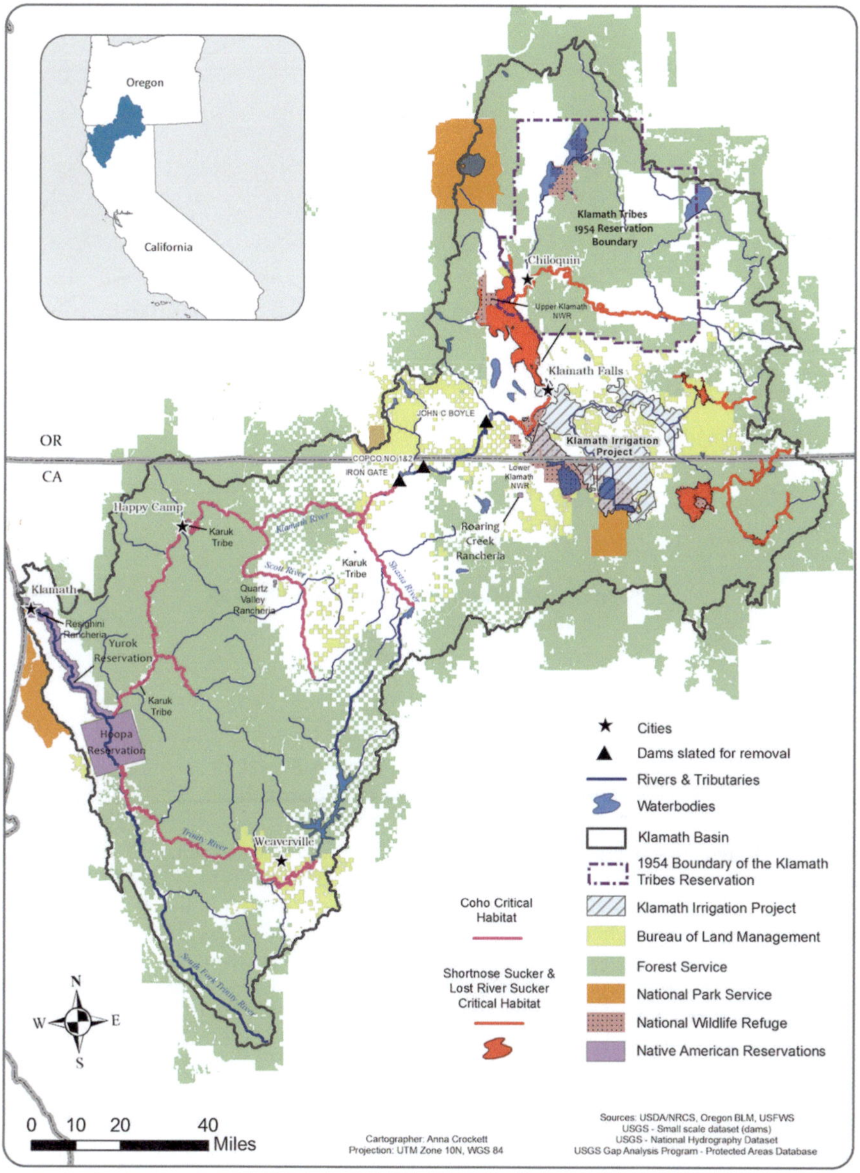

Fig. 12.1 Map of Klamath River Basin, indicating distribution of land managed by US federal agencies and designated habitat of endangered species. Inset shows location of watershed in the states of California and Oregon

salmon were listed in 1997) in anticipation of BiOp(s) that would allow them to continue operations.

During this period, NMFS and FWS issued separate BiOps. NMFS represented the needs of coho, aligning with the interests of the three Lower Basin tribes (Karuk, Yurok, and Hoopa Valley) and commercial fishermen in the development of its

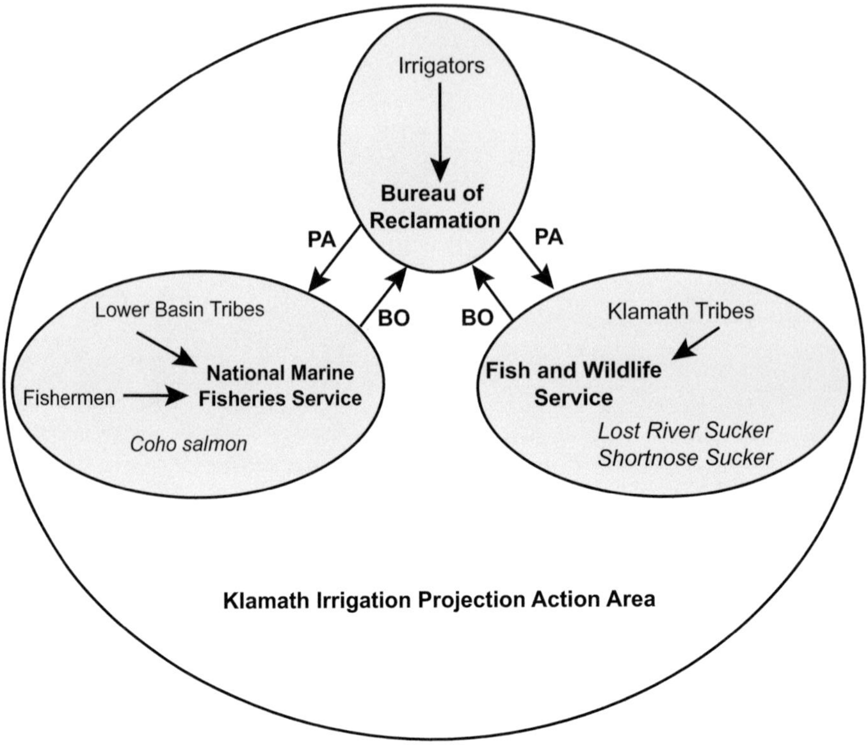

Section 7 Consultation Process Pre-2013
Upper Basin/Lower Basin Approach

Fig. 12.2 Diagram of interactions among agencies and stakeholders prior to 2013. Irrigators worked with BOR to develop a Proposed Action (PA); BOR delivered PA to FWS and NMFS separately; Lower Basin Tribes and fishermen sought to influence NMFS in development of coho BiOp; Klamath Tribes sought to influence FWS in development of suckers BiOp; FWS delivered suckers BiOp to BOR; NMFS delivered coho BiOp to BOR; process featured little or no collaboration between BOR and FWS or NMFS in development of PA, and little or no coordination between FWS and NMFS in development of BiOps for BOR

BiOps; FWS represented the needs of the suckers, aligning with the interests of the Klamath Tribes in the development of its BiOps. The Services left it to BOR to address their conflicting demands, which called for minimum Upper Klamath Lake levels for the suckers in the Upper Basin and minimum instream flows for coho in the Lower Basin (Fig. 12.2). These demands were not only at odds with each other, but individually and collectively at odds with farmers' desires to withdraw water for irrigation. BOR's task was to develop an annual operating plan for the project that would supply irrigation within the constraints of the two BiOps. The differing access of interest groups to each agency – according to alignment of interest – was compounded by the failure of the agencies to coordinate.

This dynamic resulted in a crisis that began in 2001. Acting independently, the Services each concluded that proposed operations would likely jeopardize the continued existence of both sucker species in the Upper Basin and the coho salmon in the Lower Basin. As a result of drought and low inflows to project reservoirs during the previous winter and spring, and ESA requirements for minimum lake levels and river flows, BOR was forced to limit the amount of water delivered to irrigators, leading to the economic crisis and social upheaval described earlier.

Subsequent consultations in 2002 and 2007 made it increasingly clear that the Upper Basin/Lower Basin approach was untenable (see Table 12.1 for a timeline). Figure 12.2 illustrates this segregated consultation dynamic, which had a negative effect on the fish, the water users, and other stakeholders in the Basin.

Even within apparent aligned interests, dissatisfaction surfaced. BOR's approach to operations relied on uncertain early-season runoff to forecast supply and historical water use to forecast demand. This approach frequently resulted in over-forecasting water availability and caused BOR to release too much water early in the season leading to unforeseen shortages later. Lower Basin tribes were dissatisfied with the approach to minimum downstream flows required by NMFS. They sought variability in flow timing to mimic the natural hydrologic regime, e.g., big spring pulses, in order to facilitate salmon recovery.

Interviews revealed three important dynamics related to this approach. First, the process featured little or no collaboration among BOR and the Services. As a result, it led to a reactive approach that was inefficient and involved a lot of back and forth as the BOR tried to incorporate fish needs into planned operations over multiple meetings and drafts. Second, there was little or no coordination between FWS and NMFS in the development of their respective BiOps for project operations, which resulted in a fragmented, single-species approach to ecosystem management. Third, the stakeholders' unanimous frustration with ESA implementation resulted in both fighting with each other and blaming the agencies for their problems.

The social and economic crisis in the Upper Basin initiated by Section 7 implementation in 2001 rotated to the Lower Basin in 2002. An implemented change in flow, mandated by the Bush Administration, led to a non-jeopardy opinion and to low-flow, anoxic conditions in the lower river that resulted in the death of more than 30,000 fall-run Chinook salmon (Doremus and Tarlock 2005; CAFG 2004). The 2002 fish kill dealt a catastrophic blow to Lower Basin tribes for whom salmon hold spiritual, cultural, and subsistence significance. This event was followed by a period of media warfare among a variety of adversaries in the battle for water use variously characterized as tribes vs. irrigators, Upper vs. Lower Basin, and federal agencies vs. local stakeholders.

Outside the scope of ongoing Section 7 consultation, leaders of basin tribes, irrigation groups, environmental and conservation NGOs, and federal agencies initiated a series of uncoordinated side meetings across the basin to discuss the potential for collaborative paths toward more integrated, basin-wide water and resource management (Gosnell and Kelly 2010). As side meetings became more coordinated and gained traction among a relatively representative group of stakeholder leaders in the basin, many of these leaders (on behalf of their constituent groups) intervened in the Federal Energy Regulatory Commission (FERC) relicensing process for the Klamath Hydroelectric Project around 2005. With the location of the hydroelectric

Table 12.1 Chronological list of events related to endangered species, fisheries, and water management in the Klamath River Basin

1967	The Lost River sucker (LRS) (*Deltistes luxatus*) is listed as "rare" under the California Endangered Species Act (CESA)
1971	The Shortnose sucker (SNS) (*Chasmistes brevirostris)* is listed as "rare" under CESA
1971	Both sucker species are identified as "species of concern" under CESA
1986	The Klamath Tribes close their subsistence sucker fishery to conserve the species
	Hundreds of adult suckers die during the summer months in Upper Klamath Lake (UKL).
1988	The U.S. Fish and Wildlife Service (FWS) lists both sucker species as "endangered" under the federal Endangered Species Act (ESA)
1992	The Bureau of Reclamation (BOR) prepares the first Biological Assessment (BA) for the Klamath Irrigation Project (KIP) operations; FWS issues the first Biological Opinion (BiOp) for the sucker species, imposing minimum lake levels for UKL
1995–1997	Thousands of suckers die during the summers in UKL
1997	The National Marine Fisheries Service (NMFS) lists the Southern Oregon/Northern California Coast coho salmon Evolutionarily Significant Unit (*Oncorhynchus kisutch*), the Southern Distinct Population Segment (DPS) of green sturgeon (*Acipenser medirostris*), and the Southern DPS of eulachon (*Thaleichthys pacificus*) as "threatened" under the federal ESA
2001	BOR prepares a BA for proposed KIP operations
	FWS issues a jeopardy BiOp for the two sucker species
	NMFS issues a jeopardy BiOp for the coho salmon
	BOR limits the volume of water delivered to KIP users to comply with the jeopardy BiOps, leading to social and economic hardship for irrigators
2002	BOR prepares a new BA for KIP operations spanning 2002–2012 and requests re-initiation of consultation with FWS and NMFS
	Both agencies again issue jeopardy BiOps with reasonable and prudent alternatives (RPAs) that call for reduced water deliveries
	Disregarding the BiOps, and in spite of continued drought, BOR provides full water deliveries to irrigators during summer 2002
	Curtailed flows in the Lower Basin lead to poor water quality and, in September, more than 30,000 adult salmonids (primarily unlisted Chinook) die from a bacterial pathogen (*columnaris*) associated with low flows and high stream temperatures
	Lawsuits by tribes, environmental organizations, and commercial fishermen result in a federal district court ordering the BOR to reinitiate consultation with FWS and NMFS
2005	The coho salmon is listed as "threatened" under the CESA
2005–2009	The commercial salmon fishing harvest is closed and/or restricted
2006	Settlement talks associated with the Klamath hydroelectric dam relicensing process morph into an "Extended Caucus" that creates the foundation for developing the Klamath Basin Restoration Agreement (KBRA) and the Klamath Hydroelectric Settlement Agreement (KHSA) over the next four years. (See Gosnell and Kelly 2010 for details regarding the KBRA/KHSA development and the process's influence on tribal/irrigator power relations.)

(continued)

Table 12.1 (continued)

2007	BOR reinitiates consultation with FWS and NMFS, proposing to change its operations to address concerns with its monthly time-step approach to managing UKL elevations and downstream flows
	FWS issues a nonjeopardy BiOp for suckers after working with BOR to refine the proposed action for the time period spanning 2007–2017
2008	NMFS issues a draft jeopardy BiOp for the coho which conflicts with the FWS' 2007 nonjeopardy BiOp
	BOR requests that NMFS suspend the finalization of the consultation process and BiOp until further notice
2010	BOR requests that NMFS finalize its BiOp for the KIP
	NMFS issues a final jeopardy BiOp regarding the coho for KIP operations from 2010 to 2018, with an RPA that calls for downstream flows that, when combined with KIP operations, would result in lower UKL levels than those approved in the 2007 FWS BiOp
	It becomes apparent that, under certain hydrologic conditions, BOR will be unable to meet the expectations of KIP water users and the two BiOps, putting the agency in an untenable position. It also becomes apparent that FWS will have to reinitiate consultation to evaluate the effects of the NMFS RPA on the two sucker species, scrapping much of the work that it had done to develop the 2007 BiOp
	Regional Directors for the three federal agencies – the BOR's Mid Pacific Regional Director, NMFS' Southwest Regional Administrator, and the FWS's Pacific Southwest Regional Director – meet with their respective field office managers and direct them to develop a new proposed action and the nation's first joint BiOp
	Basin stakeholders sign initial drafts of the KBRA and KHSA, which represent the culmination of four years of talks regarding a "whole basin" approach to social-ecological restoration
2012	BOR finalizes a new proposed action based on the results of its hydrologic modeling of the Klamath system, issues a new BA, and requests formal consultation with FWS and NMFS to develop a joint BiOp.
2013	FWS and NMFS issue a nonjeopardy joint BiOp
	In March, the Oregon Water Resources Department (OWRD) delivers its Findings of Fact and an Order of Determination in the Klamath River Basin Adjudication regarding water rights in the Klamath Basin (within the state of Oregon) to the Klamath County Circuit Court. Adjudication-related proceedings in the Oregon portion of the Klamath Basin had been conducted since 1975, with significant implications for Klamath Tribes' water rights and feasibility of in-stream flows to support sucker recovery. OWRD's Findings determine that the Klamath Tribes have the most senior and very large water rights in the system and quantify those rights for the first time
	In June, the Klamath Tribes "call the river" based on their newly quantified and legally enforceable water rights; OWRD enforces call and shuts off water to off-project farmers and ranchers above the lake and in, around, and above historical reservation lands. BOR, FWS, and four irrigation districts immediately follow with their own calls on the river, but because of drought conditions, these senior users' demands exceed the available flows

(continued)

Table 12.1 (continued)

	In fall, the off-project irrigators, mostly located in the basin headwaters northeast of or above the historic reservation boundary, return to the negotiating table as a result of the June call on the river by the Klamath Tribes and the subsequent economic hardship suffered by nontribal, off-project agricultural producers that year. Some of these irrigators are also instrumental leaders in the public opposition to the KBRA and KHSA that had been voiced since 2010.
2014	In April, the parties sign the Upper Klamath Basin Comprehensive Agreement (UKBCA), bringing off-project irrigators and irrigation back into the comprehensive basin water settlement agreements. This agreement is combined with the KBRA and KHSA
2015	In December, the Klamath Basin Agreements (KBRA, KHSA, UKBCA) expire when the U.S. Congress fails to authorize and fund them through federal legislation
2016	In April, the parties sign the Klamath Power and Facilities Agreement (KPFA) and an amended KHSA at a large public ceremony at the Klamath River's mouth that includes an appearance by U.S. Secretary of the Interior Sally Jewell. This agreement has very little to do with the KBRA and basin restoration; instead, it settles the outstanding Federal Energy Regulatory Commission (FERC) process that required the KHP to be either relicensed or decommissioned and facilities removed. The Agreements authorize the removal of the four mainstem Klamath dams by 2020 and do not require congressional authorization or funding.
	In June and July, the Yurok Tribes sue BOR and NMFS for mismanagement of endangered salmon in the Klamath River. A *Ceratonova shasta* parasite outbreak causes the Yurok to suspend fishing throughout 2016. The disease's prevalence is attributed to unhealthily high concentrations of salmon in the river. The Yurok suit follows the NMFS decision to reduce salmon protections rather than divert water away from the KIP and irrigation
2017	In February, a federal district court orders a review of the 2013 BiOp to re-evaluate the BOR's operating plan to protect threatened coho salmon in the Lower Klamath Basin (*Hoopa Valley Tribe v. National Marine Fisheries Service*, 230 F. Supp. 3d 1106 (N.D. Cal. 2017); *Yurok Tribe v. Bureau of Reclamation* 231 F. Supp. 3d 450 (N.D. Cal. 2017))
	In April and May, the Klamath Tribes call the river again (as in 2013) because they perceive insufficient water system recharge since the end of the drought

dams at the point of intersection between the Upper and Lower Basins, this process evolved into a venue to discuss water management problems across the entire basin, including the potential federal settlement of the Klamath Tribes claim to unquantified water rights in the Upper Basin actively being sought by the state of Oregon. These discussions would eventually culminate in a series of agreements between stakeholders, the Klamath Basin Restoration Agreement (KBRA), and Klamath Hydroelectric Settlement Agreement (KHSA), released to the public in 2010 and aimed at holistic basin restoration and water sharing (Gosnell and Kelly 2010; Chaffin et al. 2014b).

A window of opportunity (Olsson et al. 2006) opened in 2010 when NMFS issued a jeopardy BiOp that conflicted with (and essentially negated) an earlier non-jeopardy BiOp issued by FWS for the Upper Basin (Table 12.1). Frustration came to a head over the lack of agency coordination in Section 7 implementation and the need to identify a new, more coordinated approach became painfully apparent.

12.4 The 2013 Joint Biological Opinion: A SES Approach to Section 7 Consultation?

In November 2010, directors for FWS, NMFS, and BOR asked their field office managers to develop a joint BiOp. The goal was to take a more proactive and coordinated approach that would meet the conservation needs of the listed species while allowing continued operation of the project in alignment with its authorized purposes and contractual obligations. A central goal of the new approach was to address the uncertainty and instability associated with past ESA implementation, which had, at different times over the previous decade, resulted in water shortages for irrigators. Reducing uncertainty around water availability and timing for fish and farmers was a major motivator for collaboration. Federal agency leaders established an Agency Coordination Team comprised of biologists, hydrologists, managers, and support staff from each agency and, over the next 2 years, the Team met more than 25 times to develop what the agencies referred to as a new paradigm (NMFS and USFWS 2013).

The 2013 BiOp was born of a new concept for annual water sharing that addressed the needs of irrigators and listed suckers and coho and demonstrated governmental capacity for flexibility and evolution within the constraints of formal law. The new approach represented a fundamental shift from a single-species approach to ESA implementation to one more reflective of systems thinking. Specifically, it featured new levels of integration in four dimensions: between the action agency and the Services, among the Services, between biologists and hydrologists in seeking to better understand and manage the system, and between federal agencies and stakeholders in identifying innovative strategies for complying with the law that avoided putting all the risk on any one species or stakeholder (Fig. 12.3).

During the development of the joint BiOp, BOR expanded its sphere of concern from the project to include the Services and, by extension, the ESA-listed fish and the stakeholder groups advocating on their behalf (e.g., tribes, fishermen). A key variable was a new regional director in the local BOR office who had "a whole different attitude towards … the responsibility that BOR had to implement the ESA" (Interview, FWS). In a departure from the past, the new director convened a meeting of federal agency leaders and proposed a new approach to water sharing by allocating to the river and the lake first and giving the irrigation project what remained. Prioritizing the functionality and stability of the ecological system before trying to accommodate human use represented a radical shift.

The shift was made possible by changing relationships among all entities. First, water users' interest in collaborating and their willingness to make some concessions in exchange for more certainty, both of which were a result of increased trust associated with the KBRA discussions, enabled increased flexibility by BOR. Second, while BOR and FWS had developed a good working relationship over the years as a result of being in the same US Department (Interior) and both having offices in Klamath Falls (Oregon), the relations between BOR and NMFS, located in the US Department of Commerce and housed in Arcata, California (sev-

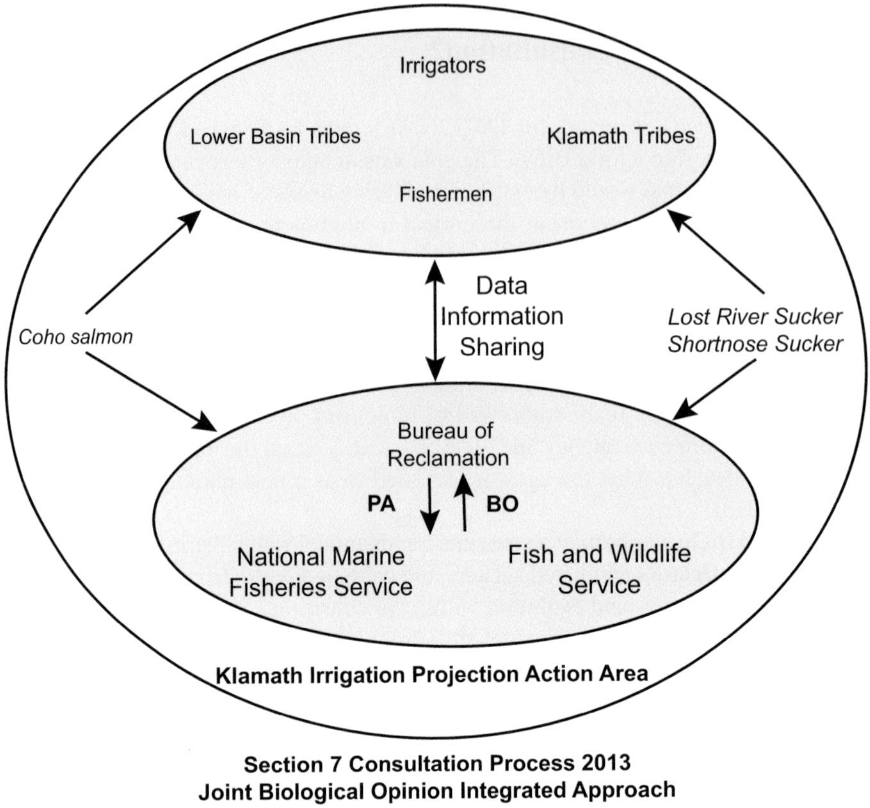

Fig 12.3 Diagram of 2013 Joint BiOp integrated approach to Section 7 compliance. Stakeholders collaborated on a water sharing plan using hydrologic modeling; stakeholders shared data and information with BOR in development of Proposed Action (PA); BOR worked with NMFS and FWS to develop the PA in a coordinated way that took into account needs of fish. NMFS and FWS hired hydrologists to facilitate data and information sharing and improve communication about the conservation role of the action area in the survival and recovery of the listed species and in the recovery support function of critical habitat; NMFS and FWS issued a nonjeopardy joint BiOp that took into account needs of irrigators; process featured collaborative, coordinated, integrated, proactive approach to development of PA and BiOp

eral hundred miles away), were more strained. The new BOR director worked to actively engage NMFS staff to overcome the physical distance and institutional divide. The new dynamic fostered a sense of shared responsibility among the three agencies for the listed species and the need for stability for water users. This approach stood in contrast to the one taken in prior consultations wherein listed fish were pitted against farmers. These new social norms around agency interactions lessened fears and uncertainty among the stakeholders. Stakeholders felt freer to engage in creative problem-solving which involved the use of hydrologic modeling and led to uncovering unexpected flexibility in the system.

To understand the importance of the hydrologic modeling, it is first important to understand that most BiOps are researched and written exclusively by biologists and ecologists from the Services. In contrast, BOR staff is dominated by hydrologists. To better support the development of the 2013 Klamath joint BiOp, NMFS and FWS hired hydrologists to facilitate data and information sharing with BOR and improve communication during the consultation process. The goal of these new hires was to help the agencies better understand a critical aspect of the ESA: the conservation role of the action area in the survival and recovery of the listed species and in the recovery support function of critical habitat.

Two aspects of the modeling effort were transformational: the greater dependence on hydrologic science to better understand the system and eliminate power imbalances between the agencies and the engagement of stakeholders in the modeling and learning about system function. By using hydrologic modeling to provide more certainty about water supply, the agencies were able to take a more creative approach to problem-solving. Focusing on the system (hydrology) rather than on a single species (biology) resulted in increased transparency about system limits (e.g., flows and water levels), which reduced tension and competitiveness between agencies. Improved understanding of the system enabled improved congruence between rules and ecological conditions and improved the institutional fit between the ESA and the basin SES (Olsson et al. 2006; Rijke et al. 2012). It also allowed the agencies to incorporate adaptive management into real-time project operations instead of trying to prescribe flows in advance. The use of hydrologic modeling also enhanced participants' ability to communicate and learn about system dynamics, resulting in a switch from a focus on optimizing water in different parts of the system for water users and individual species to fostering systems thinking.

Side meetings among leaders of the Klamath Tribes, the Lower Basin tribes, and the irrigators (those parties already involved in discussions leading to the KBRA) about the hydrologic modeling helped advance negotiations regarding sharing of scarce water resources. The meetings had the dual purpose of promoting learning about the biophysical aspects of the system and its limits while also enhancing trust and communication, as the tribes and the water users were able to ask questions of the hydrologists and of each other. The enhanced trust among the agencies and stakeholders meant that the biologists at NMFS and FWS had more flexibility to write the joint BiOp without fear of litigation. The reduced fear of litigation enabled a more creative approach to developing the BiOp that included some degree of risk taking.

Ultimately, NMFS and FWS issued a non-jeopardy joint BiOp that took into account the needs of both the fish and the irrigators. The BiOp authorized an operation model that is updated with water availability and use daily, an approach that was expected to enhance coordination between water elevations in the lake and Klamath River flows and to capture natural variations, including occasional high flows for downstream salmon. As part of the process, NMFS had to come to terms with the fact that it could not prescribe high flows for a specific time; rather it had to devolve authority to local managers to do it when possible based on real-time climatic conditions. This represented a transition toward incorporating adaptive management and feedback from natural limits into real-time operation.

12.5 Discussion and Conclusion: Tapping the Potential for Adaptive Governance

Cosens et al. (2017) note the current "dearth of legal mechanisms at any level that allow for alternatives to traditional environmental enforcement—alternatives that could enhance local adaptive capacity while maintaining stability and accountability toward achieving the goals that traditional enforcement mechanisms seek to achieve." As the Klamath case suggests, this may be more a function of implementation than the written law. The Klamath case provides insights into the ways in which implementation of the law can evolve to make room for and even begin to institutionalize adaptive governance.

This section presents key factors that made possible a more adaptive, integrated approach to Section 7 implementation in the Klamath. It describes how social change facilitated by technical and social learning had to precede institutional change and considers what led to social change in this governance context.

First, while a main focus of this chapter has been on the role of interagency relations, it is important to note that the impetus for change came from the stakeholders. Their determination to find an alternative approach to ESA implementation trickled up to influence, inspire, and motivate the agencies. Over the course of the 3-year consultation process that began in 2010, stakeholders and agency leaders began to see each other differently and developed new relationships, both personal and professional. Set against a backdrop of shifting tribal-irrigator power relations (Gosnell and Kelly 2010), some of these relationships had their origins in processes associated with the negotiation and support of the KBRA during the previous 5 years; but the interactions associated specifically with the development of the joint BiOp added a new dimension due to the collective focus on understanding the biophysical system as a whole for the purpose of balancing stability and flexibility in project operations and avoiding jeopardy for listed species.

Second, the opportunity for social and technical learning about the system was facilitated through data and information sharing and through hydrologic modeling, which allowed agency biologists and ecologists to see the system differently, shifting their focus from the life history of single species to the eco-hydrology of the whole basin, an example of transformative multiple-loop learning (Armitage et al. 2008; Pahl-Wostl 2009). The collaborative approach to hydrologic modeling helped the agencies come to terms with inherent connectivity and uncertainty in the SES as they began to consider how water use and allocation in the system could be improved. The approach also facilitated an expanded sense of place among participants. Once the eco-hydrologic system was better understood, stakeholders and government actors were empowered to improve the institutional fit or match between the social and eco-hydrologic dynamics. This suggests that ESA implementation without support for social and technical learning may not be successful in terms of public perception. Section 7 can be a trigger for social-ecological reorganization, adaptation, or transformation, but learning processes must take place; otherwise Section 7 is more likely to act like a detonator than a stabilizer. As Rogers et al. (2013) note, "[d]eep reflection leading to transformational learning is required to foster the changes in mindset and behaviors needed to adopt a complexity frame of reference."

Third, empathic leadership increased trust within federal agencies and among stakeholders and cultivated a sense of shared responsibility for ESA compliance. Agency leaders had to expand their historic spheres of responsibility and concern to include new elements. The BOR had to proactively incorporate the needs of the fish into its project operations, and the Services had to develop a greater appreciation not only for "each other's fish" but for the needs of the water users. A synergistic combination of leaders from different federal agencies and stakeholder groups came to a shared understanding that collaboration, humility, mutual respect, and empathy were necessary and desirable for progress to be made.

Essential to the Klamath Basin social dynamics was the willingness and even desire to share limited water resources as well as information and data about the system. The *willingness* to share resulted from the cultivation of trust among federal agency leaders and the different groups of stakeholders. The *desire* to share, however, was cultivated by the emergence of a culture of compassion and empathy among participants in the consultation process as well as the preceding KBRA process. Taken together, we suggest that compassion and empathy are important precursors to what many adaptive governance scholars recognize as a crucial ingredient, trust. Participants in the consultation process leading to the 2013 joint BiOp trusted that even if environmental conditions changed and Section 7 consultation had to be reinitiated, the relationships they had forged and the collective culture of compassion and empathy would ensure that they would be treated fairly. Aside from any legal documentation, the core participants felt accountability to one another to ensure a legitimate, fair approach to sharing, even if resources became scarcer.

While recognition of the important role that emotions played in Klamath Basin negotiations during the first decade of the 2000s is nothing new, scholars interested in adaptive governance and SES transformation have yet to adequately theorize the important role played by these emotional geographies. Building on theories regarding the role of trust and social entrepreneurship in adaptive governance (Westley et al. 2011; Folke et al. 2005), this case reveals the importance of a culture of compassion and empathy for cultivating enhanced cross-scale interactions and a collective sense of stability, which created space for actors to deliberately transform institutions to support adaptive governance. Although the adaptive governance literature highlights the important role of social capital, including trust, collaboration, participatory capacity, leadership, social networks, and social learning (Folke et al. 2005), it focuses on the societal or group level of this phenomenon. It generally ignores, though, the psychology of emotions, cognitive framing, empathy, and reconciliation in the creation and maintenance of social capital, which in turn supports society's adaptive capacity (Seyfarth and Cheney 2013; Arnold 2014).

In sum, the 2013 joint BiOp in the Klamath Basin was a historic first. It embodied a coordinated, place-based approach to conserving terrestrial and marine species at a bioregional scale while taking into account the socioeconomic needs of those in the basin. This approach is also partial evidence of the institutionalization of emergent adaptive governance in the Basin. The new approach both reflected and contributed to a fundamental change in the structure and function of the SES, and the resultant joint BiOp served to institutionalize this new trajectory to some degree, allowing it to gain power and legitimacy. Rather than eliciting fear and anger, as the ESA is prone to do, this approach to Section 7 implementation helped facilitate a change in behavior, procedures, customs, norms, values, and beliefs that altered environmental governance in the basin.

Acknowledgment This work was developed in part under the Adaptive Water Governance Project, funded by the US National Socio-Environmental Synthesis Center (SESYNC) under funding from the US National Science Foundation, NSF DBI-1052875.

Literature

Armitage D, Marschkem M, Plummer R (2008) Adaptive co-management and the paradox of learning. Glob Environ Chang 18(1):86–98

Arnold CA (2004) Working out an environmental ethic: anniversary lessons from Mono Lake. Wyoming Law Rev 4:1–55

Arnold CA (2014) Framing watersheds. In: Kirokawa K (ed) Environmental law and contrasting ideas of nature: a constructivist approach. Cambridge University Press, Boston, pp 271–302

Arnold CA (2015) Environmental law, episode IV: a new hope? Can environmental law adapt for resilient communities and ecosystems? J Environ Sustain Law 21(1):1–46

Arnold CA, Green OO, DeCaro DA, Chase A, Ewa JG (2014) The social-ecological resilience of an eastern urban-suburban watershed: the Anacostia River Basin Idaho Law Rev 51:29–90

Benson MH (2012) Intelligent tinkering: the endangered species act and resilience. Ecol Soc 17(4):28

California Department of Fish and Game (CDFG). 2004. September 2002 Klamath River fish-kill: final analysis of contributing factors and impacts. Northern California-North Coast Region, The Resources Agency, State of California. http://waterwatch.org/wp-content/uploads/2011/08/2002CAFinalKlamathFishKillReport.pdf

Carpenter SR, Brock WA (2008) Adaptive capacity and traps. Ecol Soc 13(2):40. http://www.ecologyandsociety.org/vol13/iss2/art40/

Chaffin BC, Craig RK, Gosnell H (2014a) Resilience, adaptation and transformation in the Klamath River basin social-ecological system. Idaho Law Rev 51:157–194

Chaffin BC, Gosnell H, Cosens BA (2014b) A decade of adaptive governance scholarship: synthesis and future directions. Ecol Soc 19(3):56. https://doi.org/10.5751/ES-06824-190356

Chaffin BC, Garmestani AS, Gunderson LH, Benson MH, Angeler DG, Arnold CA, Cosens BA, Craig RK, Ruhl JB, Allen CR (2016) Transformative environmental governance. Annu Rev Environ Resour 41:399–423

Chapin FS III, Kofinas GP, Folke C (eds) (2009) Principles of ecosystem stewardship: resilience-based natural resource management in a changing world. Springer-Verlag, New York

Cosens B (2010) Transboundary river governance in the face of uncertainty: resilience theory and the Columbia River Treaty. Univ Utah J Land Resour Environ Law 30:229

Cosens BA, Craig RK, Hirsch S et al (2017) The role of law in adaptive governance. Ecol Soc 22(1):30. http://www.ecologyandsociety.org/vol22/iss1/art30/

Cosens BA, Stow C (2014) Resilience and water governance: addressing fragmentation and uncertainty in water allocation and water quality law. In: Garmestani AS, Allen C (eds) Social-ecological resilience and law. Columbia University Press, New York

Craig RK (2014) Does the Endangered Species Act preempt state water law? Univ Kansas Law Rev 62:851

Craig RK, Garmestani AS, Allen C, Arnold CA, Benson MH, DeCaro DA, Garmestani AS, Gosnell H, Ruhl JB, Schlager E (2017) Balancing stability and flexibility in adaptive governance: an analysis of tools available in U.S. environmental law. Ecol Soc 22(2):3. https://doi.org/10.5751/ES-08983-220203

Doremus H, Tarlock AD (2005) Water war in the Klamath Basin: macho law, combat biology, and dirty politics. Island Press, Washington, DC

Folke C (2006) Resilience: the emergence of a perspective for social-ecological systems analyses. Glob Environ Chang 16(3):253–267

Folke C, Hahn T, Olsson P, Norberg J (2005) Adaptive governance of social-ecological systems. Annu Rev Environ Resour 30:441–473. https://doi.org/10.1146/annurev.energy.30.050504.144511

Folke C, Chapin FS, Olsson P (2009) Transformations in ecosystem stewardship. In: Chapin FS, Kofinas GP, Folke C (eds) Principles of ecosystem stewardship: resilience-based natural resource management in a changing world. Springer-Verlag, New York, pp 103–128

Gosnell H, Kelly E (2010) Peace on the river? Social-ecological restoration and large dam removal in the Klamath Basin, USA. Water Altern 3(2):361–383

Grantham HS, Bode M, McDonald-Madden E et al (2010) Effective conservation planning requires learning and adaptation. Front Ecol Environ 8:431–437

Gunderson LH (2013) How the Endangered Species Act promotes unintelligent, misplaced tinkering. Ecol Soc 18(1):12

Gunderson LH, Garmestani AS, Rizzardi KW, Ruhl JB, Light A (2014) Escaping a rigidity trap: governance and adaptive capacity to climate change in the Everglades social-ecological system Idaho Law Rev 51:127

Jaeger WK (2004) Conflicts over water in the Upper Klamath Basin and the potential role for market-based allocations. J Agric Resour Econ 29(2):167–184

Karkkainen BC (2008) Getting to "let's talk": legal and natural destabilizations and the future of regional collaboration. Nevada Law J 8:811–829

Moore M, Tjornbo O, Enfors E, Knapp C, Hodbod J, Baggio JA (2014) Studying the complexity of change: toward an analytical framework for understanding deliberate social-ecological transformations. Ecol Soc 19(4):54. https://doi.org/10.5751/ES-06966-190454

National Marine Fisheries Service [NMFS], U.S. Fish and Wildlife Service [USFWS] (2013) Biological opinions on the effects of proposed Klamath Project operations from May 31, 2013, through March 31, 2023, on five federally listed threatened and endangered species. https://www.fws.gov/klamathfallsfwo/news/2013%20BO/2013-Final-Klamath-Project-BO.pdf. Accessed 6 Oct 2017

Olsson P, Gunderson LH, Carpenter SR, Ryan P, Lebel L, Folke C, Holling CS (2006) Shooting the rapids: navigating transitions to adaptive governance of social-ecological systems. Ecol Soc 11:1–21

Pahl-Wostl C (2009) A conceptual framework for analyzing adaptive capacity and multi-level learning processes in resource governance regimes. Glob Environ Chang 19:354–365

Pahl-Wostl C, Becker G, Kneiper C, Sendzimir J (2013) How multilevel societal learning processes facilitate transformative change: a comparative case study analysis on flood management. Ecol Soc 18(4):58

Prokopy L, Mullendore N, Brasier K, Floress K (2014) A typology of catalyst events for collaborative watershed management in the United States. Soc Nat Resour 27(11):1177–1191

Rijke J, Brown R, Zevenbergen C, Ashley R, Farrelly M, Morison P, van Herk S (2012) Fit-for-purpose governance: a framework to make adaptive governance operational. Environ Sci Pol 22:73–84

Rogers KH, Luton R, Biggs H, Biggs R, Blignaut S, Choles AG, Palmer CG, Tangwe P (2013) Fostering complexity thinking in action research for change in social-ecological systems. Ecol Soc 18(2):31

Ruhl JB (2012) The Endangered Species Act's fall from grace in the Supreme Court. Harvard Environ Law Rev 36:487

Saundry P (ed) (2009) Endangered Species Act, United States. The Encyclopedia of Earth. https://editors.eol.org/eoearth/wiki/Endangered_Species_Act,_United_States. Accessed 6 Oct 2017

Sax JL, Keiter RB (2006) The realities of regional resource management: Glacier National Park and its neighbors revisited. Ecol Law Q 33:233

Seyfarth RM, Cheney DL (2013) Affiliation, empathy, and the origins of theory of mind. Proc Natl Acad Sci U S A 110(Supplement 2):10349–10356

Tennessee Valley Authority [TVA] v. Hill, 437 U.S. 153 (1978)

U.S. Department of the Interior (2010) Klamath Basin restoration agreement for the sustainability of public and trust resources and affected communities. http://klamathrestoration.gov/sites/klamathrestoration.gov/files/Klamath-Agreements/Klamath-Basin-Restoration-Agreement-2-18-10signed.pdf. Accessed 6 Oct 2017

Westley F, Olsson P, Folke C, Homer-Dixon T, Vredenburg H, Loorbach D, Thompson J, Nilsson M, Lambin E, Sendzimir J, Banerjee B, Galaz V, van der Leeuw S (2011) Tipping toward sustainability: emerging pathways of transformation. AMBIO J Hum Environ 40(7):762–778

Legal References

16 U.S.C. § 1531 (2012) Congressional finding and declaration of purposes and policy. Endangered species act, section 2

16 U.S.C. § 1532 (2012) Definitions. Endangered species act, section 3

16 U.S.C. § 1533 (2012) Determination of endangered species and threatened species. Endangered species act, section 4

16 U.S.C. § 1536 (2012) Interagency cooperation. Endangered species act, section 7

Part III
Social-Ecological Resilience and Adaptive Governance

Chapter 13
Cross-Basin Patterns of Systemic-Change Drivers and Adaptive Governance Features

Check for updates

Craig Anthony (Tony) Arnold, Hannah Gosnell, Melinda Harm Benson, and Robin Kundis Craig

Abstract Features of adaptive governance and drivers of systemic change were derived using qualitative textual analysis of six North American basin resilience assessments. This meta-analysis sought new knowledge that transcends each study concerning two categories of variables: (1) drivers of change in complex social-ecological-institutional water systems that affect systemic resilience and (2) features of adaptive governance. Pervasive themes, concepts, and variables from these six interdisciplinary texts were identified through inductive textual analysis and then analyzed for cross-basin patterns. Synthesis frameworks, as well as comprehensive lists of the variables that these studies uniformly or nearly uniformly addressed, are presented. These results are cross-interdisciplinary in that they identify patterns and knowledge that transcend several diverse interdisciplinary studies. The relevant and potentially generalizable insights into complex system change and adaptive governance, as well as a set of methods for synthesizing diverse interdisciplinary studies, form a foundation for future research on the dynamics of complex social-ecological-institutional systems and how they could be governed adaptively for resilience.

C. A. Arnold (✉)
Brandeis School of Law and Department of Urban and Public Affairs, University of Louisville, Louisville, KY, USA
e-mail: tony.arnold@louisville.edu

H. Gosnell
College of Earth, Ocean, and Atmospheric Sciences, Oregon State University, Corvallis, OR, USA
e-mail: gosnellh@geo.oregonstate.edu

M. H. Benson
Haub School of Environment and Natural Resources, University of Wyoming, Laramie, WYO, USA
e-mail: mhbenson@uwyo.edu

R. K. Craig
S.J. Quinney College of Law, University of Utah, Salt Lake City, UT, USA
e-mail: robin.craig@law.utah.edu

B. Cosens, L. Gunderson (eds.), *Practical Panarchy for Adaptive Water Governance*, https://doi.org/10.1007/978-3-319-72472-0_13

Keywords Adaptive governance · Interdisciplinary · Qualitative textual analysis · Resilience · Systemic-change drivers

13.1 Introduction

The Adaptive Water Governance (AWG) project was an interdisciplinary collaborative synthesis project aimed at identifying the features of adaptive governance in complex social-environmental-institutional systems, particularly governance features for managing water basin resilience under stresses caused by climate change (Cosens et al. 2014b). The AWG project brought together over two dozen researchers in many different disciplines across the natural and ecological sciences, social and policy sciences, law, engineering, and humanities.

AWG linked the concepts and science of social-ecological resilience with the theory and practice of governance, including law and policy (Cosens et al. 2014b). The project participants adopted the Walker and Salt (2012) definition of social-ecological resilience: "a measure of the amount of perturbation a social-ecological system can withstand while maintaining its structure and functions; it describes the ability of a complex system to continue to provide the full range of ecosystem services in the face of change" (Cosens et al. 2014b). Adaptive governance is governance that enables society to adapt to disturbances and changes by navigating the dynamic, multi-scalar nature of social-ecological-institutional systems (Cosens et al. 2014b).

The AWG project provided an ideal opportunity to conduct a systematic meta-analysis of the project's first set of published interdisciplinary studies in order to identify knowledge about adaptive governance and resilience that transcends each study. If clear, pervasive patterns could be identified from studies of several different basins conducted by different interdisciplinary teams using different interdisciplinary methodologies, these patterns would likely form significant generalizable insights that could improve understanding of adaptive water governance in complex social-ecological-institutional systems. This chapter uses the term "social-ecological-institutional systems," because (a) the AWG project gives particular attention to the roles of legal and governance institutions in systemic change and resilience (Cosens et al. 2014b) and (b) institutions – the rules that shape human behavior – are systems that are analytically distinct from social systems such as economies, politics, and local culture (Arnold et al. 2014; Sjostedt 2015). Institutions shape and are shaped by both social systems and ecosystems.

This chapter presents a systematic meta-analysis of the knowledge that the AWG project has generated concerning (a) the drivers of change in complex social-ecological-institutional water systems that affect their resilience and (b) the features of adaptive governance. Drivers of change include both exogenous drivers and endogenous change-producing variables as differentiated by Walker et al. (2012). The results synthesize the major themes, concepts, and variables from the published texts of resilience assessments of six North American water basins prepared by AWG teams, which are texts of interdisciplinary insights from in-depth case studies.

13.2 Collaborative Interdisciplinary Research

Collaborations across research disciplines are necessary to understand and address complex problems that have cross-scale, dynamic interactions among many environmental systems (e.g., watersheds, wetlands, climate), social systems (e.g., economies, politics, communities), and governance institutions (e.g., legal regimes, formal authorities, informal governance) (Heemskerk et al. 2003; Arnold 2004; Pohl 2008). No single discipline or even grouping of a few related disciplines possesses the theoretical frameworks and research methodologies required to comprehend fully the complexities of diverse interlinked systems.

Results from attempts at interdisciplinary collaboration have been mixed at best, though. The barriers to collaborating across disciplines are well known: lack of incentives to do so, bureaucratic barriers in organizations organized around disciplines, the usual challenges of teamwork, and the habits and patterns of rigidly thinking within one's own disciplinary paradigms (Eigenbrode et al. 2007; Gray 2008). Moreover, many different types of research fall into the broad category of interdisciplinary research, ranging from conceptual frameworks built on insights from only two disciplines to entirely new disciplines created from transdisciplinary collaboration (Fairclough 2005; Stokols et al. 2005; DeWulf et al. 2007).

Some of the most important interdisciplinary work on understanding complex systems has been in the areas of adaptive governance (Chaffin et al. 2014b) and the dynamics of social-ecological-institutional systems (Gunderson and Holling 2002). However, many studies have integrated insights from only two or three disciplines, such as law and ecology (Garmestani and Allen 2014) or ecology, economics, and sociology (Scheffer et al. 2002). Moreover, many interdisciplinary studies of adaptive governance or systemic dynamics are either primarily theoretical (Chaffin et al. 2014b) or based on empirical evidence from one particular system, such as the Florida Everglades (Light et al. 1995) or Swedish wetlands (Olsson et al. 2004). The benefits of interdisciplinary studies of complex social-ecological-institutional systems can be expanded through the systematic meta-analysis and synthesis of multiple interdisciplinary case studies (Noblit and Hare 1988; Grubert and Siders 2016). Through inductive identification of patterns of system change drivers and adaptive governance features that transcend specific case studies and their methodologies, interdisciplinary theories or frameworks of these phenomena can be developed and evaluated.

13.3 Case Studies from the Adaptive Water Governance Project

Several aspects of the AWG project make it well suited to a meta-analysis study. First, each of the resilience assessments of six North American watersheds (Part I) was done by an interdisciplinary team. The assessments were done on the Anacostia

Table 13.1 Cross-basin ranges of basin characteristics

Basin characteristic	Range
Region in the United States	Mid-Atlantic; Southeast; Great Plains; Southwest; Pacific Northwest
Basin size	From 173 square miles in the Anacostia to 258,000 square miles in the Columbia
Climate	Humid; arid; mixed
Settlement patterns	Urban; rural; mixed
Dominant hydrology	For example, surficial water flows through wetlands in the Everglades; modified riverine pulse floods in the Platte; long and high-volume snow-melt driven – but dammed – river channel flow in the Columbia
Landscapes	Various mixes of forests, plains, mountains, wetlands, urbanized areas, farmland, and deserts
Number of relevant jurisdictions	From 2 nations, 8 states, 1 province, and 15 Indian tribes in the Columbia to just 1 state but many different federal, state, and local agencies and 2 Indian tribes in the Everglades
Influence of federal laws	More influence of the Endangered Species Act in the American West; more influence of the Clean Water Act in the American East

River basin (Arnold et al. 2014), Columbia River basin (Cosens and Fremier 2014), Everglades (Gunderson et al. 2014), Klamath River basin (Chaffin et al. 2014a), portions of the Middle Rio Grande River basin (Benson et al. 2014) and the Platte River basin (Birge et al. 2014). Second, both the basins' environmental and political characteristics (Table 13.1) and the methods used by the six teams of researchers to assess the basins were diverse. Thus, any clear patterns discerned from these assessments from general knowledge about adaptive governance and resilience are likely robust to biophysical basin differences or a particular type of analytical methodology.

Despite a common research inquiry into the factors affecting the resilience and adaptive governance of each respective basin, each team developed its own methodology (Table 13.2). None of the assessments attempted to quantify or statistically test the relevance of variables influencing systemic change or adaptive governance nor was a common model that was applied and tested in each basin. The various qualitative, narrative approaches undertaken by the assessment teams reflected a belief that no single methodology is adequate to characterize resilience in every complex social-ecological-institutional system. Thus, when the basin assessments revealed shared insights, those revelations were more likely to point to significant and generalizable observations about AWG, rather than to reflect biases of a shared analytical method.

All teams employed some degree of narrative historical analysis, describing ecosystem, social system, and institutional conditions and changes in the basin over time. However, these common features did not form a standard cross-basin methodology (Table 13.2). The teams varied in the heuristics they used, ranging from more emphasis on cross-system institutional dynamics (Anacostia, Everglades), social and institutional effects on ecosystem services (Columbia, Platte), and adaptive cycle dynamics within social-ecological governance systems (Klamath, Middle Rio

Table 13.2 Cross-basin comparisons of assessment methodologies. Modified from Arnold et al. (2017)

Basin assessment	Primary analytical tools	Systemic change classifications
Anacostia	Historical-institutional analysis	Five institutional-social-ecological regimes since pre-European settlement
	Institutional-social-ecological dynamics (ISED) framework examining dynamics within and across ecological, social, and institutional systems	
Columbia	Walker and Salt list of resilience features with resilience and governance assessment of historic regime shifts	Four major historical eras since pre-European settlement as four phases in adaptive cycle (with impending possible regime shift)
	Surveys and interviews of experts re systemic resilience following Nemec et al. (2014)	
	Ecosystem services heuristics	
Everglades	Historical-institutional analysis	Fifteen governance management regimes since 1900
	Focus on legal regimes and adaptive management	
Klamath	Adaptive cycle heuristic	Four major historical eras since pre-European settlement as four phases in adaptive cycle (with impending possible regime shift)
	Social science field research	
	Historical social-ecological analysis	
Middle Rio Grande	Resilience theory analytics	Major contrasts between presettlement and current ecosystem functions, structures, and services
	Relationships among governance system, social system, and hydrologic system structures	
Platte	Ecosystem services heuristics	Major contrasts between presettlement and current ecosystem functions, structures, and services
	Panarchy theory heuristics	
	Focus on institutional effects on ecosystem services	

Grande). In addition, the Columbia and Klamath teams used social science research methods to gather empirical field data, and the Anacostia team developed a new analytical framework, the institutional-social-ecological dynamics (ISED) framework (Fig. 13.1), to describe the evolving dynamics within and across categories of ecological, social, and institutional systems (Arnold et al. 2014). Likewise, the texts of the six basin assessments contained different ways of classifying systemic changes in the basins, ranging from multiple major regime shifts (Anacostia, Everglades), movement through the phases of the adaptive cycle (Columbia, Klamath), and substantial changes in ecosystem structure and function from the presettlement era to the present (Columbia, Middle Rio Grande, Platte) (Table 13.2).

The AWG project was interdisciplinary at three different levels: the project, the basin team, and the individual scholarship. The project was designed to be an interdisciplinary collaboration and synthesis. At an early-stage project planning workshop, participants engaged in a guided group discussion to frame the project's research questions and identify the project's synthesis goals. This facilitated dialogue for cross-disciplinary communication, based on the Toolbox Project

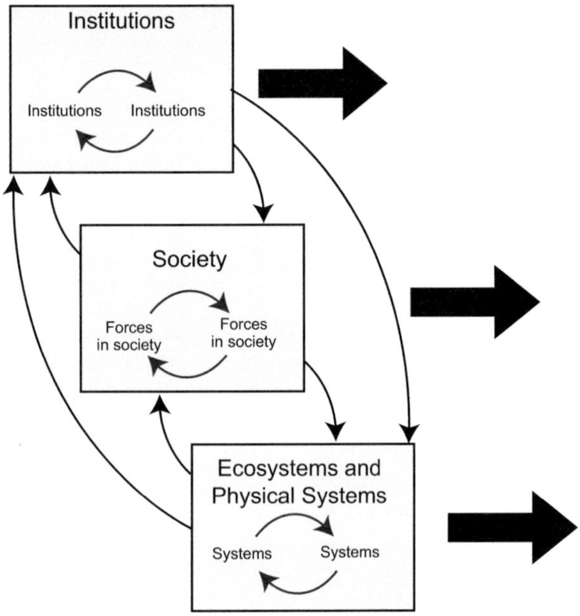

Fig. 13.1 Institutional-social-ecological dynamics framework. Depicts evolving cross-system effects across three system categories – institutions, social forces, and ecosystems – each of which is changing over time and also experiencing intra-category cross-system feedbacks (Source: Arnold et al. 2014)

(O'Rourke and Crowley 2013), explored the language, concepts, research methodologies, and even visual images that each participant brought from his or her disciplines to the group's inquiries. Comparisons, contrasts, and syntheses of the participants' contributions were made.

All of the basin assessment teams were composed of multiple researchers representing a diverse array of disciplines (Table 13.3). For example, researchers studying the Anacostia River basin brought expertise and insights from law, planning, public policy, political science, geography, geology, hydrology, sociology, anthropology, history, economics, and psychology to their work. The Platte River basin team had a different mix that includes some of the same social and policy sciences as the Anacostia team but with more representation from natural resource management and environmental sciences, including ecology, biology, and soil biochemistry. Each team collaborated to develop a resilience assessment of its respective basin that drew upon its members' diverse disciplinary backgrounds. Each assessment synthesized data and concepts from many different disciplines, weaving these diverse insights into a single analytical text.

The individual participants had engaged in interdisciplinary research and scholarship. When polled, the 25 coauthors of the basin resilience assessments had training or experience in 29 different disciplines (Arnold et al. 2017). All of the basin resilience assessment authors, except one, reported training or experience in

Table 13.3 Disciplines of basin assessment authors. The disciplines were self-assigned by each author in response to an inquiry about the disciplines in which they were systematically trained or in which they research and publish

Basin	Author	Disciplines
Anacostia	Tony Arnold	Law; planning; political science; history; public policy
Anacostia	Olivia Odom Green	Law; hydrology; geology; public policy
Anacostia	Daniel DeCaro	Psychology; political science; economics
Anacostia	Alexandra Chase	Law; public policy; anthropology; sociology
Anacostia	Jennifer-Grace Ewa	Planning; law; geography; public policy
Columbia	Barbara Cosens	Law; geology; geochemistry
Columbia	Alex Fremier	Ecology; geography; environmental science; mathematics
Everglades	Lance Gunderson	Environmental engineering; environmental science; ecology; natural resource management
Everglades	Ahjond Garmestani	Law: public policy; wildlife ecology
Everglades	Keith Rizzardi	Law; public administration
Everglades	J.B. Ruhl	Economics; law; geography
Everglades	Fred Light	Law; political science
Klamath	Brian Chaffin	Geography; environmental science
Klamath	Robin Craig	English literature and literary theory; science writing; law
Klamath	Hannah Gosnell	Geography; American civilization
Middle Rio Grande	Mindy Benson	Law; geography; environmental studies; natural resource management
Middle Rio Grande	Dagmar Llewellyn	Hydrology
Middle Rio Grande	Ryan Morrison	Civil engineering; water resource management; ecology
Middle Rio Grande	Mark Stone	Civil engineering; hydrology
Platte	Hannah Birge	Ecology; soil biochemistry; natural resource management
Platte	Craig Allen	Wildlife ecology; biology; ecology; environmental science
Platte	Robin Craig	English literature and literary theory; science writing; law
Platte	Ahjond Garmestani	Law; public policy; wildlife ecology
Platte	Joseph Hamm	Law; psychology; natural resource management
Platte	Christina Babbitt	Natural resource management; environmental science; international relations; public policy
Platte	Kristine Nemec	Natural resource management; biology
Platte	Edella Schlager	Political science; public policy; anthropology; economics

more than one discipline. The disciplines include the humanities, social sciences, natural sciences, policy systems design and management, and physical systems design and management. Of the 29 reported disciplines, 12 were in law, 7 in public policy, 6 in geography or natural resource management, and 5 in environmental science or ecology (Table 13.3).

The broad interdisciplinary perspectives of the AWG project researchers helped to make the basin resilience assessments themselves interdisciplinary in nature. The basin assessment scholars regularly use their backgrounds in multiple disciplines to engage in interdisciplinary research, as their previous work shows (Arnold 2004; Cosens et al. 2014a; Chaffin et al. 2014b; Garmestani and Allen 2014). Moreover, research shows that systematic and sustained experience in a discipline, whether through formal training or repeated research, influences how a scholar will perceive and frame a phenomenon and then go about researching it (Fry 2001; Fairclough 2005; O'Rourke and Crowley 2013). According to epistemic network theory, humans form new, networked frameworks of knowledge, thinking, and analysis from mental connections that they make among the ideas and methods that they use from multiple disciplines, both individually and through group work (Heemskerk et al. 2003; Shaffer et al. 2009). Interdisciplinary research collaborations with others affect a scholar's capacity to see a phenomenon from multiple perspectives and to synthesize these perspectives (Fry 2001; Heemskerk et al. 2003; Stokols et al. 2005). The following section describes the methods used to synthesize the case studies using an analysis of the texts.

13.4 Cross-Basin Textual Analyses

The six basin resilience assessments were analyzed to identify drivers of systemic change and features of adaptive governance. This was done by coding the text of each assessment for themes, concepts, and variables and then synthesizing the patterns that emerged from the textual analysis. Basically, coding is a process of identifying each concept or variable that is present in a written text and that is relevant to the analyst's research question (i.e., system change drivers, adaptive governance features). Synthesis involves identifying common relationships among coded concepts and variables (e.g., farming, agriculture, and conversion of forests to cropland are all coded as agriculture, but conversion of forests to cropland is also coded as land use, as is conversion of cropland to suburban housing). Synthesis also involves tracking how many different texts used or identified the same concepts and variables, in other words, looking for patterns of the concepts and variables that are present in multiple studies.

The published assessments' texts were coded in NVivo, a qualitative data analysis software program commonly used in social science research (Bringer et al. 2004; Bazeley and Jackson 2013). Qualitative analysis of the six basin resilience assessment for common themes and patterns is an appropriate method for developing cross-disciplinary theory, because the assessments are both texts containing verbal data about common research questions on basin resilience and adaptive water governance (Strauss 1987; Fairclough 2005; Thomas and Harden 2008; Bazeley and Jackson 2013) and rich empirical case studies using diverse data of basin conditions, historical narratives of systemic changes, and critical analyses of basin resilience and governance (Eisenhardt and Graebner 2007).

Instead of looking for specific features of a preconceived theory or model in the texts, multiple iterations of analysis and synthesis were used to construct theoretical frameworks from the language and ideas of the texts themselves and from deep, critical thinking about the themes, concepts, and variables communicated by each basin team (Strauss 1987; Corbin and Strauss 1990). The texts and coded nodes in this project went through seven iterations of analysis and synthesis.

Five important checks on coder bias were used: (1) an intentionally iterative inductive coding process; (2) a process of disciplined and critical self-reflectivity to identify and control personal and disciplinary biases (Bettez 2015), including recoding, annotation, and journaling processes; (3) intentional avoidance of predictive models, theories, or synthesis works when coding; (4) careful analysis of each assessment text in the second and third iterations of coding for each concept and variable coded in any of the assessment texts in the first iteration, thus allowing insights from later-coded texts to inform the coding of earlier-coded texts; and (5) review and feedback from at least one coauthor of each basin assessment.

Coding used nodes, which are references in data to concepts that may be either terminal variables or points of connection in networked concepts (Bazeley and Jackson 2013). Annotations and journal notes were made throughout the process, explaining the coding, analyzing relationships among nodes, and exploring the ideas behind the coded text (Bazeley and Jackson 2013; Bringer et al. 2004).

After all coding was completed, synthesis activities focused on nodes that appeared in at least five of the six assessments: a total of 100 nodes appearing in all six assessments and 23 nodes appearing in five of the six assessments. By using a five-out-of-six rule for including a node, the results were based on features appearing in several different types of basin assessments, yet avoided excluding a feature that may be present in every basin but overlooked by just one basin assessment team or may be implicit in that team's assessment but missed in the coding process. These 123 nodes were synthesized through several iterations that explored the relative importance of and relationships among each theme, concept, or variable to the assessments. The coding outcomes were subjected to critical analyses, questioning and revealing assumptions behind them and identifying gaps and weaknesses in the assessments. The final iterations of analysis involved aggregation and synthesis of these nodes into two forms: a framework of broad categories and relationships and a list of relevant variables. This was done for both systemic change drivers and adaptive governance features. The broad frameworks are more interpretive than descriptive, and the lists are more descriptive than interpretive, but both forms remain faithful to the assessment texts (Thomas and Harden 2008).

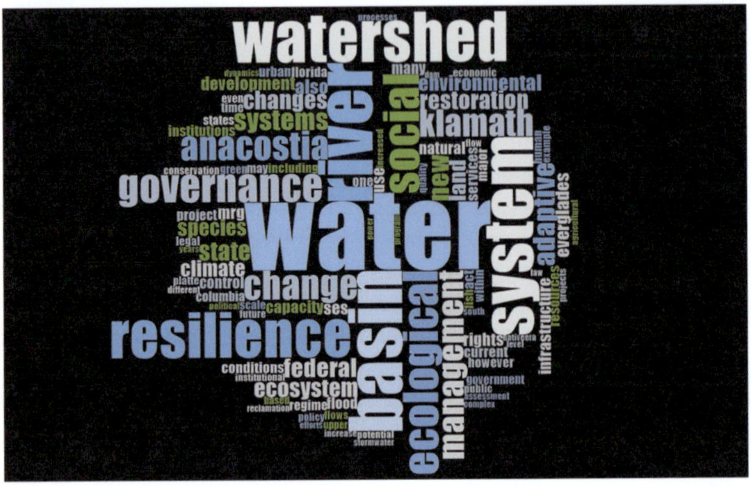

Fig. 13.2 Word cloud generated from exact word frequency analysis of the text from all basin assessments. The size of the word is proportional to the number of times the word was used in the text. For example, water was the most frequently used in all of the assessments

Fig. 13.3 Word cloud generated from frequency analysis of the concepts in the text from all basin assessments. The size of the word is proportional to the number of times the word was used in the text. For example, change was the most frequently used concept

13.5 Results of Textual Analyses

Word clouds (Figs. 13.2 and 13.3) graphically depict the words and concepts mostly commonly used in the case studies. The diagrams were generated using the NVivo software's word frequency function (Bazeley and Jackson 2013), and the size of the word or phrase in the cloud corresponds to frequency of use. The most frequently used exact words appearing in all assessments' text in aggregate all involved some type of system description (Fig. 13.2). Hydrologic system descriptors, such as basins, watersheds, and rivers, were the most frequently used words across all of the assessments, followed by phrases of ecological, social, and governance systems and system properties of resilience and adaptive capacity (Fig. 13.2). The most frequently used concepts and word stems appearing in all assessments' text in aggregate were change (Fig. 13.3), particularly the relationships among change, conditions, actions, and systemic states. The changes, conditions, actions, and states that were being studied occurred within ecosystems, governance organizations and institutions, and communities. The word clouds reveal general themes but not patterns of systemic change drivers and adaptive governance features; coding and synthesis were necessary to discover these patterns.

13.6 Drivers of Systemic Change

The results of textual analysis and synthesis of the cross-assessment drivers of systemic change are presented in both graphic (Fig. 13.4) and tabular form (Table 13.4). A lengthy, detailed list of specific change drivers within each category was

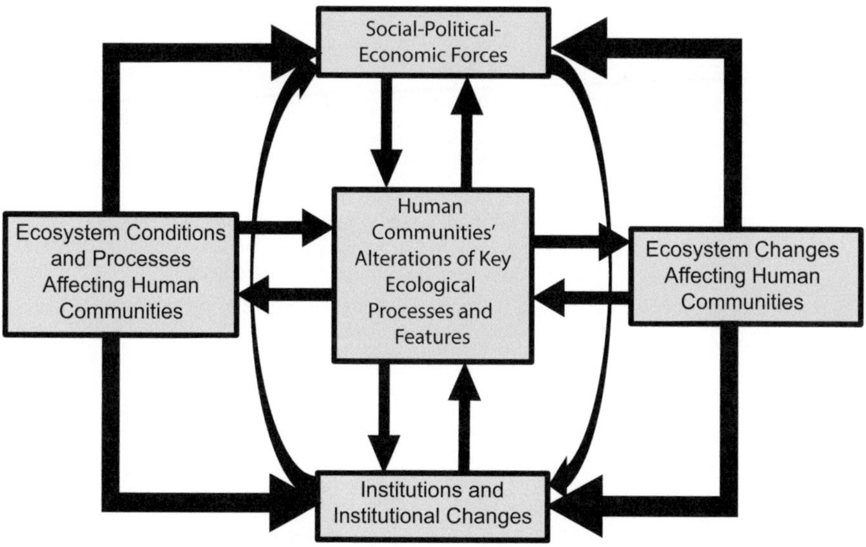

Fig. 13.4 Framework of drivers of systemic change. Diagram of systemic change drivers affecting one another across systems. Human communities' alterations of key ecological processes and features are the central drivers of change in water basins. They influence and are influenced by ecosystem conditions and processes, social-political-economic forces, evolving governance institutions, and ecosystem services feedbacks to society (Source: Arnold et al. 2017)

Table 13.4 Results of textural analyses, indicating drivers of change

Category	Drivers	
Ecosystem conditions and processes affecting human communities	Hydrological conditions and processes	General hydrological conditions and processes
		Drought (e.g., ➔ water storage, irrigation)
		Flooding (e.g., ➔ flood control projects)
		Groundwater levels and dynamics (e.g., ➔ dependence and uncertainty)
		Geological or topographical conditions and changes (e.g., ➔ settlement patterns)
		Fire (e.g., ➔ fire suppression)
		Extreme or traumatic events; disasters
		Ecosystem services and feedbacks (e.g., ➔ exploitation)
Human communities' alterations of key ecological features and processes	Land use activities	Agricultural activities
		Land use and development
	Alteration of hydrologic structure and processes	Engineered infrastructure and systems
		Altered stream and riparian structure(s)
		Altered stream flows
		Water demand and supply-demand mismatches
	Alteration of land-water nexus	Pollution
		Runoff
		Wetland alteration or loss
	Climate change	
	Conservation practices	Use or conservation of green infrastructure
		Land conservation practices
		Ecological restoration
Social-political-economic forces	Economic valuation and exploitation of nature	Treatment of environmental features and services as commodifiable resources for exploitation and consumption
		Forces of economic development and production
		Cross-scale economic forces
		Change in public values and norms over time and with respect to the psychology of human connection to place and framing/reframing of watersheds
		Technology and engineered infrastructure as tools for exploiting nature
		Business organization behavior
	Societal growth and development	Settlement, resettlement, and migration
		Urbanization
		Population growth
		Land use and development
		Technology

(continued)

Table 13.4 (continued)

Category	Drivers	
	Sociopolitical activity	Power including interest group power and conflict
		Public values and norms: economic, health, environmentalism, framing/reframing, change over time
		Social inequity: oppression and marginalization, indigenous peoples, racial and ethnic minorities, structural racism, socioeconomic class
		Mobilization and activism: community-based mobilization and activism, political mobilization and activism, environmentalism
		Sociopolitical change
	Social interaction dynamics	Conflict and collaboration
		Trust and distrust
		Culture and spirituality
Institutions and institutional changes	Institutionalization of social-political-economic forces	Examples: economic goals and values, oppression and marginalization (indigenous communities, racial and ethnic minorities), environmentalism
	Governance structure	Cross-scale governance
		Fragmentation of governance
		Mixed and hybrid governance institutions (e.g., formal-informal)
		Federal resource ownership and management
		State and local governments (polycentric structure)
		Government-created infrastructure
		Government-provided financial resources
		Planning
		Changes in governance agency mission
		Emergence of adaptive methods and tools (e.g., adaptive management)
	Legal activity and legal institutions	Legislation
		Regulation
		Enforcement
		Litigation
		Judicial decisions
		Law as disturbance (e.g., altering status quo)
		Law as facilitator (e.g., grant of authority)
		Law as resistance and entrenchment (e.g., rigid rules)
		Changes in legal institutions
	Institutional change	Emergence: gradual and organic development of new regimes
		Evolution: incremental modifications to existing regimes
		Hybridization: new regimes from synthesis of 2+ existing regimes
		Transformation: complete shift from one regime to another regime
		Revolution: sudden, radical change in regimes

(continued)

Table 13.4 (continued)

Category	Drivers	
Ecosystem changes affecting human communities: ecosystem services feedbacks	Extreme or traumatic events or disturbances	Drought
		Flooding
		Fire
	Decline or loss of ecosystem features on which humans depend	Species decline or loss
		Alteration or loss of wetlands
		Alteration of stream flows and water flows
	Major systemic changes	Altered hydrology
		Climate change
		Green infrastructure performance

developed from variables identified in at least five of the six basin assessments, similar to Ostrom's SES framework (Ostrom 2009).

The strong influence of ecosystem conditions, processes, and changes on entire system change was apparent in all of the case studies (Fig. 13.4). All of the assessments described presettlement and post-settlement ecosystem conditions and processes that strongly affected human efforts to inhabit, obtain food from, build economic activity within, relate to, and govern the basins. The ecosystem conditions and processes have affected how humans have altered the ecosystem, as well as the structure of society (social-political-economic forces) and institutions. As humans have altered ecosystems, the changes have had ecosystem services feedbacks to society, affecting human communities, as well as ecosystems' structure and function. These feedbacks affect social-political-economic forces and governance institutions, as well as the ways in which humans continue to alter ecosystems. For example, negative influences by humans on environmental systems have prompted efforts to restore watershed features and processes or manage the basin for ecological function, including changes in laws, governance goals, and public environmental values. The separate boxes for ecosystems in Fig. 13.4 reflect the assessments' substantial and distinct attention to both ecosystems' structures/functions and ecosystems' services to society.

Central to systemic change in North American water basins is alteration of key ecological features and processes by human communities. Humans have changed ecosystem structures, functions, and processes through land use activities, reengineering of hydrological structure and processes, alteration of the land-water nexus, climate change, and conservation practices (Table 13.4).

The social components of linked social-ecological-institutional systems can be decomposed into three parts: (1) human communities' alteration of key ecological features and processes, (2) social-political-economic forces, and (3) institutions and institutional changes. All basin assessments gave special attention to institutions and their dynamics, which are certainly part of society and interconnected with societal forces yet also analytically different than societal forces.

Four major categories of change or driving forces with intertwined social, political, and economic dimensions were identified (Table 13.4). The first category is economic valuation and exploitation of nature, especially societal and political treatment of environmental features and services as commodifiable resources for exploitation and consumption. The second category is societal growth and development, particularly settlement and migration trends, urbanization, population growth, land development, and technology. The third category is sociopolitical activity. This includes the exercise of and conflict over power, public values and norms, social inequity (especially the oppression and marginalization of indigenous peoples and racial and ethnic minorities), mobilization and activism, and sociopolitical change. The fourth category is social interaction dynamics. Conflict and collaboration, trust and distrust in communities, and cultural and spiritual beliefs and practices are aspects of social interaction dynamics that received substantial cross-basin analysis.

Institutions and institutional change received substantial attention as a major change driver category (Fig. 13.4). The effects of social-political-economic forces on the basins were strengthened through their institutionalization. For example, economic goals were incorporated into basin governance policies to support agriculture and development; forces of oppression and marginalization were institutionalized in slavery, segregation, and Native American conquest and displacement; and environmentalism became codified in environmental statutes and regulations. In addition, the structure of governance has influenced systemic change in the basins. Federal, state, local, and hybrid formal-informal governance entities, organized in fragmented, cross-scale governance systems, have influenced change, particularly by providing infrastructure, financial resources, and planning. The institutionalized missions of government agencies have changed over time, and adaptive methods and tools have emerged.

All six basin assessments gave much attention to legal activity and legal institutions, including the roles of legislation, regulation, litigation, enforcement, and judicial decision-making. Legal institutions have contributed to change by disturbing the status quo and granting legal authority to governance entities. They have resisted change through rigid rules and entrenchment of the status quo, eventually triggering systemic change when governance institutions proved too brittle and inflexible. Changes in legal institutions have contributed to broad systemic changes. For example, Congress adopted land disposal, river management, and water development laws in the nineteenth and early twentieth centuries that altered the Columbia, Klamath, and Platte basins, and new watershed governance institutions emerged in the Anacostia in the shadow of the Clean Water Act's evolving enforcement. Institutional change generally was a major theme in all basin assessments.

This analysis of the basin assessments also revealed the significant cross-category effects that social-political-economic forces, human alterations of ecosystems, and institutions have had on one another. These effects are cumulative, synergistic, and iterative. Cumulative effects result from the aggregation of multiple drivers. For example, in the Klamath basin, the combination of the basin's hydrological processes, emergent agricultural activities, economic forces, human settlement and population growth, legal regimes, governance institutions for irrigation-oriented basin management, and drought condition feedbacks to local communities and economies produced a reengineering and reallocation of the basin's hydrological system to support agriculture. Synergistic effects result from the dynamic interaction of multiple drivers that produce effects greater or different than the mere sum of the parts. In the Anacostia, for example, urbanization, land development, population growth, pollution generation, stormwater runoff, river flow conditions, deforestation, wetland loss, private property rights, and institutionalized economic development policies have intersected and reinforced one another to produce shallow, sluggish, sediment-filled, and polluted river conditions and the risk of hydrological collapse. Likewise, in the Middle Rio Grande basin, the intersecting effects of deforestation, fire suppression norms and practices, invasive bark beetle infestation, drought, climate change, and land use activity in the wildland-urban interface have changed forest and basin systems in ways that are more than merely cumulative, resulting in frequent, intense, and system-changing fires. Iterative effects occur when one set of changes produces another set of changes that leads to yet another set of changes, and so forth. In the Platte River basin, economic-driven alterations of basin hydrology had imperiled species and produced negative ecosystem services feedbacks by the early 1970s. Emergent environmentalist values, activism, and interest group power led to congressional enactment of the Endangered Species Act, which in turn begat a series of disruptive regulatory and enforcement actions over time. A dam relicensing requirement brought together Nebraska, Colorado, and Wyoming to negotiate a restoration and adaptive management agreement. A change in Nebraska law authorized integrated water management, in turn removing a major barrier to Nebraska's ability to enter into the tri-state agreement. From the agreement have emerged adaptive management methods, ecological restoration projects, land and water conservation, and new plans and policies for the basin. Through multiple iterations, the Platte River basin's governance system has changed.

According to this meta-analysis of cross-basin data, many relevant variables contribute to systemic change in water basins (Table 13.4). This surprising result is inconsistent with panarchy models that posit only a few major variables drive systemic change across adaptive cycles (Holling et al. 2002; Walker et al. 2012; Walker and Salt 2012).

There are four possible explanations for this numerosity. First, not all variables contribute to all changes. Each basin assessment analyzed several major systemic changes over time, ranging from 3 to 15 depending on how the team defined the system's core structure and the nature of change. Thus, some of the variables driving change toward environmental exploitation, basin reengineering, and minority group oppression were different than some of the variables driving change toward environmentalism, ecosystem restoration, and intergroup cooperation.

Second, most assessments did not clearly categorize the types of change-contributing variables that they described. Some of what has been included in the list may have been triggers or disturbances, mere contributors to change, or sub-drivers of change, instead of standing alone as major drivers per se. In addition, most assessments did not clearly classify drivers as slow variables or fast variables (Holling et al. 2002; Walker et al. 2012; Walker and Salt 2012) or differentiate between exogenous drivers and endogenous variables (Walker et al. 2012). The basin assessments consistently described many variables working together – cumulatively, synergistically, and iteratively – to cause systemic change. Some variables that may have started out slowly at large geographic scales seem to have ended up operating quickly at small geographic scales (e.g., urbanization in the Anacostia; climate change in the Middle Rio Grande). If researchers were to attempt to narrow the major drivers of change over the post-settlement history of a complex social-ecological-institutional basin system to a small number, they would either have created a list at the broad category level of generality reflected in Fig. 13.4 or have a list that ignored relevant variables and therefore sacrificed truth.

Third, social-ecological systems are complex. Water basins are dynamic social-ecological-institutional systems with many features and processes that are interconnected through many linear and nonlinear cross-system feedbacks that are changing over time. Lengthy lists may be needed to describe the many interacting variables that have caused changes in these complex systems over the historical periods studied by the basin assessment teams. In attempting to describe how complex social-environmental systems (SES) work and how institutions for SES governance emerge, Elinor Ostrom developed similarly long lists of relevant variables for analysis: 6 categories and 47 variables in her social-ecological systems (SES) framework and 8 design principles, 21 categories, and over 300 terms and concepts in her institutional analysis and development (IAD) framework (Ostrom 2009; McGinnis 2011). According to Sjostedt (2015), institutions in complex SES should be studied for both exogenous and endogenous drivers of change that do not have simple ecosystem or social system proxies.

Fourth, this study's lengthy list of change drivers is a starting point for further research on the variables affecting water basins and other SES. Now that cross-basin patterns have been identified, the list of relevant variables can and should be refined through further empirical studies of systemic changes, including both qualitative case studies and the development and testing of quantitative models using statistical analysis tools.

13.7 Features of Adaptive Governance

The analysis and synthesis of the six basin assessments also produced a relatively lengthy list of features of adaptive governance, which were categorized and summarized in Table 13.5. These features were identified in all or all but one of the assessments as characteristics that made the basin governance systems adaptive or would be needed in order to make the basin governance systems adaptive. They fall into four major categories: the goals, structure, methods, and resources of the governance system.

Table 13.5 Features of adaptive governance derived from the textual analysis of the case studies

Goals	Structure	Methods	Resources
Poly-resilience social goals	Polycentric	Flexibility	Sociopolitical capacity
	Networked	Learning	Social change
Ecosystems	Scaled	Resource management for ecosystem functions and processes	Government resources
Social systems	Modular		Use of law and institutions for adaptation and transformation
Institutions	Participatory		
Changing public values	Legitimate	Integrated problem-solving	
		Accountability	

The governance system's goals affect its adaptive character. According to the basin assessments, adaptive governance aims to enhance the resilience of desired ecosystems, social systems, and institutional systems, not just the resilience of one system or subset of systems. This is known as poly-resilience (Arnold and Gunderson 2013). In contrast, water governance in each basin has, at times, maladaptively made the basin more vulnerable to undesirable transformation by favoring the resilience of some systems, such as exploitative economies or property rights institutions, over the resilience of other systems, such as natural flow watersheds or marginalized human communities. Given the strong influence of public values and norms on governance decisions, social norms must evolve so that the public values poly-resilience as a societal goal if governance is to become adaptive. Resource problems must be reframed so that multiple goals match the multidimensional nature of the problems. Poly-resilience also requires that resources be managed for ecosystem functions and processes in integrated ways.

The basin assessments identified several features of governance structure that facilitate adaptive governance. Adaptive governance is polycentric and modular, yet connected through multi-stakeholder and multi-government networks. Adaptive governance has nested scales of governance that are matched to the problem scale, the government system's capacity, ecological scale (i.e., the basin), and the smallest appropriate level for the function (i.e., subsidiarity). The public and multiple stakeholders participate meaningfully in water governance and accept the legitimacy of the governance system, because they have robust opportunities for participation and deliberation.

Adaptive governance uses a variety of methods for adaptively managing complex social-ecological-institutional systems, such as multi-scenario planning, ecological restoration, green infrastructure, land use management, adaptive ecosystem management, and many others. The overarching themes of adaptive governance methods are flexibility, learning, resource management for ecosystem functions and processes, integrated problem-solving, and accountability. In particular, flexibility is embedded in governance, and adaptive capacity is developed. Both expert/scientific and public/social learning occur through experimentation, monitoring, feedback loops, and adaptive management. Ecosystem functions and processes are served by both conservation of key ecosystem features and governance or management actions that deliberately transform the ecosystem toward a desirable state.

These ecosystem functions and processes must be assessed regularly for resilience in the context of the basin's interlinked systems. Integrated problem-solving is also a method used in adaptive governance. Furthermore, accountability is a critical feature, achieved through the use of identified performance measures, regulations and laws, and cross-system feedbacks, which signal which governance decisions or societal behaviors undermine the resilience and functioning of the basin.

Finally, resources are essential to adaptive governance. One set of resources centers on sociopolitical capacity, including social capital (especially cooperation and trust), communication, social and institutional innovation, and the power to decide and act. A second set of resources is about social change, including social mobilization, political action, evolving public values, and government facilitation of social change. Law and institutions form a third set of resources for adaptive governance. Law serves as a disturbance that stimulates adaptation and/or transformation, provides the authority to act, and is a tool to achieve accountability or transformation. Legal flexibility, reform, and change are common features of adaptive governance systems. Institutional changes trigger other institutional changes and build adaptive capacity in basin systems.

This synthesis framework of cross-assessment adaptive governance features overlaps considerably with, yet differs somewhat from, conceptualizations by Walker and Salt (2012), Chaffin et al. (2014b), or Cosens et al. (2014b). Differences between this framework and other frameworks appear in both their conceptual structures and their specific features. The advantages of this study's synthesis framework are the following: (a) it is based on broadly cross-interdisciplinary patterns that pervade assessments of water governance in six diverse basins; and (b) it is built from coding, analysis, and synthesis of the language and ideas of the assessment texts themselves through an inductive coding process, instead of prior syntheses based on both empirical studies and theory about the nature of adaptive governance.

13.8 Conclusions

This study has resulted in cross-interdisciplinary knowledge about systemic change drivers and adaptive governance features from syntheses of patterns in the texts of six diverse interdisciplinary assessments of basin resilience. Four key insights from this study should inform research on complex social-ecological-institutional systems.

First, the alteration of key ecological features and processes by human communities is the central driver of system change, affecting the structures and functions of ecosystems, social systems, and institutions. The adaptive governance of water basins must pay particular attention to land use activities (including both land development and agricultural activities), the alteration of hydrologic structures and processes, the alteration of the land-water nexus, climate change, and emergent conservation and restoration practices. This is consistent with much of the social-ecological resilience literature (Gunderson and Holling 2002; Light et al. 1995; Olsson et al. 2006). It cautions against too much emphasis on abstract principles of adaptive governance that are disconnected from the core problem of human alteration of key ecological features and processes in social-ecological-institutional systems. Furthermore, this study shows that the central focus on human change to ecosystems has increasingly come to include restoration of ecological functions and features in water basins and management for natural processes, such as flows, biotic infrastructure, forest succession, and fire regimes.

Second, more research should explore the many different drivers of change in complex social-ecological-institutional systems, instead of seeking to identify a few major drivers of change as is often urged in the literature (Holling et al. 2002; Walker et al. 2012; Walker and Salt 2012). The "few drivers" imperative may be more theoretically driven than empirically supported, as suggested by the very large number of variables in Table 13.4, which were identified as drivers of change in at least five of the six basin assessments. For example, where the broad category of societal growth and development drove systemic change, it often involved settlement and migration, urbanization, population growth, land use and development, and technology. Likewise, ecosystem feedbacks to human communities that drove changes in policies and actions included drought, flooding, fire, species decline or loss, wetland loss, altered hydrology, and climate change, among others. Legal system processes are many and diverse and can serve as change-causing disturbances, facilitators of transformation, or resistance to change. At the very least, the Ostrom-style table of relevant variables (Table 13.4) should serve as a reference for the study of system change drivers in water basins and encourage others to gather thorough cross-basin lists of relevant variables before determining which are the most significant.

Third, the six basin assessments showed that institutions matter and that the term "social-ecological-institutional systems" is often more appropriate than merely "social-ecological systems." Institutions are systems that are distinct from the social, political, and economic forces and systems that affect and are affected by institutions and ecosystems. This meta-analysis study supports the increasingly intentional analysis of institutions in complex system dynamics and adaptive governance (Cosens et al. 2014a; Garmestani and Allen 2014; Sjostedt 2015). However, it also encourages researchers to study institutions, social systems, and ecosystems as cross-dynamic, each continuously changing and affecting each other across time. Even though resilience theory is based on nonlinear cross-system dynamics (Gunderson and Holling 2002), many social-ecological-institutional systems tend to be described in linear cause-and-effect ways, such as when economic and political interests demand flood control and stable water supplies, which leads to institutionalization of major dam-building on rivers, which in turn causes a decline in natural flow regimes and aquatic biodiversity (Birge et al. 2014). This historical narrative is practically unavoidable, yet fails to represent accurately the ongoing instabilities and evolutions of each (sub)system throughout the basin's transformation. Both the ISED framework (Fig. 13.1) and the framework of drivers of systemic change (Fig. 13.4) help to focus systemic analysis on the continuous evolutions and transformations in systems and the effects of these changes on other systems. Moreover, the emergence of the institutional and social system characteristics to support adaptive governance can serve as drivers of change, and drivers of change can form the features of adaptive governance (Olsson et al. 2006). For example, in the six basin assessments, policy innovation, socio-cognitive reframing processes, multi-stakeholder networks, and improved social capital, among others, were both drivers of change and features of adaptive governance.

Fourth, despite the concern of much of the adaptive governance literature with facilitating adaptive management of complex systems (Chaffin et al. 2014b), this meta-analysis of six water basin resilience assessments suggests that more attention should be given to the role of adaptive governance in facilitating social change and changes in social values. More change to basin functions occurred when public values and sociopolitical forces changed than when administrative methods or scientific approaches to basin management changed. For example, public attitudes and social forces for both the consumptive exploitation of natural systems and the marginalization and oppression of minority groups strongly shaped the transformation of the river basins' environmental conditions, societal functions, and governance institutions. Environmentalism, the activism of historically marginalized groups, socio-cognitive reframing processes, and shifts from conflict to collaboration and distrust to trust have stimulated efforts at ecosystem restoration in many basins, as well as moves toward adaptive governance systems with sufficient sociopolitical capacity to support social, ecological, and institutional transformations. The lack of explicit attention to social justice and activism in many studies of adaptive governance is troubling in the light of critics' charges that resilience scholarship fails to address structural inequities in society and the power arrangements behind social-ecological-institutional systems (Cote and Nightingale 2012; MacKinnon and Derickson 2013). Resilience analysis can be used to expose structural inequities in the social benefits and vulnerabilities of ecosystems and the roles of social, political, and economic forces that oppress marginalized groups (Langridge et al. 2006; Ernstson 2013; Walsh-Dilley et al. 2013; Arnold et al. 2014). More systematic and explicit attention to social justice, equitable conditions, and community-based activism as features of adaptive governance systems is needed in collaborative interdisciplinary research on social-ecological-institutional resilience.

Acknowledgments This work was developed in part under the Adaptive Water Governance project, funded by the US National Socio-Environmental Synthesis Center (SESYNC) under funding from the US National Science Foundation, NSF DBI-1052875.

References

Arnold CA (2004) Working out an environmental ethic: anniversary lessons from Mono Lake. Wyoming Law Rev 4:1–55

Arnold CA, Gunderson LH (2013) Adaptive law and resilience. Environ Law Reporter 43:10426–10443

Arnold CA, Green OO, DeCaro DA, Chase A, Ewa JG (2014) The social-ecological resilience of an eastern urban-suburban watershed: the Anacostia River Basin Idaho Law Rev 51(1):29–90
Arnold CA, Gosnell H, Benson MH, Craig RK (2017) Cross-interdisciplinary insights into adaptive governance and resilience. Ecol Soc 22(4):14. https://doi.org/10.5751/ES-09734-220414
Bazeley P, Jackson K (2013) Qualitative data analysis with NVIVO, 2nd edn. Sage, Los Angeles
Benson MH, Llewellyn D, Morrison R, Stone M (2014) Water governance challenges in New Mexico's Middle Rio Grande Valley: a resilience assessment Idaho Law Rev 51(1):195–228
Bettez SC (2015) Navigating the complexity of qualitative research in postmodern contexts: assemblage, critical reflexivity, and communion as guides. Int J Qual Stud Educ 28(8):932–954
Birge HE, Allen CR, Craig RK, Garmestani AS, Hamm JA, Babbitt C, Nemec K, Schlager E (2014) Social-ecological resilience and law in the Platte River Basin. Idaho Law Rev 51(1):229–256
Bringer JD, Johnston LH, Brackenridge CH (2004) Maximizing transparency in a doctoral thesis: the complexities of writing about the use of QSR NVIVO within a grounded theory study. Qual Res 4(2):247–265
Chaffin BC, Craig RK, Gosnell H (2014a) Resilience, adaptation, and transformation in the Klamath River Basin social-ecological system. Idaho Law Rev 51(1):157–193
Chaffin BC, Gosnell H, Cosens BA (2014b) A decade of adaptive governance: synthesis and future directions. Ecol Soc 19(3):56
Corbin J, Strauss A (1990) Grounded theory research: procedures, canons, and evaluative criteria. Qual Sociol 12:3–21
Cosens BA, Fremier A (2014) Assessing system resilience and ecosystem services in large river basins: a case study of the Columbia River Basin. Idaho Law Rev 51(1):91–125
Cosens BA, Gunderson LH, Allen CR, Benson MH (2014a) Identifying legal, ecological and governance obstacles, and opportunities for adapting to climate change. Sustainability 6:2338–2356
Cosens BA, Gunderson LH, Chaffin BC (2014b) The adaptive water governance project: assessing law, resilience and governance in regional socio-ecological water systems facing climate change. Idaho Law Rev 51(1):1–27
Cote M, Nightingale AJ (2012) Resilience thinking meets social theory: situating social change in socio-ecological systems (SES) research. Prog Hum Geogr 36(4):475–489
DeWulf A, Francois G, Pahl-Wostl C, Tailieu T (2007) A framing approach to cross-disciplinary research collaboration: experiences from a largescale research project on adaptive water management. Ecol Soc 12(2):14
Eigenbrode SD, O'Rourke M, Wulfhorst JD, Althoff DM, Goldberg CS, Merrill K, Morse W, Nielsen-Pincus M, Stephens J, Winowiecki L, Bosque-Perez NA (2007) Employing philosophical dialogue in collaborative science. Bioscience 57(1):55–64
Eisenhardt KM, Graebner ME (2007) Theory building from cases: opportunities and challenges. Acad Manag J 50(1):25–32
Ernstson H (2013) The social production of ecosystem services: a framework for studying environmental justice and ecological complexity in urbanized landscapes. Landsc Urban Plan 109:7–17
Fairclough N (2005) Critical discourse analysis in transdisciplinary research. In: Wodak R (ed) New agenda in (critical) discourse analysis: theory, methodology and interdisciplinarity. John Benjamins, Philadelphia, pp 53–70
Fry GLA (2001) Multifunctional landscapes – towards transdisciplinary research. Landsc Urban Plan 57:159–168
Garmestani AS, Allen CR (eds) (2014) Social-ecological resilience and law. Columbia University Press, New York
Gray B (2008) Enhancing transdisciplinary research through collaborative leadership. Am J Prev Med 35((2):S124–S132
Grubert E, Siders A (2016) Benefits and applications of interdisciplinary digital tools for environmental meta-reviews and analyses. Environ Res Lett 11(9):093001
Gunderson LH, Holling CS (eds) (2002) Panarchy, understanding transformations in human and natural systems. Island Press, Washington, DC

Gunderson LH, Garmestani A, Rizzardi KW, Ruhl JB, Light A (2014) Escaping a rigidity trap: governance and adaptive capacity to climate change in the everglades social ecological system Idaho Law Rev 51(1):127–156

Heemskerk M, Wilson K, Pavao-Zuckerman M (2003) Conceptual models as tools for communication across disciplines. Conserv Ecol 7(3):8

Holling CS, Gunderson LH, Peterson GD (2002) Sustainability and panarchies. In: Gunderson LH, Holling CS (eds) Panarchy, understanding transformations in human and natural systems. Island Press, Washington, DC, pp 63–102

Langridge R, Christian-Smith J, Lohse KA (2006) Access and resilience: analyzing the construction of social resilience to the threat of water security. Ecol Soc 11(2):18

Light SS, Gunderson LH, Holling CS (1995) The Everglades: evolution of management in a turbulent system. In: Gunderson LH, Holling CS, Light SS (eds) Barriers and bridges to the renewal of ecosystems and institutions. Columbia University Press, New York, pp 103–168

MacKinnon D, Derickson KD (2013) From resilience to resourcefulness: a critique of resilience policy and activism. Prog Hum Geogr 37:253–270

McGinnis MD (2011) An introduction to IAD and the language of the Ostrom Workshop: a simple guide to a complex framework for the analysis of institutions and their development. Policy Stud J 39:169–183

Nemec KT, Chan J, Hoffman C, Spanbauer TL, Hamm JA, Allen CR, Hefley T, Pan D, Shrestha P (2014) Assessing resilience in stressed watersheds. Ecol Soc 19 (1)

Noblit GW, Hare RD (1988) Meta-ethnography: synthesizing qualitative studies. Sage, Newbury Park

O'Rourke M, Crowley SJ (2013) Philosophical intervention and cross-disciplinary science: the story of the Toolbox Project. Synthese 190(11):1937–1954

Olsson P, Folke C, Hahn T (2004) Social-ecological transformation for ecosystem management: the development of adaptive co-management of a wetland landscape in southern Sweden. Ecol Soc 9(4):2

Olsson P, Gunderson LH, Carpenter SR, Ryan P, Lebel L, Folke C, Holling CS (2006) Shooting the rapids: navigating transitions to adaptive governance of social-ecological systems. Ecol Soc 11(1):18

Ostrom E (2009) A general framework for analyzing sustainability of social-ecological systems. Science 325:419–422

Pohl C (2008) From science to policy through transdisciplinary research. Environ Sci Policy 11(1):46–53

Scheffer M, Westley F, Brock WA, Holmgren M (2002) Dynamic interaction of societies and ecosystems – linking theories from ecology, economy, and sociology. In: Gunderson LH, Holling CS (eds) Panarchy, understanding transformations in human and natural systems. Island Press, Washington, DC, pp 195–239

Shaffer DW, Hatfield D, Svarovsky GN, Nash P, Nulty A, Bagley E, Frank K, Rupp AA, Mislevy R (2009) Epistemic network analysis: a prototype for 21st century learning. Int J Learn Media 1(2):1–21

Sjostedt M (2015) Resilience revisited: taking institutional theory seriously. Ecol Soc 20(4):23

Stokols D, Harvey R, Gess J, Fuqua J, Phillips K (2005) In vivo studies of transdisciplinary scientific collaboration. Am J Prev Med 28(2S2):202–213

Strauss AL (1987) Qualitative analysis for social scientists. Cambridge University Press, Cambridge

Thomas J, Harden A (2008) Methods for the thematic synthesis of qualitative research in systematic reviews. BMC Med Res Methodol 8:45

Walker B, Salt D (2012) Resilience practice: building capacity to absorb disturbance and maintain function. Island Press, Washington, DC

Walker BH, Carpenter SR, Rockstrom J, Crepin AS, Peterson GD (2012) Drivers, "slow" variables, "fast" variables, shocks, and resilience. Ecol Soc 17(3):30

Walsh-Dilley M, Wolford W, McCarthy J (2013) Rights for resilience: bringing power, rights and agency into the resilience framework. Oxfam America, Washington, DC

Chapter 14
Trajectories of Change in Regional-Scale Social-Ecological Water Systems

Lance Gunderson, Barbara Cosens, and Brian C. Chaffin

Abstract The six North American case studies described in Part I of this volume provide the opportunity to explore patterns of change over time associated with the development and management of social-ecological systems. The historical assessments show the trajectories that have led to the current heavily regulated and developed social-ecological systems nested within a hierarchical governmental system. This chapter uses Panarchy theory as a general framework to evaluate the interactions between societal and ecological regime shifts and the governance regimes that mediate those interactions. The resilience assessments indicate that complex interactions among the governance and ecosystem components of these systems can produce different trajectories, which include patterns of (a) development and stabilization, (b) cycles of crisis and recovery which include lurches in adaptation and learning, and (c) periods of innovation, novelty, and transformation. Exploration of cross-scale interactions (Panarchy) among levels and sectors of government and society reveals that larger-scale processes and structures may constrain development and growth, but may also provide resources for recovery and renewal following crises or create windows or opportunities for system change. Smaller-scale processes provide opportunities for innovation and novelty, but may also be the source of revolts or crises that lead to broader system transformations. The case studies illustrate different ways that adaptive governance may be triggered, facilitated, or constrained by ecological and social (and particularly legal) processes.

Keywords Adaptive governance · Cross scale dynamics · Social ecological system · Transformation · Panarchy

L. Gunderson (✉)
Department of Environmental Sciences, Emory University, Atlanta, GA, USA
e-mail: lgunder@emory.edu

B. Cosens
College of Law and Waters of the West Program, University of Idaho, Moscow, ID, USA
e-mail: bcosens@uidaho.edu

B. C. Chaffin
W. A. Frank College of Forestry and Conservation, University of Montana, Missoula, MT, USA
e-mail: brian.chaffin@mso.umt.edu

B. Cosens, L. Gunderson (eds.), *Practical Panarchy for Adaptive Water Governance*, https://doi.org/10.1007/978-3-319-72472-0_14

14.1 Introduction

Water management in the United States and Australia is largely centered on hydrologic units, particularly the drainage basin. The cases in this volume (Chaps. 3, 4, 5, 6, 7, 8 and 9) discuss how water is managed in basins that range in size from a few hundred square kilometers (the Anacostia River basin – Chap. 3) to ones that cover and drain areas over one million square kilometers (the Lake Eyre basin – Chap. 9). These hydrologic basins sit in climatic zones that vary from the wet to dry, hot to cold, with different resultant timings and volumes of water flows (Chap. 2). Prior to intensive development, these systems were dynamic sets of ecosystems – riverine, riparian, wetland, aquatic, and terrestrial that supported a diversity of life, including humans.

In order to meet expanding human populations, the twentieth century witnessed an acceleration in the technology and resources used to exert increasing control over these dynamic ecosystems. Stone and concrete dams replaced wooden ones built by beavers. Levees were constructed to constrain flood effects and provide water and energy for human activity, replacing levees naturally created by sediment transportation. Channelization and other constructs allowed for the movement of water to meet social demands for navigation, agriculture, urban development, and economic growth. Land use changes in the drainage basins resulted in shifts in water quantity and quality, which in turn has altered ecosystem structures and functions. In short, development of water resources has led to ecosystems that are currently highly controlled and managed to meet specific social goals. Management of these systems has largely centered on controlling and stabilizing key biophysical processes to achieve multiple social objectives. Such social objectives or expectations are often defined in legal terms, but also play key roles in ethical, religious, aesthetic, and economic frameworks. The success in meeting management objectives has led to dramatic shifts in the social and ecological components of these systems, as discussed in the next section.

14.2 Assessing Resilience in the Case Studies

A key component of resilience as introduced by Holling (1973) is the proposition that ecosystems can exist in specific configurations or regimes. Since that time, ecological studies have documented and described the existence of alternative ecosystem states or regimes (Gunderson and Pritchard 2002; Folke et al. 2004). Examples of such regime shifts from the case studies include landscape level changes in the Everglades from sawgrass marshes to cattail marshes (Gunderson and Pritchard 2002; Chap. 5), changes in riparian and upland vegetation along the middle Rio Grande river (Benson et al. 2014; Chap. 7), the Platte River (Birge et al. 2014; Chap. 8), Columbia River (Cosens and Fremier 2014; Chap. 3), and Klamath River (Chaffin et al. 2014a; Chap. 6). Another shift in ecosystem form and function occurred in those systems where large-scale dams were built. Dams have altered the movement of water, sediments, and migratory fish and contributed to the endangerment of aquatic species in the western rivers (Klamath, Columbia, Platte, and Rio Grande). Dams can also change the temperature and flow regimes in the river system, changing sediment movement and deposition critical to biotic habitats (Platte River, Chap. 8).

Alternative ecological regimes can provide different sets of ecosystem goods and services. For example, undammed or uncontrolled rivers provide many provisioning, regulatory, aesthetic, and supporting services, such as flood abatement, nutrient and sediment transport, food production, and recreational experiences (Gunderson et al. 2016). Damming of rivers has often switched the bundles of ecosystem services by trading off services such as aesthetic and biodiversity supporting services, in order to provide dependable water and energy supplies as well as reduce vulnerability to flood risk. How to reconcile such trade-offs is an ongoing issue for management and governance (Chelleri et al. 2015; Craig et al. 2017). Since prior management activities have tended to reduce ecological resilience (Holling and Meffe 1996), scenarios of climate change and other global drivers indicate that ecological regime shifts that can impact and alter the bundle of ecosystem goods and services are more likely to occur in the future.

Resilience is the systemic property that mediates transitions among or between alternative or alternating ecosystem configurations (Holling 1973), while adaptive governance is the societal process that mediates transitions among alternative societal configurations. As such, there is a distinction between system state and resilience. Each of the ecological states or regimes described in the previous paragraphs is characterized by a different set of self-organized processes that generate structural features of that particular ecological state (Folke et al. 2004; Walker and Salt 2006; Scheffer et al. 2001). Unlike the concept of sustainability, resilience is nonnormative or value neutral. The social components of coupled systems do make value judgments about alternative states; sawgrass marshes in the Everglades are desired over cattail marshes. As discussed elsewhere (Dietz et al. 2003; DeCaro et al. 2017), much of adaptive governance is about contrasting and weighing alternative values of different ecological states. A system can be highly resilient either because it is quite adaptable or quite resistant to change (Walker et al. 2004). An overgrazed field taken over by invasive weeds may be resistant to returning to its original state if livestock are removed – the current regime is therefore resilient but not necessarily desirable. A brutal military dictatorship may be highly resistant to change – also resilient but not necessarily desirable. Hence, it is important to distinguish between resilience, regime shifts, and the human values placed on a system state.

A second bridging concept from the resilience literature, Panarchy, provides a dynamic cross-scale lens through which both social-ecological systems and their systems of governance can be viewed (Gunderson and Holling 2002; Benson and Garmestani 2011; Garmestani and Allen 2014; Chaffin and Gunderson 2016). Panarchy describes an interconnected hierarchy of systems, with systems at each level defined by spatial and temporal organizational patterns. While spatial scales remain constant for systems within the Panarchy, the systems exhibit temporal

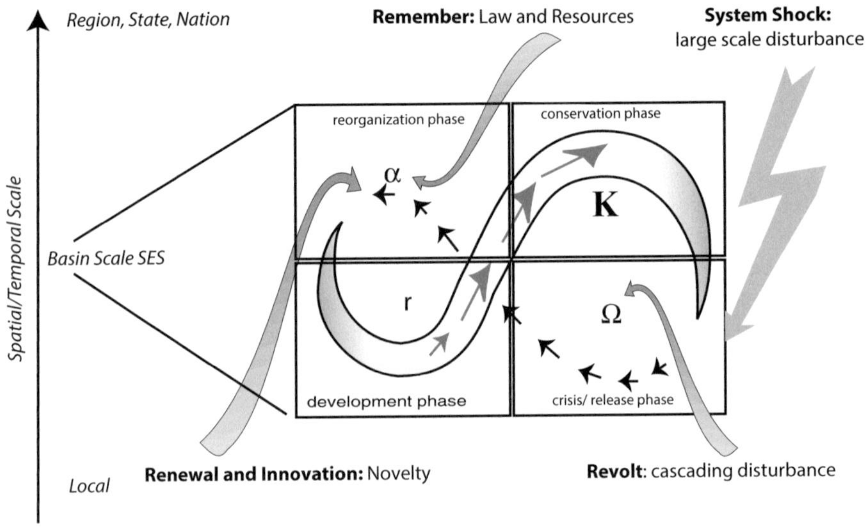

Fig. 14.1 Panarchy theory emphasizes key features of change observed in the basin scale social-ecological systems described in Chaps. 3, 4, 5, 6, 7, 8, and 9. All of the systems have undergone cycles of growth, development, crisis, and reorganization in their history, as represented by the four phases of system dynamics: growth, conservation, crisis/release, and reorganization (Holling 2001). External influences can influence the system from larger and smaller scales. Large-scale disturbances such as floods and droughts can trigger a crisis followed by a reorganization in which broader-scale social factors such as laws and fiscal resources or ecological factors such as biodiversity or climate can lead to recovery of the system to a pre-disturbance status. The Everglades SES has remained in the same configuration for decades, as legal actions and large infusions of resources have suppressed revolts and innovation, indicative of a rigidity trap. SES transformation has been triggered by instabilities either from cascades of small-scale revolts (such as lawsuits) or system shocks followed by reorganization phase. The transformation occurs when small-scale efforts interject novel solutions to old problems, or innovation in the form of new technology or institutions restructure the system into a new development trajectory

changes, going through stages of growth, collapse, innovation, and reorganization (Fig. 14.1). Panarchy expands the concept of resilience in a number of ways. One is that ecological resilience of a system is defined at a particular level within the Panarchy. For that particular or focal scale, system resilience is a dynamic property that waxes and wanes over time. The dynamics of resilience are mediated by the interaction between larger (slower)- and smaller (faster)-scale processes. The cross-scale interactions appear to play a role in system transformation, as focal systems morph into new regimes in both ecological and social system configurations. The next section discusses these distinctions in the context of several aspects of law, management, and environmental governance with respect to the case studies and different types of system trajectories.

14.3 Trajectories of Change

At least three trajectories or patterns of change were gleaned from the historical resilience assessments of the seven case studies. One is a growth and development trajectory, in which increasing control over the water resources is a primary management goal. Water resource infrastructure and institutions grow, as do the laws and rules that govern the developing SES. Through water control, economic development and human population growth are both fostered. The second caricature is a trajectory characterized by cycles, in which periods of growth and development were interrupted by periods of crises, destruction, or instabilities and quickly followed by reorganization, recovery, and renewal. These cycles can occur because of many mechanisms, including declining ecological resilience and shifting social values (Gunderson et al. 1995), and provide opportunities for adaptation. The third trajectory of change can involve transformative change in many characteristics that define social-ecological systems. That is, new ecosystem configurations, new institutional arrangements, and new physical infrastructure all interact to create a new system identity (Gunderson and Holling 2002). Each of these trajectories can be abstracted from the case studies, as described in the following sections.

14.3.1 Growth and Development

A common trajectory can be described as a growth and development pathway, in which infrastructure is built and operated to achieve particular societal goals (Holling and Meffe 1996). In the water basin case studies, these pathways involved the construction of dams, levees, and canals to control and constrain water movement to meet social goals of flood control and water supply. During the periods of growth and development, many formal governance structures were devised to oversee construction and implementation of infrastructure. Also, multiple authorities for resource allocation were specified. Much of the governmental aspect of governance needed during these periods focuses on coordination among redundant, overlapping management loci, multiple nodes of decision-making, and rules for participation by stakeholders (Table 14.1). Among the case studies, both the small and the mighty rivers were tightly controlled and regulated during these eras of development. As a result, the social objectives of flood control and water diversion for human use were achieved. During these phases, governance becomes focused on efficiency and cost control, and economies become dependent on continued growth (Chaps. 4, 5 and 6).

Development and growth in all of the North American case studies relied on resources and capacity building from the federal level, as well as management that is at the same time redundant, overlapping, and contested among federal, state, and local levels. As water management systems developed over time, policies and actions have been largely successful in meeting social objectives. This is a period or time of formal structures of governance or institutionalization of social objectives in

Table 14.1 Relationships among components of the role of law in adaptive governance and Panarchy theory, emphasizing how structures, capacities, and processes of government need to vary with phases of system development and cross-scale interactions

Component	Facet	Development/implementation phases	Instability/reorganization phases	Panarchy/cross-scale considerations
Structure	Polycentricity	Multiple centers of authority	Connections across loci activate to respond to ecological surprise	Adaptive governance provides bridge across multiple loci of government
	Redundancy	Overlapping management and multiple decision-making functions	Increases capacity for unexpected ecological dynamics	Within and cross-scale functional checks and balance
	Complementarity	Multiple arenas for decisions		
	Subsidiarity	Authority at scale of resource issue		Resources/stability from larger scales local levels innovate
	Integration			Across scales and between sectors
	Persistence	Formal networks established	Informal networks emerge/disappear	Cross-scale networks are available to respond at the scale of the problem rather than jurisdictional scale
Capacity	Adaptive	Adaptive management	Adaptive planning, adaptive assessment	Provide resources/capital for responding to change
	Participatory	Determination of who participates	Question of new participants	Rules and resources for participation
Process	Legitimacy	Authority for exercise and perception	Provides opening for reestablishment of or new legitimacy	Modes of decision-making –science, accountability, transparency
	Procedural justice	Maintains social stability	Maintains trust and prevents corruption when responding to surprise	Higher levels provide forums to prevent local marginalization of minority or disenfranchised groups
	Problem solving	Allows accumulation of knowledge about system response	Problem reframed the face of uncertainty	Scale matching – Networked governance and problem
	Reflection/learning	Policy as hypothesis, single-loop learning	Multiple hypotheses, double-loop learning	Memory and wisdom
	Balance between stability and flexibility	Resources from growth are used to facilitate local capacity building	Increased flexibility at the scale of the change	Higher levels provide stability while local levels innovate

law, government, or social norms (Chaffin and Gunderson 2016). But it is also a period in which the growth and stability of higher levels of government might have facilitated preparation and development of tools to navigate change. Among these are cross-scale and cross-sector networks and the use of resources to build local capacity as well as to reengineer local water infrastructure to provide space for adaptation.

One key role of law is that it attempts to implement a collective value for a given ecosystem regime or state (Cosens et al. 2014, 2017). It does so by protecting the characteristics or identity of the regime but also by establishing rules associated with thresholds that separate a defined or desired state from other regimes. The basin case studies in this volume highlight several examples of this. The establishment of water quality standards under the Clean Water Act in the United States sets the bounds for clean water for purposes of both human health and ecosystem function. Legislation passed in the Everglades following the lawsuit of 1988 (John 1994), in which the federal government sued the state government (Chap. 5), established water quality standards for phosphorus discharge. These standards were set to prevent ecosystem flips from sawgrass to cattail marshes. In all of the US basins, various plants and animals are listed and protected by the Endangered Species Act (ESA). Federal listing defines an SES regime for that population of a species. Once a species is listed, it is subject to legal protections intended to prevent the population from crossing a threshold toward extinction. In many of the basins, some sort of legal instrument exists that defines a minimum threshold for water allocation. For example, a federal law (PL 91-181) defines the minimum water delivery requirements to Everglades National Park (Chap. 5).

In the case studies, science and law have combined to identify these critical thresholds. While the thresholds have been identified, there remains large scientific uncertainties due to the complexity of these systems, partial understanding of system dynamics, and ability to predict threshold transgression (Gunderson and Pritchard 2002). Furthermore, the actual threshold in question is likely to change over time and in response to other changes in the system. The process is further complicated by the fact that setting the (often arbitrary) bounds of a system and thresholds implicates the interests of the basin's society including the water-based economic system. Once set, current management assumes a linear path from the current state of the system to crossing a threshold toward a new state.

The examples from the case studies illustrate that not only is the system trajectory likely to be nonlinear, but that once a threshold is crossed, it will be much more difficult to return the system to its prior configuration, making the common societal objective of restoration that much more challenging (Chaps. 4, 5 and 8). In a system in the growth trajectory that is in a state desired by society, with that preference articulated and maintained by laws, social norms, or institutions, management actions tend to focus on understanding the latitude of the system for absorbing disturbance and understanding ecosystem thresholds (Gunderson et al. 1995). For example, water quality standards in the Anacostia River (Chap. 3) or the Everglades (Chap. 5) focus on nutrient thresholds in light of the types of disturbances such as droughts or floods in exceeding the thresholds. Adaptive management was highlighted as a way to monitor and detect dynamic thresholds, but has had limited application or success in these case studies, perhaps Platte (Chap. 8) River system. Legal authority for implementation of adaptive management may theoretically help threshold detection, yet institutional fragmentation, power asymmetries, and lack of institutional knowledge and learning present barriers to its implementation (Lee 1993; LoSchiavo et al. 2013; Craig and Ruhl 2014). In the complex water management systems, adaptive management alone is generally insufficient without the mechanisms for adaptive governance (Gunderson and Light 2006; Chaffin et al. 2014b; Chaffin and Gunderson 2016; Chaffin et al. 2016).

Perhaps one unintended consequence of a management trajectory focused on achieving growth and development has been the emergence of a perverse system resilience (Zellmer and Gunderson 2009). Control and resources from larger scales appear to have multiple influences and may constrain subsequent system trajectories in ways that have been described as maladaptive or as rigidity traps (Holling 2001). Case study regions, such as the Everglades Social Ecological System (SES), appear to be in a rigidity trap and are quite resilient to change (Gunderson and Light 2006; Zellmer and Gunderson 2009). Trapped systems have high institutional diversity (numerically and functionally) yet can only appear to change (for better or worse) following crises. While polycentric, the Everglades governance system is hierarchical, rigid, and inflexible. Another indication is the inability to negotiate (or even discuss) many policy changes, much less attempt them. The result of large influxes of capital has sustained existing power relations in the system, leading to the current governance and management system that keep the system stable and rigid (Gunderson et al. 2014).

14.3.2 Building Adaptive Capacity Through Cycles of Change

Many of the environmental issues identified in the case studies suggest that management legacies have eroded system resilience (Holling and Meffe 1996; Folke et al. 2004). In order to maximize certain social objectives for water management, other ecosystem goods and services were degraded (Gunderson et al. 2016). The legacy of engineered optimization for a limited number of services is twofold. One is that economic growth enabled by water management, such as hydropower in the Columbia River Basin (Cosens and Fremier, Chap. 4) or irrigation in the Platte River Basin (Birge et al. Chap. 8), creates not only a financially and politically powerful constituency for maintenance of the system state but also a regional economic dependency on its continuation. Development of services to an optimum level of provisioning leaves little latitude for adaptation to a changing water balance, leaving the basin vulnerable to any disturbance. In particular, change in the water balance due to climate change may destabilize the economic system and move the basin over a threshold from which it will be difficult to return.

Once an approaching threshold is signaled, experimentation may pose unacceptable risk (Lee 1993). For example, the rigidity of the ESA following species listing reflects a societal value that once the possibility of irreversible extinction is identified, flexibility and experimentation pose too great a risk. Management in this system trajectory should focus on restoration of resilience. Rather than gamble with a system close to a threshold, restoration may not only move the system away from the threshold, but increase latitude for adaptation. Arguably, this is the essence of the ultimate focus of the ESA, embodied in the goal of species recovery.

Each of the basins studied has certain aspects of this second major trajectory and potential for active management to increase resilience. For example, restoration in the Columbia River to increase the latitude for adaptation might include measures such as reconnecting rivers to some of their former floodplain, altering dam operation to mimic natural hydrographs, restoring riparian habitat, altering release points at dams to reduce instream water temperatures, and altering hatchery operation for careful selection of genetic stock and layout to mimic natural rearing. In addition, diversification of the source of ecosystem services from the system can be critical to managing resilience. For example, increasing local flood control measures in the Columbia River can free up some water storage space for other purposes as well as provide capacity to handle unexpected flows (Chap. 4).

Sustainability of economic and cultural systems in the face of cycles of change often requires an incremental broadening of the range of services provided through a targeted reduction in optimization and the restoration of key natural services. Restoration measures increase resilience latitude in the face of change and uncertainty, but require both adaptive governance and financial investment to navigate the trade-offs associated with reduction in optimization. The role of law is in preparation of the formal system of government to allow emergence of adaptive forms of governance and in forcing governmental provision of financial resources including government assistance for local capacity building (Cosens et al. 2017; Garmestani and Allen 2014). Law played a role in preparation for the emergence of adaptive processes including the empowerment of indigenous communities through providing a forum and basis for litigation in both the Columbia and Klamath basins. The Clean Water Act triggered and sometimes funded the development of local watershed groups in the Anacostia River (Arnold et al. 2014).

Following periods of instabilities, systems reorganize and start new phases of growth and development. It is during the period of reorganization that system resilience is tested and the period in which a new regime (as described above) can come into being. Such new regimes are characterized by a different set of processes and structures (system controlling variables). These periods are when adaptive governance may emerge through formal and informal networks of response to the disturbance. This emergence must be proceeded, however, by the development of appropriate structure, capacity, and processes to, at best, facilitate its emergence and, at a minimum, not create barriers (Table 14.1). This is also the period in which cross-scale interactions are critical.

During phases of instability and reorganization, new connections across loci of governance emerge or are strengthened. Examples include the formation of National Academy of Science committees in the Columbia River or the Klamath Basin. Such emergent groups tend to be epistemic and focus on resolving uncertainties that contributed to the resource surprise and what are possible responses and adaptations to the unforeseen system dynamics. Cross-scale interactions may facilitate these connections through the provision of resources including technical support from higher levels of government.

In addition, new forms of management or new forms of government may emerge separately or to institutionalize those that have informally arisen. One example is the creation of the South Florida Water Management District, following a severe drought in the Everglades (Light et al. 1995). Another example is the establishment of the Northwest Power and Conservation Council in the US portion of the Columbia River, an interstate council authorized by congress to engage with the public in regional electric power planning and enhancement of fish and wildlife within the basin (Lee 1993; NWPCC n.d.). These entities generally formalize cross-scale and cross-jurisdictional networks and are useful when the need to respond at the scale of the new entity is likely to recur.

14.3.3 Transformative Change

The third management trajectory of transformation occurs under two circumstances. First, when either through overdevelopment, a disturbance, or a combination of both, the system has crossed a threshold and is in an alternative, generally stable configuration. Second, when the system is in an undesirable state and society seeks to navigate to a more desirable alternative state, a major change in ecosystem management and significant financial investments may be needed to traverse the threshold and achieve the desired state. In complex systems, regime shift is not a simple matter of a single variable (e.g., river temperature or sediment supply) being out of balance. Given climate change, basin response may include changes in precipitation and temperature with complex secondary impacts on everything from biota to water demand. It is difficult to imagine a smooth navigation of regime shift without assistance in both leadership and financing from a larger-scale entity or entities. In the Klamath Basin, crisis led to the emergence of local leadership and a local collaborative process, yet it was facilitated through federal and NGO funding to provide capacity for participation by Native American tribes and a federal and state process for participation in a water settlement (Chaffin et al. 2014a; Chap. 6). A key variable in the Klamath Basin identified as necessary to potentially catalyze social-ecological transformation, the removal of major dams, will require substantial federal funding to accomplish and, at the time of this writing, remains uncertain. In addition, the collaborative governance that led to these proposed changes will need to be institutionalized if it is to be sustained in the long term. In the Florida Everglades, the collaborative efforts of local water management districts and key scientists led to passage of a federal act with funding for restoration, yet the fragmentation of water allocation and water quality law combined with entrenched interests may stand in the way of the transformation needed to prepare the social-ecological system for climate change (Gunderson et al. 2014).

Adaptive governance alone may not be sufficient to navigate a regime shift (Bruner et al. 2005; Chaffin et al. 2014b, 2016). It must be coupled with changes in the law that allow for cross-sector and cross-scale integrated water management. It must be coupled with leadership and funding from outside the basin. It must be coupled with the political and personal will to transform water-based economies to new livelihoods. Such transformative changes are evaluated through a different theoretical lens, one that involves patterns of stability and instability over time within the social-ecological system and examines cross-scale linkages. The characteristics of governance needed for this more complex and systemic change have been called transformative governance (Chaffin et al. 2016).

By using this framework to connect the understanding of complex system response in ecological systems to an understanding of the complex governance systems that mediate social-ecological system interaction, it becomes possible to chart a course more likely to assist society in the navigation of change. Moving from identification of the role of system trajectories and cross-scale interactions – i.e., Panarchy – in the basins studied to synthesis of the key lessons this framework and other theoretical constructs provide for understanding the barriers and opportunities for enhancing the adaptability of regulated water systems is the goal of this special issue.

14.4 Summary

The six North American water basins that were chosen as cases to investigate the interaction among ecosystems, legal systems, and adaptive governance all represent heavily regulated and developed social-ecological systems. Reviews of the historical development or trajectories of these systems reflect complex interactions among adaptive governance, ecosystem regimes, and the legal systems. The basin assessments show different ways that adaptive governance may be triggered, facilitated, or constrained by ecological and/or legal processes. The basin assessments indicated that complex interactions among the legal, governance, and ecosystem components of these systems can produce different trajectories, which include patterns of (a) development and stabilization, (b) cycles of crisis and recovery which builds the adaptive capacity of the system, and (c) periods of innovation, novelty, and transformation.

Acknowledgments This work was developed in part under the Adaptive Water Governance Project, funded by the US National Socio-Environmental Synthesis Center (SESYNC) under funding from the US National Science Foundation, NSF DBI-1052875.

References

Arnold CA, Green OO, Decaro DA, Chase A, Ewa JG (2014) The social-ecological resilience of an eastern urban-suburban watershed: the Anacostia River Basin Idaho Law Rev 51:29–90

Benson MH, Garmestani A (2011) Embracing Panarchy, building resilience and integrating adaptive management through a rebirth of the National Environmental Policy Act. J Environ Manag 92:1420–1427

Benson M, Llewellyn D, Morrison R, Stone M (2014) Water governance challenges in New Mexico's Middle Rio Grande River Valley: a resilience assessment Idaho Law Rev 51:195–228

Birge H, Allen CR, Craig RK, Hamm JA, Babbitt C, Nemec K, Schlager E (2014) Social-ecological resilience and law in the Platte River Basin Idaho Law Rev 51:229–256

Bruner R, Steelman TA, Coe-Juell L, Cromley C, Edwards C, Tucker D, (2005) Adaptive governance: integrating science, policy, and decision making. Columbia University Press, New York

Chaffin BC, Gunderson LH (2016) Emergence, institutionalization and renewal: rhythms of adaptive governance in complex social-ecological systems. J Environ Manag 165:81–87

Chaffin B, Craig R, Gosnell H (2014a) Resilience, adaptation, and transformation in the Klamath River Basin docial-ecological system. Idaho Law Rev 51:157–183

Chaffin BC, Gosnell H, Cosens BA (2014b) A decade of adaptive governance scholarship: synthesis and future directions. Ecol Soc 19(3):56. http://dx.doi.org/10.5751/ES-06824-190356

Chaffin BC, Garmestani AS, Gunderson LH, Benson MH, Angeler DG, Arnold CA, Cosens BA, Craig RK, Ruhl JB, Allen CR (2016) Transformative environmental governance. Annu Rev Environ Resour 41:399–423. https://doi.org/10.1146/annurev-environ-110615-085817

Chelleri L, Waters JJ, Olazabal M, Minucci G (2015) Resilience tradeoffs: addressing multi-scale and temporal aspects of urban resilience. Environ Urban J 27(1):181–198

Cosens B, Fremier A (2014) Assessing system resilience and ecosystem services in large river basins: a case study of the Columbia River Basin. Idaho Law Rev 51:91–126

Cosens B, Gunderson L, Chaffin B (2014) The adaptive water governance project: assessing law, resilience and governance in regional social ecological water systems facing a changing climate. Idaho Law Rev 51:1–27

Cosens BA, Craig RK, Hirsch SL, Arnold CA, Benson MH, DeCaro DA, Garmestani AS, Gosnell H, Ruhl JB Schlager E (2017) The role of law in adaptive governance. Ecol Soc 22:30

Craig RK, Ruhl JB (2014) Designing administrative law for adaptive management. Vanderbilt Law Rev 67:1–86

Craig RK, Garmestani AS, Allen CR, Arnold CA, Birgè H, DeCaro DA, Fremier AK, Gosnell H, Schlager E (2017) Balancing stability and flexibility in adaptive governance: an analysis of tools available in U.S. environmental law. Ecol Soc 22:3

DeCaro DA, Chaffin BC, Schlager E, Garmestani AS, Ruhl JB (2017) Legal and institutional foundations of adaptive environmental governance. Ecol Soc 22:32

Dietz T, Ostrom E, Stern PC (2003) The struggle to govern the commons. Science 302:1907–1912

Folke C, Carpenter S, Walker B, Scheffer M, Elmqvist T, Gunderson L, Holling C.S (2004) Regime shifts, resilience, and biodiversity in ecosystem management. Annu Rev Ecol Evol Syst 35:557–581

Garmestani AS, Allen CR (2014) Socio-ecological resilience and law. Columbia University Press, New York

Gunderson LH, Holling CS (2002) Panarchy: understanding transformations in systems of humans and nature. Island, Washington, DC

Gunderson L, Light SS (2006) Adaptive management and adaptive governance in the everglades ecosystem. Policy Sci 39:323–334

Gunderson LH, Pritchard L (2002) Resilience and the behavior of large scale systems. Island, Washington, DC

Gunderson LH, Holling CS, Light SS (1995) Barriers and bridges to the renewal of ecosystems and institutions. Columbia University Press, New York

Gunderson L, Garmestani A, Rizzardi KW, Ruhl JB, Light A (2014) Escaping a rigidity trap: governance and adaptive capacity to climate change in the everglades social ecological system Idaho Law Rev 51:127–156
Gunderson L, Cosens B, Garmestani A (2016) Adaptive governance of riverine and wetland ecosystem goods and services. J Environ Manag 183(1):353–360
Holling CS (1973) Resilience and stability of ecological systems. Annu Rev Ecol Syst 4:1–23
Holling CS (2001) Understanding the complexity of economic, ecological, and social systems. Ecosystems 4(5):390–405
Holling CS, Meffe GK (1996) Command and control and the pathology of natural resource management. Conserv Biol 10:328–337
John D (1994) Civic environmentalism: alternatives to regulation in states and communities. CQ Press, Washington, DC
Lee KN (1993) Compass and gyroscope. Island, Washington, DC
Light SS, Gunderson LH, Holling CS (1995) The Everglades: evolution of management in a turbulent ecosystem. In: Gunderson LH, Holling CS, Light SS (eds) Barriers and bridges to the renewal of ecosystems and institutions. Columbia University Press, New York
LoSchiavo AJ, Best RG, Burns RE, Gray S, Harwell MC, Hines EB, McLean AR, St. Clair T, Traxler S, Vearil JW, (2013) Lessons learned from the first decade of adaptive management in comprehensive Everglades restoration. Ecol Soc 18:70
Northwest Power and Conservation Council [NWPCC] (n.d.) Mission and strategy. https://www.nwcouncil.org/about/mission/. Accessed 6 Oct 2017
Scheffer M, Carpenter S, Foley JA, Folke C, Walker B (2001) Catastrophic shifts in ecosystems. Nature 413:591–596
Walker B, Salt D (2006) Resilience thinking: sustaining ecosystems and people in a changing world. Island, Washington, DC
Walker B, Holling CS, Carpenter SR, Kinzig A (2004) Adaptability and transformability in social-ecological systems. Ecol Soc 9:5
Zellmer S, Gunderson L (2009) Why resilience may not always be a good thing: lessons in ecosystem restoration from Glen Canyon and the Everglades. Nebraska Law Rev 87:893

Chapter 15
Uncertainty and Trade-Offs in Resilience Assessments

Craig R. Allen, Hannah Birgé, David G. Angeler, Craig Anthony (Tony) Arnold, Brian C. Chaffin, Daniel DeCaro, Ahjond S. Garmestani, and Lance H. Gunderson

Abstract Several frameworks have been developed to assess the resilience of social-ecological systems, but most are time consuming and require substantial time and technical expertise. Stakeholders and practitioners often lack the resources for such intensive efforts. Furthermore, most resilience assessments end with problem framing and fail to explicitly address trade-offs and uncertainty inherent in any assessment of resilience. This chapter reports on a rapid assessment of survey responses to compare the relative resilience across four North American social-ecological watershed systems with respect to a number of proposed resilience properties. Responses were compared among four stakeholder categories: (1) government (policy, regulation, management), (2) end users (farmers, ranchers, landowners, industry), (3) agency/public science (research, university, extension), and (4) non-governmental organizations (environmental, citizen, social justice) in each of the watersheds. Conceptually, social-ecological systems are comprised of components

C. R. Allen (✉)
U.S. Geological Survey, Nebraska Cooperative Fish and Wildlife Research Unit, School of Natural Resources, University of Nebraska, Lincoln, NE, USA
e-mail: callen3@unl.edu

H. Birgé
Nebraska Cooperative Fish and Wildlife Research Unit, School of Natural Resources, University of Nebraska, Lincoln, NE, USA

The Nature Conservancy, Omaha, NE, USA

D. G. Angeler
Swedish University of Agricultural Sciences, Uppsala, Sweden
e-mail: david.angeler@slu.se

C. A. Arnold
Louis D. Brandeis School of Law, Department of Urban and Public Affairs, University of Louisville, Louisville, KY, USA
e-mail: tony.arnold@louisville.edu

B. C. Chaffin
W. A. Frank College of Forestry and Conservation, University of Montana, Missoula, MT, USA
e-mail: brian.chaffin@mso.umt.edu

B. Cosens, L. Gunderson (eds.), *Practical Panarchy for Adaptive Water Governance*, https://doi.org/10.1007/978-3-319-72472-0_15

ranging from strictly human to strictly ecological, but that relate directly or indirectly to one another in complex ways. They have soft boundaries and several important dimensions or axes that together describe the nature of social-ecological interactions (e.g., variability, diversity, modularity, slow variables, feedbacks, capital, innovation, redundancy, and ecosystem services). There is no absolute measure of resilience, so our design takes advantage of comparisons across watersheds and therefore focuses on relative resilience. Our approach quantifies and compares the relative resilience across watershed systems and the potential trade-offs among different aspects of the social-ecological system (e.g., among social, economic, and ecological contributions). This approach permits explicit assessment of several types of uncertainty (e.g., self-assigned uncertainty for stakeholders; uncertainty across respondents, watersheds, and subsystems) and subjectivity in perceptions of resilience among key actors and decision-makers and provides an efficient way to develop the mental models that inform stakeholders and stakeholder categories.

Keywords Coupled human-natural system · Resilience assessment · Social-ecological system · Stressed watersheds · Resilience metrics

15.1 Introduction

Resilience is a measure of the amount of perturbation or disturbance a system can withstand without crossing a critical threshold (Holling 1973). When such a threshold is exceeded, the system collapses and reorganizes. Reorganization can occur with the same driving variables and processes, in which case the original system renews, or the system can reorganize around a new set of variables and drivers in which case a new organization emerges and quite often a very different system (Chaffin et al. 2016). When reorganization occurs around new drivers, the new system may be less desirable to humankind than the former system in terms of the provision of goods and services (and relative predictability). Therefore, it is often in humankind's interest to maintain systems in desirable states, avoid critical thresholds, and enhance resilience. Alternatively, when a system is undesirable, it may be

D. DeCaro
Department of Urban and Public Affairs, Department of Psychological and Brain Sciences, and Center for Land Use and Environmental Responsibility, University of Louisville, Louisville, KY, USA
e-mail: daniel.decaro@louisville.edu

A. S. Garmestani
School of Natural Resources, University of Nebraska, Lincoln, NE, USA
e-mail: garmestani@hotmail.com

L. H. Gunderson
Department of Environmental Sciences, Emory University, Atlanta, GA, USA
e-mail: lgunder@emory.edu

in humankind's interest to erode resilience and purposely transform the system to a more desirable state (Chaffin et al. 2016). In either case, it is important to be able to assess, at least relatively, how resilient a system is, so that humankind can either foster or erode resilience, depending upon the desirability of the current system state.

Resource managers often rely on the resilience, or the capacity, of a social-ecological system to absorb and respond to a disturbance while maintaining its essential structure and functions (Holling 1973; Folke et al. 2002). An improved understanding of the boundaries of resilience – the thresholds that separate one state of a system from alternative regimes – may help resource managers avoid or facilitate regime shifts so that desired ecosystem services are maintained or restored. Resilience in a complex system of people and nature has both ecological and social dimensions (Folke et al. 2002). Whereas ecological resilience is the magnitude of disturbance that an ecosystem can absorb before it shifts into a new regime, social resilience is the capacity of social systems to withstand and adapt to disturbances that result from social, political, or environmental changes (Adger 2000). While enhancing and maintaining resilience is increasingly identified as a management goal or strategy for projects focused on either ecosystems (Kareiva et al. 2008; Benson and Garmestani 2011; WWF 2012) or social systems (Godschalk 2003; Norris et al. 2008; Longstaff et al. 2010), resilience is perhaps better understood as a product of reinforcing social and ecological interactions.

As resilience is an emergent property of complex social and ecological system dynamics, quantifying resilience remains a developing field (Angeler and Allen 2016). Some resilience assessment approaches are based on a structured set of discussions among stakeholders and experts who envision a system of interest and determine dominant drivers, including influences from scales above and below the focal scale of interest (Angeler and Allen 2016; Quinlan et al. 2016). Advancing resilience theory toward a quantifiable science is important if the concept is to have utility and application for navigating a rapidly changing Anthropocene (Biggs et al. 2012, 2015).

Although several frameworks have been proposed for conducting resilience assessments (Walker et al. 2002, 2009; Resilience Alliance 2010; Biggs et al. 2015; Quinlan et al. 2016), the application of these techniques to real-world systems continues to be a challenge because of the highly dynamic and multidimensional nature of linked social-ecological systems (Berkes and Folke 1998; Walker et al. 2002). Such techniques involve multiple workshops or other long-term projects (Resilience Alliance 2015). Many of these methods involve detailed and time-consuming activities to winnow the complexity inherent in social-ecological systems. Since these activities are largely qualitative, measures of uncertainty are rarely pursued. Uncertainty takes many forms (Williams 2001, 2011), as does its quantification, and identifying key uncertainties through resilience assessments may provide insight into how assessments can be improved, how the system itself functions, and how structured experimentation and learning can reduce uncertainties (Allen et al. 2011, 2016a, b; Birge et al. 2016). Therefore, methods to assess uncertainty are critical for forwarding resilience theory and for the application of resilience approaches to particular systems or challenges.

Table 15.1 The nine properties of a resilient social-ecological system from Walker and Salt (2006), along with their definitions (adapted from Allen et al. 2016a, b), and a survey question designed to address that property (asterisks represent definitions given to participants via hyperlinks)

Property	Definition	Example of survey question
Biological diversity	The number and evenness of species, functional groups, and response to disturbance in the ecosystem	How would you assess the ecological diversity* of the basin?
Ecological variability	Natural variability and fluctuations in ecological processes, structures, and populations	How would you assess the water flow variability of the basin?
Modularity	System components are connected to one another so that information is transferred effectively, but not so overly connected that shocks cause disproportionate damage	What would the impact be of a random loss of an institution on the basin as a whole?
Acknowledging slow variables	Incorporation of information about long-term outcomes that result from near-term decision-making	How does the basin's governance acknowledge slow variables*?
Tight feedbacks	Feedbacks among critical system components respond quickly allowing practitioners to avoid dangerous thresholds	How do decision-makers in the basin understand information about the ecosystem?
Social capital	It is comprised of the net sum of benefits generated from relationships among components in a system's social network	How would you assess trust among stakeholders in the basin?
Innovation	Degree of learning, experimentation, education, and locally developed rules in order to embrace change and creatively improve conditions	How would you assess grassroots governance efforts in the basin?
Overlap in governance	Institutions have redundancy in their roles and responsibilities	How evenly is institutional power* distributed in the system?
Ecosystem services	Essential and nonessential benefits people obtain from ecosystems	How would you assess the production of ecosystem services* from the basin?

An absolute measure of resilience has not been developed, is not likely to be developed, and may not be useful (Quinlan et al. 2016). Rather, there have been at least two approaches to developing resilience metrics. The first involves quantification of specific resilience, that is, the resilience of what, to what, for whom (Carpenter et al. 2001), which should also involve quantifying uncertainty. The second type attempts to measure relative resilience by comparing attributes across similar systems relative to one another (Nemec et al. 2014). Such a simplified approach to resilience assessment incorporates Walker and Salt's (2006) nine facets of resilience: ecological variability, diversity, modularity, acknowledgement of slow variables, tight feedbacks, social capital, innovation, overlap in governance, and ecosystem services (Table 15.1). This chapter examines the relative resilience of these facets across a subset of the case studies described in Part I of this volume, as a way to compare and contrast trade-offs among facets of resilience and uncertainty.

15.1.1 Study Areas

In addition to the largely historical resilience assessments done in the Anacostia (Chap. 3), Columbia (Chap. 4), Middle Rio Grande (Chap. 7), and Platte River Basin (Chap. 8), a rapid prototype assessment of relative resilience was undertaken. The issues associated with resilience varied among the systems.

The Anacostia River Basin is the smallest (1140 km^2) of the river drainage basins. The basin is the most urban, flowing through rural, suburban, and urban areas around the District of Columbia. Water quality regimes dominate the ecological questions surrounding the resilience of the Anacostia River Basin, especially on the system's ability to withstand ongoing or increased pollution, runoff, and flooding. This is especially true in light of the complex institutional interactions that define the ability of social and ecological components to respond to disturbances related to water quality (Arnold et al. 2014; Chap. 3).

The Columbia River Basin covers an area of roughly 670,000 km^2 and extends across large parts of Idaho, Washington, Oregon, British Columbia, and smaller parts of Wyoming, California, Nevada, and Utah. The basin includes largely rural cover, but also encompasses a handful of metropolitan areas including Portland, OR; Boise, ID; and Spokane, WA, where most of the six million human inhabitants of the basin reside (Cosens and Fremier 2014; Chap. 4). In the Columbia River Basin, issues emerging from the intersection of climate change and water scarcity – characteristic of western river systems – include competition among hydropower, irrigation, flood control, ecological integrity, and other valuable social-ecological goods and services for water flow and supply (Cosens and Fremier 2014; Chap. 4).

The Middle Rio Grande Basin, defined as the portion of the Rio Grande between Cochiti Dam to Elephant Butte Reservoir in Central New Mexico, extends roughly 8000 km^2 and is contained entirely in the US state of New Mexico (Benson et al. 2014). While sparsely populated through most of its extent, it contains the city of Albuquerque and a total population of 690,000 (MRGBS 2012). The resilience of the Middle Rio Grande River Basin generally concerns the ability of the social-ecological system to withstand ongoing human population growth and urbanization, biodiversity loss, and cyclical drought in the face of ongoing water appropriation and climate change (Benson et al. 2014; Chap. 7). Despite the definitional focus on the portion of the river in New Mexico, with its headwaters in Colorado and its lower reaches forming the boundary between Texas and Mexico, much like the other western water systems in our study (Columbia and Platte), interstate water compact obligations and treaties strongly interact with social-ecological components in the system.

The Platte River Basin drains an area of more than 23,000 km^2 extending across the US states of Colorado, Nebraska, and Wyoming (Palmer 1996). In the Platte Basin, the population is disproportionately distributed among a few dozen medium (e.g., Kearney, Nebraska) to very large (e.g., Denver, Colorado) metropolitan areas, many of which are focused in the South Platte Basin, where population is expected to double in the next 40 years, putting additional pressure on already overappropriated water sources (Garner 2010). Much like the other western basins, interstate water compacts and arid, low-flow environments constrain the amount of water available for appropriation to various components in the social-ecological system, including endangered species, riverine wetlands, irrigation projects that generate power and recreation opportunities, and downstream users entitled to flows. Resilience in this system is therefore largely focused on the ability of the social-ecological system as a whole to withstand flow variability – specifically droughts, but also major floods – without any integral system component "losing" access to water and undermining the rest of the system.

15.2 Methods

We surveyed governments, researchers, end users, and NGOs within each of the four watersheds of interest to determine relative resilience of the watersheds, trade-offs among social and ecological components of resilience, and uncertainty in the assessments. Although social and ecological resilience are intertwined, we evaluated the properties for social and ecological resilience separately as well as in combination to provide a clearer assessment of the elements of resilience (Allen et al. 2003). Because not all of the properties applied to both kinds of resilience, we assessed both social and ecological resilience utilizing a survey consisting of 26 questions (Table 15.1). We concurred with Walker and Salt's (2006) creation of a property for ecological variability but not social variability because, as they defined it, variability refers to variability in the occurrence and magnitude of ecological phenomena such as flooding and wildfire that do not have a social equivalent. Similar aspects of a system such as social and cultural heterogeneity are incorporated into the social diversity and social modularity variables. Likewise, the social properties of social capital, innovation, and overlapping governance do not have an ecological equivalent. Slow ecological variables themselves and tight ecological feedbacks are so closely related to ecological variability that these properties are encompassed by the ecological variability property, and ecosystem services are a social and not an ecological construct because they refer to the benefits that humans obtain from nature.Within each of these five basins, we invited 40 survey participants that we identified as representatives of end users (farmer, rancher, and/or private citizen), state or federal government, nongovernmental organizations, or research/extension stakeholder groups, but allowed respondents to reassign themselves at the outset of the survey to the stakeholder category with which they most closely identified. This, along with the survey hyperlink being nonspecific to user, and the collection of no identifiable information beyond stakeholder group and basin, meant that anonymity among participants was carefully preserved.

We sent consenting participants a hyperlink to an online questionnaire. The questions we included in the survey were designed to assess stakeholder perception and uncertainty concerning various aspects of their system's social-ecological resilience. Specifically, we designed our questionnaire to analyze stakeholders' assessments of Walker and Salt's (2006) nine properties of a resilient world, but within individual contexts (Table 15.1). We used a Likert scale of five points for the questionnaire, which included 26 content questions each followed by a question asking respondents to rank their level of certainty in their response to the previous question (see Table 15.2 for the full list of questions). Some definitions (e.g., "modularity") were provided to respondents via hyperlinks embedded in the online survey.

Our questions addressed how different stakeholders from the five basins assessed the level of (1) biological (species, response, and trait) diversity, (2) ecological variability, (3) modularity, (4) acknowledgement of slow variables, (5) feedback length, (6) social capital, (7) innovation, (8) overlap in critical social roles, and (9) ecosystem services provisioning in their social-ecological system. We asked stakeholders explicitly to assess these properties (i.e., "assess the ecological diversity in your system") but also designed questions to address the properties for a stressed river basin context.

These nine resilience variables likely represent an incomplete and subjective list, but their use in prior assessments (Nemec et al. 2014; Allen et al. 2016a, b; Cosens and Fremier 2014), ease of interpretation, and applicability across different systems lend them well to a rapid comparative approach and to uncertainty assessments. The selection of particular variables (survey questions) is important and necessarily varies based on the systems of interest and the aspects of those systems that are important. A balanced design may be most useful (i.e., the same or similar number of questions for different aspects of resilience such as ecological, social, or economic), but we did not employ that approach here. Our empirical results are for illustrative uses only because our sample sizes are small. Thus, we explore the utility of this approach to assess relative resilience, trade-offs, and uncertainty.

We utilized a rapid prototyping approach (Nicolson et al. 2002) to score relative resilience by having stakeholders individually respond to a series of survey questions meant to probe different aspects of the social-ecological systems inferred to confer resilience, on a Likert scale of 1–5, with 1 being least resilient and 5 being most resilient (Table 15.1). Additionally, we rated the uncertainty of each resilience score on a scale of 1–5, i.e., low to high. The choice of metric for resilience assessment is important. Here, we focus on social and ecological components, but incorporating components of critical infrastructure, and explicitly incorporating economics, may also be useful. Organizations such as the Resilience Alliance (http://www.resalliance.org/) have spent considerable time and resources in determining relevant aspects of resilience. Although it would be appropriate for assessments to develop particular aspects of resilience relevant for their analyses, we utilize a modification of those presented in Nemec et al. (2014), because they are both useful and illustrative. Our method is based on surveys of stakeholders in the social-ecological systems of interest (described below) and based on 26 questions focused on social and ecological aspects of resilience and governance.

Table 15.2 Resilience questionaire

Q1	To which stakeholder group do you most closely belong (please check only one)?
	□ Government (policy, regulation, management)
	□ End user (farmer, rancher, private landowner, industry)
	□ Agency/public service (research, university, extension)
	□ NGO (environmental, citizen, social justice)
Q2	To which river basin do you belong?
	□ Anacostia
	□ Columbia
	□ Klamath
	□ Middle Rio Grande
	□ Platte
Q3a	How would you assess the ecological diversity of the basin?
	□ There is **very low** ecological diversity
	□ There is **low** ecological diversity
	□ There is **moderate** ecological diversity
	□ There is **high** ecological diversity
	□ There is **very high** ecological diversity
Q3b	How certain are you in your answer to question 3a?
	□ Very uncertain
	□ Uncertain
	□ Somewhat uncertain
	□ Somewhat certain
	□ Certain
	□ Very certain
Q4a	How would you assess the ecological variability in the basin?
	□ There is **very low** variability
	□ There is **low** variability
	□ There is **moderate** variability
	□ There is **high** variability
	□ There is **very high** variability
Q4b	How certain are you in your answer to question 4a?
	□ Very uncertain
	□ Uncertain
	□ Somewhat uncertain
	□ Somewhat certain
	□ Certain
	□ Very certain

(continued)

Table 15.2 (continued)

Q5a	How often are decisions made collectively in in the basin?
	□ Management decisions are **never** made collectively
	□ Management decisions are **rarely** made collectively
	□ Management decisions are **sometimes** made collectively
	□ Management decisions are **often** made collectively
	□ Management decisions are **always** made collectively
Q5b	How certain are you in your answer to question 5a?
	□ Very uncertain
	□ Uncertain
	□ Somewhat uncertain
	□ Somewhat certain
	□ Certain
	□ Very certain
Q6a	How would you assess stakeholder involvement in decision-making?
	□ **No** stakeholder interests/perspectives are incorporated into decision-making
	□ **Few** stakeholder interests/perspectives are incorporated into decision-making
	□ **Some** stakeholder interests/perspectives are incorporated into decision-making
	□ **Many** stakeholder interests/perspectives are incorporated into decision-making
	□ **All** stakeholder interests/perspectives are incorporated into decision-making
Q6b	How certain are you in your answer to question 6a?
	□ Very uncertain
	□ Uncertain
	□ Somewhat uncertain
	□ Somewhat certain
	□ Certain
	□ Very certain
Q7a	How would you assess the diversity of livelihoods in the basin?
	□ The social system has a **very low** diversity of livelihoods
	□ The social system has a **low** diversity of livelihoods
	□ The social system has a **moderate** diversity of livelihoods
	□ The social system has a **high** diversity of livelihoods
	□ The social system has a **very high** diversity of livelihoods
Q7b	How certain are you in your answer to question 7a?
	□ Very uncertain
	□ Uncertain
	□ Somewhat uncertain
	□ Somewhat certain
	□ Certain
	□ Very certain

(continued)

Table 15.2 (continued)

Q8a	How would you assess the water flow variability of the rivers and wetlands in the basin?
	□ The rivers and wetlands in the basin have **very low** variability
	□ The rivers and wetlands in the basin have **low** variability
	□ The rivers and wetlands in the basin have **moderate** variability
	□ The rivers and wetlands in the basin have **high** variability
	□ The rivers and wetlands in the basin have **very high** variability
Q8b	How certain are you in your answer to question 8a?
	□ Very uncertain
	□ Uncertain
	□ Somewhat uncertain
	□ Somewhat certain
	□ Certain
	□ Very certain
Q9a	How would you assess the vulnerability of streams and wetlands to further groundwater depletion?
	□ Groundwater inputs to the streams and wetlands have **very low vulnerability** to groundwater depletion
	□ Groundwater inputs to the streams and wetlands have **low vulnerability** to groundwater depletion
	□ Groundwater inputs to the streams and wetlands have **moderate vulnerability** to groundwater depletion
	□ Groundwater inputs to the streams and wetlands have **high vulnerability** to groundwater depletion
	□ Groundwater inputs to the streams and wetlands have **very high vulnerability** to groundwater depletion
Q9b	How certain are you in your answer to question 9a?
	□ Very uncertain
	□ Uncertain
	□ Somewhat uncertain
	□ Somewhat certain
	□ Certain
	□ Very certain
Q10a	How would you assess the social modularity of the basin?
	□ Economic, political, and social institutions have **very low modularity**
	□ Economic, political, and social institutions have **low modularity**
	□ Economic, political, and social institutions have **moderate modularity**
	□ Economic, political, and social institutions have **high modularity**
	□ Economic, political, and social institutions have **very high modularity**

(continued)

Table 15.2 (continued)

Q10b	How certain are you in your answer to question 10a?
	□ Very uncertain
	□ Uncertain
	□ Somewhat uncertain
	□ Somewhat certain
	□ Certain
	□ Very certain
Q11a	How would you assess the social impact on other institutions/the system as a whole when an institution is lost from the basin?
	□ The elimination of any one institution **has a very low impact** on other institutions and/or the system as a whole
	□ The elimination of any one institution **has a low impact** on other institutions and/or the system as a whole
	□ The elimination/impairment of any one institution **has a moderate impact** on other institutions and/or the system as a whole
	□ The elimination/impairment of any one institution **has a high impact** other institutions and/or the system as a whole
	□ The elimination/impairment of any one institution **has a very high impact** on other institutions and/or the system as a whole
Q11b	How certain are you in your answer to question 11a?
	□ Very uncertain
	□ Uncertain
	□ Somewhat uncertain
	□ Somewhat certain
	□ Certain
	□ Very certain
Q12a	Does the basin's governance acknowledge slow variables?
	□ Slow variables are **never acknowledged** in the governance of the system
	□ Slow variables are **rarely acknowledged** in the governance of the system
	□ Slow variables are **sometimes acknowledged** in the governance of the system
	□ Slow variables are **often acknowledged** in the governance of the system
	□ Slow variables are **always acknowledged** in the governance of the system
Q12b	How certain are you in your answer to question 12a?
	□ Very uncertain
	□ Uncertain
	□ Somewhat uncertain
	□ Somewhat certain
	□ Certain
	□ Very certain

(continued)

Table 15.2 (continued)

Q13a	Is information about the ecosystem incorporated into the basin's governance?
	□ Information about the ecosystem is **never incorporated** into the basin's governance
	□ Information about the ecosystem is **rarely incorporated** into the basin's governance
	□ Information about the ecosystem is **sometimes incorporated** into the basin's governance
	□ Information about the ecosystem is **often incorporated** into the basin's governance
	□ Information about the ecosystem is **always incorporated** into the basin's governance
Q13b	How certain are you in your answer to question 13a?
	□ Very uncertain
	□ Uncertain
	□ Somewhat uncertain
	□ Somewhat certain
	□ Certain
	□ Very certain
Q14a	Do decision-makers in the basin understand information about the ecosystem?
	□ Decision-makers **never** understand ecosystem feedbacks, identify potential thresholds, or respond to change
	□ Decision-makers **rarely** understand ecosystem feedbacks, identify potential thresholds, or respond to change
	□ Decision-makers **sometimes** understand ecosystem feedbacks, identify potential thresholds, or respond to change
	□ Decision-makers **often** understand ecosystem feedbacks, identify potential thresholds, or respond to change
	□ Decision-makers **always** understand ecosystem feedbacks, identify potential thresholds, or respond to change
Q14b	How certain are you in your answer to question 14a?
	□ Very uncertain
	□ Uncertain
	□ Somewhat uncertain
	□ Somewhat certain
	□ Certain
	□ Very certain
Q15a	Is information about the social system incorporated into the basin's governance?
	□ Information about the social system is **never incorporated** into the governance of the system
	□ Information about the social system is **rarely incorporated** into the governance of the system
	□ Information about the social system is **sometimes incorporated** into the governance of the system
	□ Information about the social system is **often incorporated** into the governance of the system
	□ Information about the social system is **always incorporated** into the governance of the system

(continued)

Table 15.2 (continued)

Q15b	How certain are you in your answer to question 15a?
	□ Very uncertain
	□ Uncertain
	□ Somewhat uncertain
	□ Somewhat certain
	□ Certain
	□ Very certain
Q16a	Do decision-makers understand social information in the basin?
	□ Decision-makers **never** understand social system feedbacks, identify potential thresholds, or respond to change
	□ Decision-makers **rarely** understand social system feedbacks, identify potential thresholds, or respond to change
	□ Decision-makers **sometimes** understand social system feedbacks, identify potential thresholds, or respond to change
	□ Decision-makers **often** understand social system feedbacks, identify potential thresholds, or respond to change
	□ Decision-makers **always** understand social system feedbacks, identify potential thresholds, or respond to change
Q16b	How certain are you in your answer to question 16a?
	□ Very uncertain
	□ Uncertain
	□ Somewhat uncertain
	□ Somewhat certain
	□ Certain
	□ Very certain
Q17a	How many non-governmental groups are there in the basin?
	□ **Very few** groups exist
	□ **Few** groups exist
	□ **Some** groups exist
	□ **Many** groups exist
	□ **Very many** groups exist
Q17b	How certain are you in your answer to question 17a?
	□ Very uncertain
	□ Uncertain
	□ Somewhat uncertain
	□ Somewhat certain
	□ Certain
	□ Very certain

(continued)

Table 15.2 (continued)

Q18a	How would you assess trust among stakeholders in the basin?
	□ There is **very little** trust among stakeholders
	□ There is **little** trust among stakeholders
	□ There is **moderate** trust among stakeholders
	□ There is **high** trust among stakeholders
	□ There is **very high** trust among stakeholders
Q18b	How certain are you in your answer to question 18a?
	□ Very uncertain
	□ Uncertain
	□ Somewhat uncertain
	□ Somewhat certain
	□ Certain
	□ Very certain
Q19a	How would you assess trust in government institutions in the basin?
	□ There is **very little** trust in government institutions
	□ There is **little** trust in government institutions
	□ There is **moderate** trust in government institutions
	□ There is **high** trust in government institutions
	□ There is **very high** trust in government institutions
Q19b	How certain are you in your answer to question 19a?
	□ Very uncertain
	□ Uncertain
	□ Somewhat uncertain
	□ Somewhat certain
	□ Certain
	□ Very certain
Q20a	How would you assess access to social services in the basin?
	□ A **very small subset** of society has access to social services
	□ A **small subset** of society has access to social services
	□ A **moderate subset** of society has access to social services
	□ A **large subset** of society has access to social services
	□ **All** of society has access to social services
Q20b	How certain are you in your answer to question 20a?
	□ Very uncertain
	□ Uncertain
	□ Somewhat uncertain
	□ Somewhat certain
	□ Certain
	□ Very certain

(continued)

Table 15.2 (continued)

Q21a	How would you assess grassroots governance efforts in the basin?
	□ There is **very little** grassroots governance
	□ There is **little** grassroots governance
	□ There is **moderate** grassroots governance
	□ There is **high** grassroots governance
	□ There is **very high** grassroots governance
Q21b	How certain are you in your answer to question 21a?
	□ Very uncertain
	□ Uncertain
	□ Somewhat uncertain
	□ Somewhat certain
	□ Certain
	□ Very certain
Q22a	How would you assess the degree of <u>innovation</u> in the basin?
	□ There is **very little** innovation
	□ There is **little** innovation
	□ There is **moderate** innovation
	□ There is **high** innovation
	□ There is **very high** innovation
Q22b	How certain are you in your answer to question 22a?
	□ Very uncertain
	□ Uncertain
	□ Somewhat uncertain
	□ Somewhat certain
	□ Certain
	□ Very certain
Q23a	How does the governance of the system respond to unexpected change in the basin?
	□ Change is **never** embraced
	□ Change is **infrequently** embraced
	□ Change is **sometimes** embraced
	□ Change is **often** embraced
	□ Change is **very often** embraced
Q23b	How certain are you in your answer to question 23a?
	□ Very uncertain
	□ Uncertain
	□ Somewhat uncertain
	□ Somewhat certain
	□ Certain
	□ Very certain

(continued)

Table 15.2 (continued)

Q24a	How evenly is institutional power distributed in the system?
	□ Power is **very unevenly** distributed
	□ Power is **unevenly** distributed
	□ Power is **moderately evenly** distributed
	□ Power is **evenly** distributed
	□ Power is **very evenly** distributed
Q24b	How certain are you in your answer to question 24a?
	□ Very uncertain
	□ Uncertain
	□ Somewhat uncertain
	□ Somewhat certain
	□ Certain
	□ Very certain
Q25a	How would you assess overlap in essential government services?
	□ There is **very little overlap** in essential governance services
	□ There is **little overlap** in essential governance services
	□ There is **moderate overlap** in essential governance services
	□ There is **high overlap** in essential governance services
	□ There is **very high overlap** in essential governance services
Q25b	How certain are you in your answer to question 25a?
	□ Very uncertain
	□ Uncertain
	□ Somewhat uncertain
	□ Somewhat certain
	□ Certain
	□ Very certain
Q26a	How would you assess the production of ecosystem services from the basin?
	□ The river and the wetlands produce a **very low** amount of ecosystem services
	□ The river and the wetlands produce a **very low** amount of ecosystem services
	□ The river and the wetlands produce a **moderate** amount of ecosystem services
	□ The river and the wetlands produce a **high** amount of ecosystem services
	□ The river and the wetlands produce **a very high** amount of ecosystem services
Q26b	How certain are you in your answer to question 28a?
	□ Very uncertain
	□ Uncertain
	□ Somewhat uncertain
	□ Somewhat certain
	□ Certain
	□ Very certain

(continued)

Table 15.2 (continued)

Q27a	How often is the generation of ecosystem services incorporated into management decision?
	□ The generation of ecosystem services is **never** incorporated into management decisions
	□ The generation of ecosystem services is **rarely** incorporated into management decisions
	□ The generation of ecosystem services is **sometimes** incorporated into management decisions
	□ The generation of ecosystem services is **often** incorporated into management decisions
	□ The generation of ecosystem services is **always** incorporated into management decisions
Q27b	How certain are you in your answer to question 29a?
	□ Very uncertain
	□ Uncertain
	□ Somewhat uncertain
	□ Somewhat certain
	□ Certain
	□ Very certain

15.3 Results and Analyses

We received completed responses (n = 30) from representatives of all four user groups and four watersheds. Total resilience scores (the sum of average response to questions) were even among NGO, end user, and government user groups, but markedly lower for research/extension respondents (Table 15.3). Among river basins, respondents from the Anacostia reported the highest scores, followed by Columbia, Platte, and Middle Rio Grande (Table 15.3).

To assess resilience and uncertainty, we calculated the mean scores reported for each question by assigning values corresponding with the Likert score for each question (i.e., very high = 5, high = 4, moderate = 3, low = 2, very low = 1, etc.). To assess uncertainty, we calculated mean uncertainty across individual stakeholders as reported in the questionnaires ("explicit uncertainty") as well as the degree of variance across respondents and the explicit uncertainty ("implicit uncertainty"). Only completed surveys were included in our analysis.

A comparative measure of resilience could be inferred from the total area under the curve in the radar plots of results (Figs. 15.1, 15.2, 15.3, 15.4, 15.5, 15.6, 15.7, 15.8 and 15.9). Ecological and social areas can be compared to assess trade-offs among different components of resilience. Relative resilience scores and degree of uncertainty can be compared among user groups and basins.

No individual questions from the survey appeared to drive or diverge from total resilience trends across user groups, with research/extension consistently reporting lower average scores (Fig. 15.2). Variance in our assessment is meant as an index of

Table 15.3 Total resilience scores by user group and watershed

	Total resilience
User Group	
NGO	78
End user	81.7
Government	80.8
Research/extension	66.5
SES Watershed	
Anacostia	82.5
Columbia	78.5
Middle Rio Grande	69.2
Platte	75.2

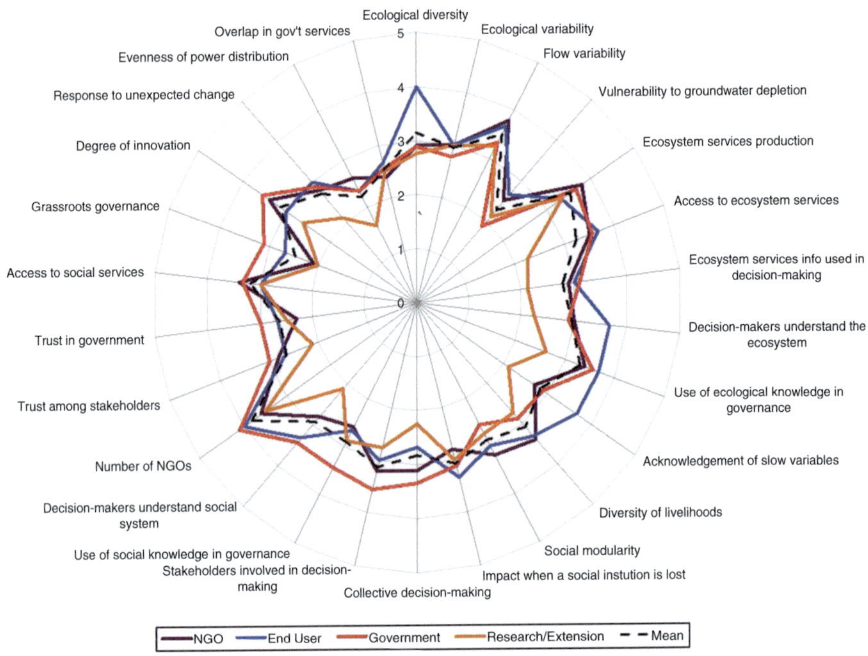

Fig. 15.1 Spider diagram showing average resilience scores for individual survey questions, by user group

implicit, group-level uncertainty. While most user groups had relatively low variance relative to their mean resilience scores, there are departures from this trend for the end users in their responses to questions on flow variability, ecosystem service production, and trust among stakeholders.

In addition to this group-level, implicit certainty, we asked respondents to report their individual level of certainty in their responses (Fig. 15.3). End users reported higher mean certainty in their responses, but this was not clearly reflected in vari-

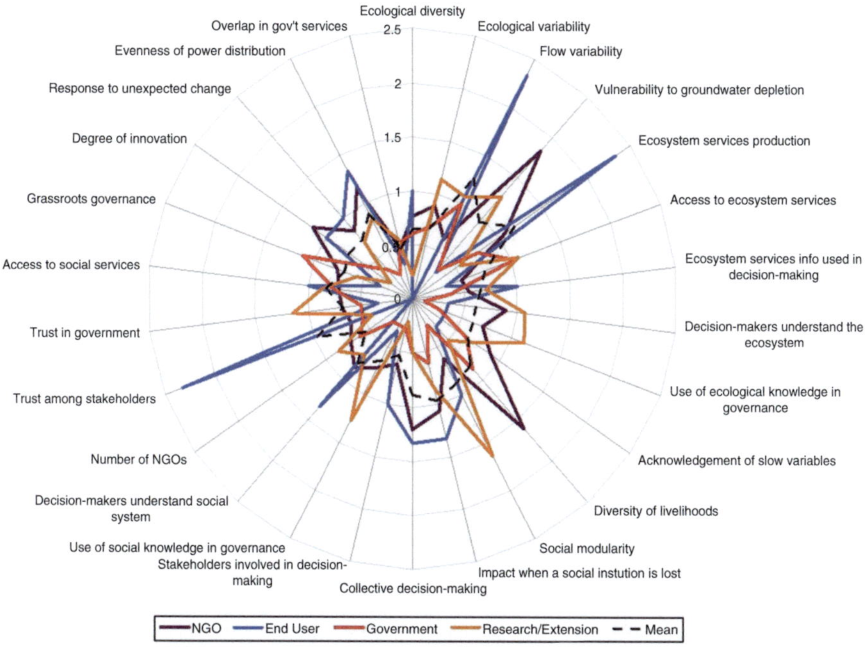

Fig. 15.2 Spider diagram showing variance in resilience scores for individual survey questions, by user group

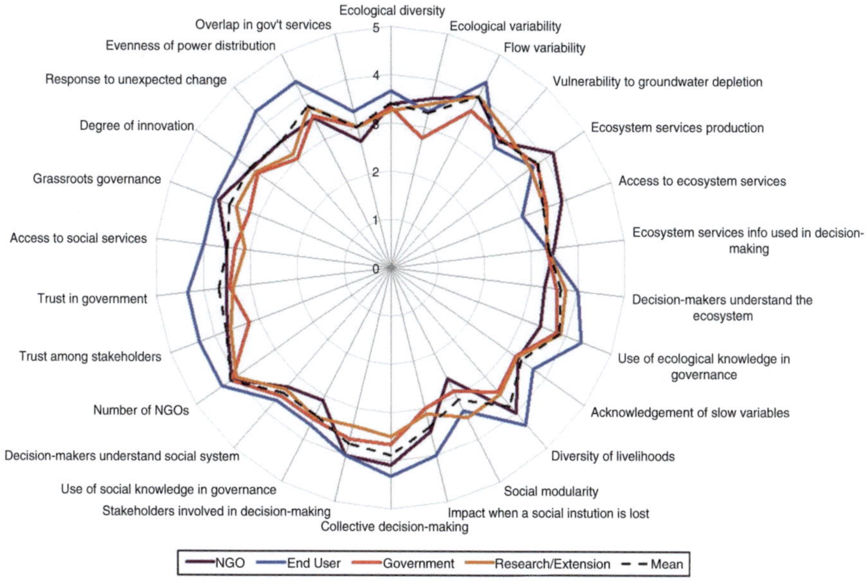

Fig. 15.3 Spider diagram showing average certainty in response for individual survey questions, by user group

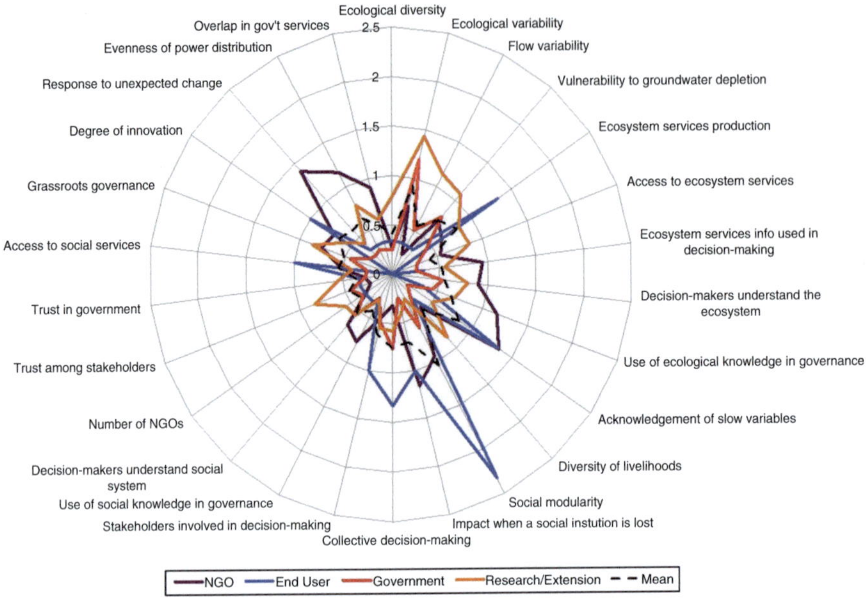

Fig. 15.4 Spider diagram showing variance in level of certainty for individual survey questions, by user group

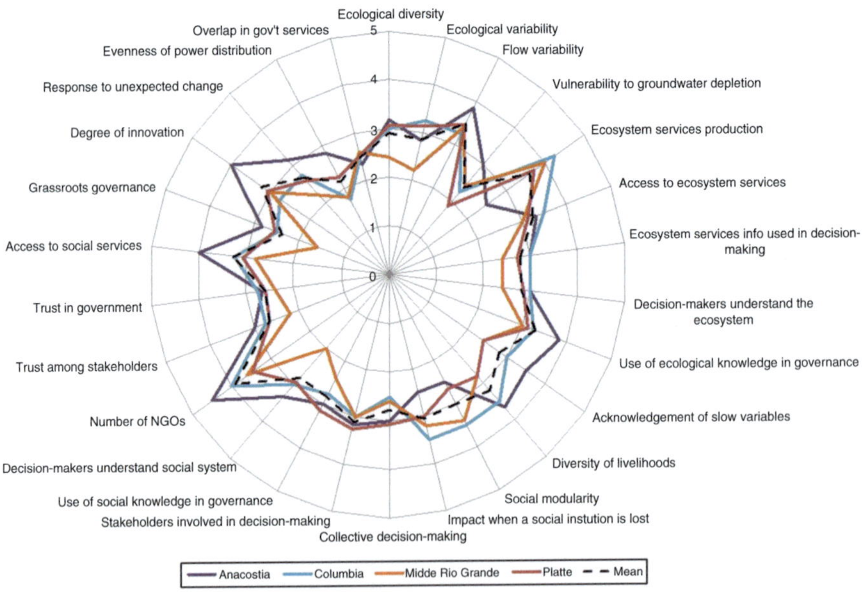

Fig. 15.5 Spider diagram showing average resilience scores for individual survey questions, by watershed basin

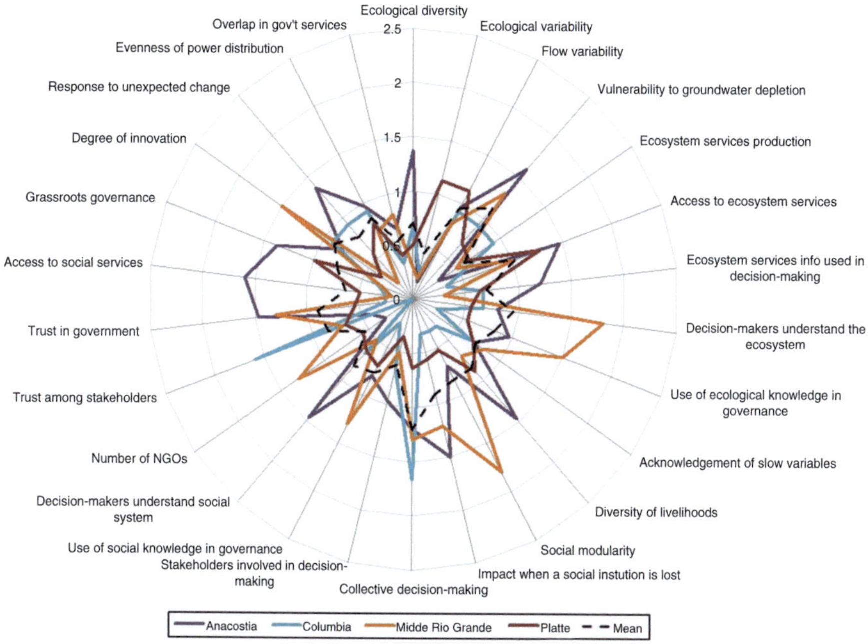

Fig.15.6 Spider diagram showing variance in resilience scores for individual survey questions, by watershed basin

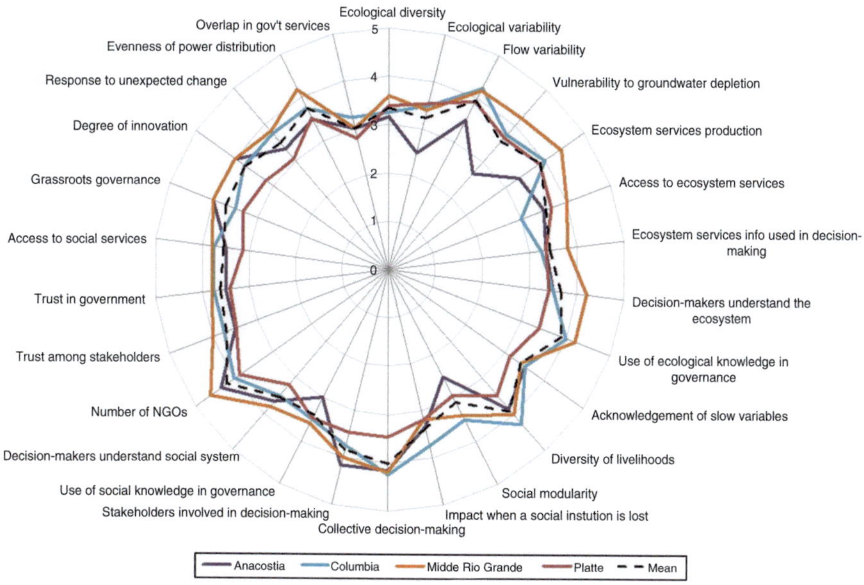

Fig. 15.7 Spider diagram showing average certainty in response for individual survey questions, by watershed basin

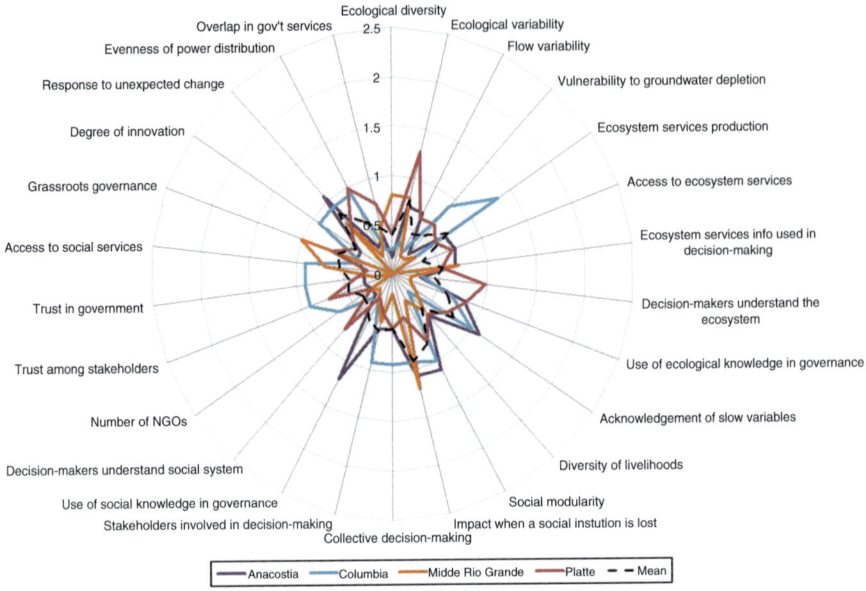

Fig. 15.8 Spider diagram showing variance in level of certainty for individual survey questions, by watershed basin

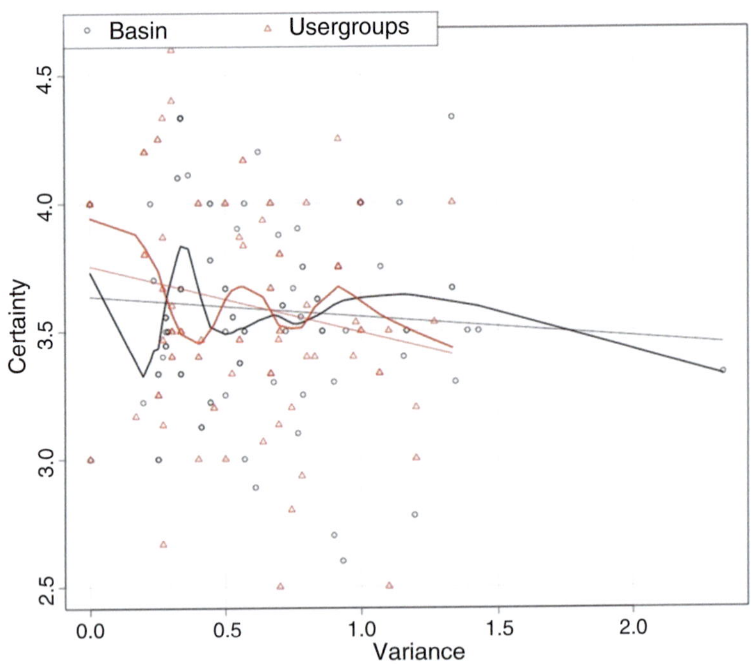

Fig. 15.9 Relationship between observed variance and user-identified level of certainty

ance of level of certainty (Fig. 15.4), which again shows no discernable trend aside from a spike in variance for end users regarding a question concerning social modularity.

Resilience scores for each question across basins (Fig. 15.5) do not provide any additional insight into the total resilience scores (Table 15.3). However, it is noteworthy that the shape of the resilience scores in Fig. 15.1 is rounder than that of Fig. 15.5, even though they draw from the same source data. This is perhaps reflected in the significantly higher variance, or implicit group-level uncertainty, in resilience scores across basins (Fig. 15.6). Individual uncertainty was also higher across basins (lower certainty scores; Fig. 15.7) than across user groups, but this difference is nominal and likely attributed to the redistribution of end users across basins in the former analysis. Similar to the pattern of variance across user groups, there was no discernable pattern of variance in level of response certainty by basin (Fig. 15.8).

When the variance of mean response to questions is plotted against certainty (i.e., group-level uncertainty versus individual certainty), we expected a negative relationship. However, this relationship is weak (Fig. 15.9) and only significant ($P<0.043$) when comparing variance versus certainty of user groups.

15.4 Discussion

Quantifying absolute, general resilience of complex systems of people and nature is problematic, but assessing relative resilience and specific resilience (resilience of what to what) is a realistic goal that will provide useful tools for managers and policy makers, as well as other stakeholders. Furthermore, the search for better methods to quantify resilience will lead to greater understanding of the drivers of resilience in complex systems. This study expanded upon the properties of resilience forwarded by Walker and Salt (2006) and modified by Nemec et al. (2014) to develop a straightforward survey of stakeholders in SES and approaches to quantify relative resilience (Nemec et al. 2014), trade-offs among social and ecological (or economic or infrastructure) components, and uncertainty. Such uncertainty was evaluated in two ways, user self-assessment and analysis of variance within responses and across watersheds and user groups.

It is clear that resilience in complex systems of people and nature encompasses both social and ecological components. Furthermore, purely social components can be differentiated from economic components. Among social, ecological, and economic components, clear trade-offs are often apparent. For example, the Platte River Basin, encompassing nearly all of Nebraska and parts of Colorado and Wyoming, is a complex system of people and nature that has been heavily altered by human activities, primarily agriculture in the uplands, and hydrological alteration of surface and groundwater in support of agriculture and development. In support of agriculture, and for other reasons including energy production, the Platte River itself has been dammed and hydrological variability greatly reduced. The riparian corridor has undergone a regime shift. Prior to damming, pulsing floods created bare sandbar habitat necessary for (currently endangered) least terns and piping plovers. With damming, sandbars became vegetated and eventually armored by herbaceous and woody vegetation, and channels reduced and incised. The riparian corridor, ecologically, is in an undesirable state with deep hysteresis. However, the social and economic aspects of the Platte Basin are highly desirable, and, unconsciously at least, a decision has been made to sacrifice the ecological component in favor of the social and economic components. Such imbalance is likely to lessen overall resilience. The described methods allow for assessing trade-offs in system subcomponents, which, in addition to social and ecological, could include economic and infrastructure.

Identifying individual components of resilience that are weak, and or highly uncertain, either at gross levels (ecological versus social, example above) or at the survey question level, should be beneficial for practitioners and managers. Explicitly considering trade-offs and identifying the areas of highest uncertainty will allow for focus on those aspects or components in most need of intervention or further understanding. These types of results can also provide baseline information against which the success of interventions can be assessed over time.

Acknowledgments This work was developed in part under the Adaptive Water Governance Project, funded by the US National Socio-Environmental Synthesis Center (SESYNC) under funding from the US National Science Foundation, NSF DBI-1052875. The Nebraska Cooperative Fish and Wildlife Research Unit is jointly supported by a cooperative agreement between the US Geological Survey, the Nebraska Game and Parks Commission, the University of Nebraska–Lincoln, the US Fish and Wildlife Service, and the Wildlife Management Institute. The views set forth by contributors to this volume represent their own and do not represent the views of any public or private entity the contributor is affiliated with.

References

Adger WN (2000) Social and ecological resilience: are they related? Prog Hum Geogr 24:347–364

Allen CR, Pope K, Fontaine TJ (2011) Adaptive management for a turbulent future. J Environ Manage 92:1339–1345

Allen CR, Angeler DG, Cumming FS, Folke C, Twidwell D, Uden DR (2016a) Quantifying spatial resilience. J Appl Ecol 53:625–635

Allen CR, Birge HE, Bartelt-Hunt SL, Bevans RA, Burnett JL, Cosens BA, Cai X, Garmestani AS, Linkov I, Scott EA, Solomon MD, Uden DR (2016b) Avoiding decline: fostering resilience and sustainability in midsize cities. Sustain For 8(9):844

Allen TFH, Tainter JA, Hoekstra T (2003) Supply-side sustainability. Columbia University Press, New York

Angeler DG, Allen CR (2016) Quantifying resilience. J Appl Ecol 53:617–624

Arnold CA, Green OO, DeCaro DA, Chase A, Ewa JG (2014) The social-ecological resilience of an eastern urban-suburban watershed: the Anacostia River Basin. Idaho Law Rev 51:29–90

Benson MH, Garmestani AS (2011) Can we manage for resilience? The integration of resilience thinking into natural resource management in the United States. Environ Manage 48:392–399

Benson MH, Llewellyn D, Morrison RR, Stone M (2014) Water governance challenges in New Mexico's Middle Rio Grande Valley: a resilience assessment Idaho Law Rev 51:195–228

Berkes F, Folke C (1998) Linking social and ecological systems for resilience and sustainability. In: Berkes F, Folke C (eds) Linking social and ecological systems: management practices and social mechanisms for building resilience. Cambridge University Press, New York, pp 1–25
Biggs R, Schlüter M, Biggs D, Bohensky EL, BurnSilver S, Cundill G, Dakos V, Daw TM, Evans LS, Kotschy K, Leitch AM, Meek C, Quinlan A, Raudsepp-Hearne C, Robards MD, Schoon ML, Schultz L, West PC (2012) Toward principles for enhancing the resilience of ecosystem services. Annu Rev Environ Resour 37:421–448
Biggs R, Schlüter M, Schoon M (2015) Principles for building resilience: sustaining ecosystem services in social-ecological systems. Cambridge University Press, Cambridge
Birge H, Pope K, Garmestani AS, Pope KL (2016) Adaptive management for ecosystem services. J Environ Manag 183:343–352
Carpenter SR, Walker BH, Anderies JM, Abel N (2001) From metaphor to measurement: resilience of what to what? Ecosystems 4:765–781
Chaffin BC, Garmestani AS, Gunderson LH, Benson MH, Angeler DG, Arnold CA, Cosens BA, Craig RK, Ruhl JB, Allen CR (2016) Transformative environmental governance. Annu Rev Environ Resour 41:399–423. https://doi.org/10.1146/annurev-environ-110615-085817
Cosens B, Fremier A (2014) Assessing system resilience and ecosystem services in large river basins: a case study of the Columbia. Idaho Law Rev 51:91–125
Folke C, Carpenter S, Elmqvist T, Gunderson L, Holling CS, Walker B (2002) Resilience and sustainable development: building adaptive capacity in a world of transformations. Ambio 31(5):437–440
Garner E (2010) Colorado population forecast. Colorado Department of Local Affairs. https://raqc.org/postfiles/board_meetings/2010/may07/StateDemographer050710.pdf. Accessed 6 Oct 2017
Godschalk DR (2003) Creating resilient cities. Nat Hazard Rev 4:136–143
Holling CS (1973) Resilience and stability of ecological systems. Annu Rev Ecol Syst 4:1–23
Kareiva P, Enquist C, Johnson A, Julius SH, Lawler J, Petersen B, Pitelka L, Shaw R, West JM (2008) Synthesis and conclusions. In: Julius SH, West JM (eds) Preliminary review of adaptation options for climate-sensitive ecosystems and resources: final report, synthesis and assessment product 4.4. U.S. Environmental Protection Agency, Washington, DC pp 9-1 to 9-66
Longstaff PH, Armstrong NJ, Perrin K, Parker WM, Hidek MA (2010) Resilient communities: a preliminary framework for assessment. Homel Secur Aff 6(3):1–23
Middle Rio Grande Basin Study [MRGBS] (2012) New Mexico Water Science Center. http://nm.water.usgs.gov/projects/middleriogrande/. Accessed 6 Oct 2017
Nemec KT, Chan J, Hoffman C, T. L. Spanbauer TL, Hamm JA, Allen CR, Hefley T, Pan D, Shrestha P (2014) Assessing resilience in stressed watersheds. Ecol Soc 19(1):34. https://doi.org/10.5751/ES-06156-190134
Nicolson CR, Starfield AM, Kofinas GP, Kruse JA (2002) Ten heuristics for interdisciplinary modeling projects. Ecosystems 5:376–384
Norris FH, Stevens SP, Pfefferbaum B, Wyche KF, Pfefferbaum RL (2008) Community resilience as a metaphor, theory, set of capacities, and strategy for disaster readiness. Am J Community Psychol 41:127–150
Palmer T (1996) America by rivers. Island, Washington, DC
Quinlan AE, Berbés-Blázquez M, Haider LJ et al (2016) Measuring and assessing resilience: broadening understanding through multiple disciplinary perspectives. J Appl Ecol 53:677–687
Resilience Alliance (2010) Assessing resilience in social-ecological systems: workbook for practitioners. Version 2.0. https://www.resalliance.org/files/ResilienceAssessmentV2_2.pdf. Accessed 6 Oct 2017
Resilience Alliance (2015) Assessment projects. https://www.resalliance.org/assessment-projects. Accessed 6 Oct 2017
Walker B, Salt D (2006) *R*esilience thinking: sustaining ecosystems and people in a changing world. Island, Washington, DC

Walker B, Carpenter S, Anderies J et al (2002) Resilience management in social-ecological systems: a working hypothesis for a participatory approach. Conservation Ecology 6(1):14

Walker BH, Abel N, Anderies JM, Abel N, Cumming G, Janssen M, Lebel L, Norberg J, Peterson GD, Pritchard R (2009) Resilience, adaptability, and transformability in the Goulburn-Broken Catchment, Australia. Ecol Soc 14(1):12

Williams BK (2001) Uncertainty, learning, and the optimal management of wildlife. Environ Ecol Stat 8:269–288

Williams BK (2011) Passive and active adaptive management: approaches and an example. J Environ Manage 92:1371–1378

World Wildlife Federation [WWF] (2012) Resilience-building: a new approach to protecting arctic ecological values. http://wwf.panda.org/who_we_are/wwf_offices/denmark/index.cfm?uProjectID=9Z1414. Accessed 6 Oct 2017

Chapter 16
Theory and Research to Study the Legal and Institutional Foundations of Adaptive Governance

Daniel A. DeCaro, Brian C. Chaffin, Edella Schlager, Ahjond S. Garmestani, and J. B. Ruhl

Abstract Adaptation to major social and ecological changes requires the participation, innovation, social learning, and political deliberation of many stakeholders, doing many different governance activities at different scales. Legal and institutional systems set the ground rules for this governance activity, establishing boundaries and opportunities for widespread innovation and cooperation. However, the enabling conditions for adaptive governance are poorly understood, making it difficult to facilitate. Candidate design principles that describe enabling conditions for adaptive environmental governance are proposed. Research opportunities are outlined to study the effects of these factors in different social-ecological systems and to further refine the principles.

Keywords Adaptive governance · Climate change · Design principles · Environmental law · Social-ecological resilience · State-reinforced self-governance · Water governance

---

D. A. DeCaro (✉)
Department of Urban and Public Affairs, Department of Psychological and Brain Sciences, University of Louisville, Louisville, KY, USA
e-mail: daniel.decaro@louisville.edu

B. C. Chaffin
W. A. Frank College of Forestry and Conservation, University of Montana, Missoula, MT, USA
e-mail: brian.chaffin@mso.umt.edu

E. Schlager
School of Government and Public Policy, The University of Arizona, Tucson, AZ, USA
e-mail: Schlager@email.arizona.edu

A. S. Garmestani
School of Natural Resources, University of Nebrasksa, Lincoln, NE, USA

J. B. Ruhl
Vanderbilt University Law School, Nashville, TN, USA
e-mail: jb.ruhl@vanderbilt.edu

B. Cosens, L. Gunderson (eds.), *Practical Panarchy for Adaptive Water Governance*, https://doi.org/10.1007/978-3-319-72472-0_16

16.1 Introduction

Environmental governance systems must embrace a broader set of stakeholders and become more innovative and flexible to cope with major changes in social-ecological systems (SESs). Vital social and natural resource systems are collapsing worldwide because many governance systems are based on outdated conceptions of nature (Garmestani and Allen 2014), antiquated assumptions about political engagement (Bingham et al. 2005; Ostrom 2010), rigid and poor-fitting policies (Arnold and Gunderson 2013), and traditional modes of governance that fail to accommodate uncertainty and changing conditions (Ruhl 2011; Cosens et al. 2014). These shortcomings contribute to poverty, conflict, and degraded societal conditions, threatening human welfare (IUCN 1980, 2008).

In theory, environmental governance systems can be designed to facilitate adaptation to social and environmental stressors, like economic downturn and climate change, helping society navigate important transitions (Chaffin et al. 2016; Olsson et al. 2006). Most stakeholders in these systems have some inherent capacity for self-organization and decision-making (Ostrom 1990, 2010). Furthermore, the general characteristics of adaptive governance (Dietz et al. 2003; Folke et al. 2005; Ruhl 2011), self-organization (Ostrom 1990, 2014), and the adaptive management of environmental resource systems (Holling 1978; Allen and Garmestani 2015) are well documented (Chaffin et al. 2014b). Such adaptive processes are flexible by being open to revision, employing iterative decision-making, and engaging in experimentation. They are innovative and participatory. They are polycentric, meaning they are spread across multiple centers of activity, social networks, and environmental stakeholders in a pluralistic decision-making context (Armitage 2007; Pahl-Wostl 2009). With sufficient knowledge, societies may be able to intentionally design governance systems to support these attributes, facilitating adaptation (Clarvis et al. 2014a).

Scholarship about adaptive environmental governance (AEG) is still relatively new, and there are large gaps in our understanding. Existing research has focused primarily on defining AEG and describing its core characteristics (Chaffin et al. 2014b). Importantly, AEG is context specific, self-organizing, and emergent (Folke et al. 2005; Ostrom 2014). Governance systems designed to facilitate AEG must therefore provide structure and guidance, but also enable localization and support stakeholder autonomy (Ostrom 2007, 2010). However, very few scholars have examined how AEG may be facilitated by legal frameworks and other institutional (rule-governed) systems or how to design governance systems to enable AEG (Shivakumar 2005).

This knowledge gap is addressed by introducing legal and institutional design principles of AEG (Table 16.1). These principles represent legal, economic, and democratic elements intended to create enabling conditions favorable to self-organization, collaboration, and adaptive problem solving among multiple centers of governance in society. The design principles are hypothesized to create sufficient resources, opportunity, guidance, and authority for AEG. Possible applications are

Table 16.1 Descriptions and examples of candidate legal and institutional design principles

Legal design principle	Description	Example
Reflexive	Does not rely on static rules (e.g., fixed water allocations) when flexibility is needed. Standards, priorities, and procedures guide decision-making	National Civic League (2013) amendments to city charters and US administrative procedure act, defining standards of "adequate" public participation and outlining mechanisms to allow agency discretion in meeting those standards
Legal sunsets	Planned periods of evaluation, in which environmental policies and agreements can be periodically examined, renegotiated, and potentially modified	Craig and Ruhl (2014) proposed amendments to US administrative procedure act, outlining periods when environmental agencies may test alternative policies, programs, without external review
Legally binding authority	Authority to make decisions and implement chosen solutions is institutionalized in binding legislation, for multiple centers of governance activity in polycentric society (e.g., agencies, communities, regional governments)	Land-use laws, giving individual farmers authority to self-organize in increasingly complex collective decision-making structures, parallel to formal government at local, state, and national levels (Sarker 2013)
Legally binding responsibility	Formal definition and assignment of responsibility to resolve, or contribute to resolution, of a social-ecological dilemma or element thereof	US state of Oregon requiring local and regional governments to align their development plans with 1970 statewide planning goals (Wheeler 2000)
Tangible support	Assistance from national, state, local government, etc., in the form of dedicated and sufficient funds, technology, information, and training to meet one's legally binding responsibilities and pursue one's legally granted authority	Individual farmers (Sarker 2013) and local and regional governments (Wheeler 2000) given tangible assistance from state and national authorities; coral reef conservation act, providing funding for research, restoration, and technical training for coral reef management (Green et al. 2014)

(continued)

Table 16.1 (continued)

Institutional design principle (Ostrom 1990, 2010)	Description	Example
Well-defined boundaries	Political, institutional, and ecosystem boundaries of the social-ecological dilemma are well-defined, clarifying legal and institutional jurisdiction	River compacts (e.g., Platte River, recovery and implementation program, Birge et al. 2014) and water management agreements (e.g., NYC watershed MOA 1997), which clarified sociopolitical, legal, and institutional boundaries and refined sector or subsystem jurisdiction (see also Schlager and Heikkila 2011)
Participatory decision-making	Variety of participatory methods and processes used so that affected stakeholders can influence design and implementation of governance systems that affect them	National Civic League (2013) and Bingham (2009, 2010) amendments to US administrative procedure act
Internal enforcement	In addition to external monitoring and enforcement systems, organizations and collectives (e.g., partners, compacts, communities) have internal mechanisms to monitor and enforce compliance	NYC watershed memorandum of agreement (2012), South Platte River compact in context of the Platte River recovery and implementation program (Birge et al. 2014) and traditional community-based irrigation governance (e.g., Cox 2014), which have external and internal mechanisms to monitor and regulate stakeholder compliance (e.g., water usage and deliveries) (see also Heikkila et al. 2011)
Internal conflict resolution	Internal mechanisms for neutral and transparent conflict resolution	Traditional community-based irrigation institutions (Ostrom 1994, Cox 2014), contemporary polycentric, multi-stakeholder farmer irrigation collectives (Sarker 2013)

demonstrated in case examples drawn from multiple sources, including case studies in Part I of this book. Research directions and opportunities are also outlined.

16.2 Conceptual Frameworks

To explore legal and institutional foundations of AEG, this chapter focuses on its emergent and self-organizing properties. The idea of an emergence frontier is used to conceptualize how legal and institutional systems shape and constrain emergence in SESs. The emergence frontier aligns with theories of self-governance (Ostrom 1990), state-reinforced self-governance (Sarker 2013), adaptive law (Craig 2010; Ruhl 2011), and resilience (Holling 1986, 2001). These conventions help us understand how legal and institutional systems influence the evolution of stakeholder cooperation and self-organization in different SESs (for example, Larson et al.

2013). We were also inspired by several disciplinary traditions: e.g., dynamic federalism and reflexive law (Engel 2006; Garmestani and Benson 2013), political economy (Schlager et al. 2012), and participatory democracy (Cosens 2013; DeCaro and Stokes 2013). This chapter describes emergence frontiers and self-organization (DeCaro et al. 2017b).

16.2.1 Emergence Frontiers

Many stakeholders and centers of governance activity are involved in the governance of complex SESs. These centers fulfill many different, potentially complementary roles (Larson et al. 2013). For example, in US water governance, government agencies carry out various legal mandates, enforcing the law and managing official government resources. Many are also involved in transboundary—interstate or international—agreements, governing shared water problems with different stakeholders (Heikkila et al. 2011). Governments also create policies and work within various legal frameworks that broadly influence water governance (Cosens et al. 2017). Stakeholder groups, such as grassroots watershed organizations are actively engaged in environmental monitoring, advocacy, and restoration projects. Such groups often monitor government agencies and local actors to ensure compliance with water regulations (Sabatier et al. 2005) and advocate for better participatory democracy (Arnold et al. 2014; Chap. 3). In addition, some communities organize governance systems that independently manage a resource, address a watershed problem, or complement formal government activities by filling management gaps (Ostrom 1990, 2010). These groups often use the law itself (e.g., lawsuits), and public advocacy, and other methods to influence legal systems (Wheeler 2000). Finally, formal and informal bridging structures including partnerships or compacts and organizations such as NGOs, or universities often serve as liaisons among different governance centers, helping to reconcile scales, share information, and coordinate activities (Garmestani and Benson 2013).

Chaffin and Gunderson (2016) introduce the term emergence to describe the relatively spontaneous, or self-organizing, nature of adaptive processes within and among centers of activity in AEG. Every center has some capacity for self-organization, in terms of semiautonomous decision-making, rulemaking, and implementation (Ostrom 2005). The phrase emergence frontiers (Fig. 16.1) is introduced here to describe how the expansion and contraction of capacity, authority, and resources provide opportunities or space for adaptation within a particular center of activity. This space is characterized by flexibility, decision-making authority, polycentricity, cooperation, and social-ecological fit (Armitage 2007). Emergence frontiers are bounded by existing legal conditions: legal systems influence the space, or latitude, for adaptation by establishing the authority, formal rules, and procedures for governance activity. For example, the US Congress delegates substantive authority to environmental agencies such as the US Fish and Wildlife Service to manage environmental resources. However, the Administrative Procedure Act (APA) and pre-decision assessment statutes like the National Environmental Policy Act effectively limit agency adaptation by requiring feasibility assessments before acting (Craig and Ruhl 2014).

Such opportunities and constraints are represented as a solid box in Fig. 16.1 to illustrate how existing governance conditions shape the space in which adaptation can emerge in a center of activity, such as a local watershed association. Overly rigid legal provisions may confine emergence, constricting an emergence frontier and constraining adaptation. Legal systems and traditional governments can also expand or broaden emergence frontiers (Fig. 16.1). Social-ecological factors, such as politics, economy, culture, or environmental events, can additionally affect emergence.

A focal center of activity is often linked to other centers; it may be embedded in broader institutional arrangements, involving multiple stakeholders and legal/institutional systems (Larson et al. 2013). For example, New York City water quality is governed through a Memorandum of Agreement involving the city, multiple jurisdictions in the Catskill and Delaware watersheds, and New York State (Galusha 2002). Such linkages and their effect on emergence in a focal activity center often exist at different scales (Fig. 16.1). Such linked centers also have emergence frontiers that can adapt and influence other activity centers.

System dynamics, such as new policies or a crisis, affect adaptation within and among different governance centers of activity. For example, larger-scale dynamics, such as modification of federal law, can lead to changing rules and norms within a

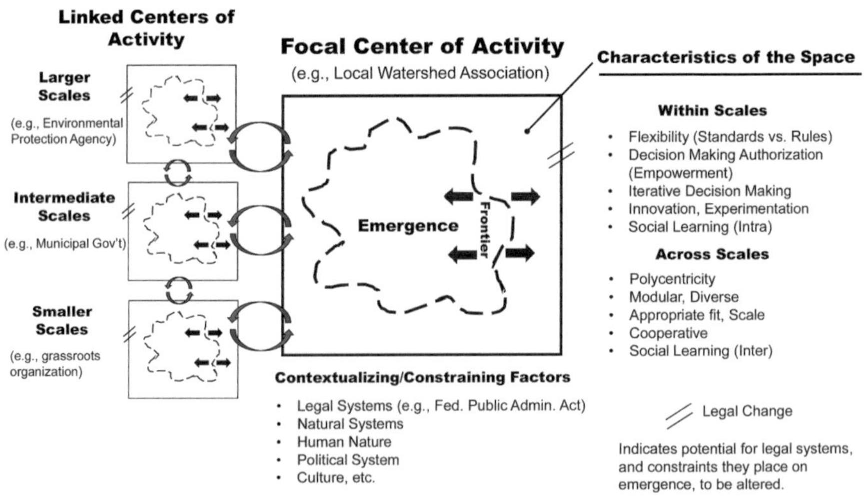

Fig. 16.1 Conceptual diagram of emergence frontiers within a focal center of activity (*large box in center*) and cross-scale linkages (*smaller boxes to the left*). Emergence frontiers describe how the expansion and contraction of capacity, authority, and resources provide arenas or space for adaptation within a particular center of activity, such as a local watershed association. Characteristics of the frontier spaces are listed in the *right hand column*; constraints on the space are listed *below* the large box. *Circular arrows* represent reciprocal polycentric connections to larger, intermediate, and smaller scales

smaller scale governance center or sets of centers. As one example, the APA (federal scale) establishes standards for public involvement in formal government activities (focal center). Bingham (2009) and NCL (2013) found such standards are limited and poorly defined and may lead to superficial public involvement that ultimately places government agencies in crisis because of public opposition or undermines public-government collaboration. The interaction of various centers at any scale may trigger system failure, resulting in either beneficial adaptation and broadening of emergence frontiers as shown for the Anacostia River (Chap. 3) or lead to further rigidity and entrenchment, as demonstrated in the Everglades case study (Chap. 5).

Emergence and self-organization are contextual properties of adaptive governance (Folke et al. 2005) and probably cannot be replicated in a top-down fashion (Ostrom 1990, 2014). Traditional centers of authority may be able to create enabling conditions for adaptation, using a suite of legal, economic, and democratic tools that facilitate self-organization and promote coordination among multiple (polycentric) activity centers. Sarker (2013) refers to this as state-reinforced self-governance.

In the next section, candidate design principles are proposed for AEG based on the central guiding concept of state-reinforced self-governance. The institutional design principles of Ostrom (1990, 2010) are expanded to include legal considerations, such as APA and NEPA, which determine the process for governmental and nongovernmental stakeholders to make certain decisions and carry out actions in a self-organizing way. Aspects of substantive law, which deal with compliance, and aspects of administrative law that influence how members of the general public collaborate with traditional government, are also presented. Hence, the chosen principles encompass a broad range of stakeholders and governance activities but are not exhaustive. Although important, aspects of law like common law, property law, and litigation, which do not focus explicitly on governance scope and authority, are not discussed.

16.2.2 Candidate Design Principles

In this chapter, design principles are features of legal, and other rule-governed (institutional), systems that may be important enabling conditions for self-organization and adaptation (Table 16.1). Legal design principles pertain to (a) any elements of official legal systems that determine structure, authority, function, and guidelines for government agencies or private centers of activity. For example, administrative law is an important determinant for government agencies (Cosens et al. 2017; Chap. 10), and nongovernmental organizations, like charitable organizations, are likewise influenced by various laws and legal processes, including ones that limit their financial and political activities (US State Department 2017). Legal design principles also pertain to (b) rules and regulatory systems that address compliance like the Endangered Species Act (Clarvis et al. 2014a). Institutional design principles refer more broadly to features of rule-governed systems that help groups and other collectives solve problems cooperatively. They originated from Ostrom's

(1990, 2010) analyses of self-governing communities and may also apply to larger-scale, formal governance systems (Schlager and Heikkila 2011; Sarker 2013).

It is important to address both legal and institutional principles because most complex governance problems involve both formal and informal governance systems and stakeholders. Altogether, the principles complement one another, creating sufficient resources, opportunity, guidance, and authority for robust cooperation and emergence in AEG. These design principles are important aspects of AEG as derived from many case studies (DeCaro et al. 2017a, b), including those in Part I of this volume. The following section provides a brief introduction to legal design principles, including reflexive law, legally binding authority and responsibility, and tangible support.

16.2.3 Legal Design Principles

Reflexive Law and Legal Sunsets: Iterative Decision-Making

One of law's primary roles is to establish ground rules for society, creating a sense of security and stability, which enables diverse societal stakeholders to cooperate, solve complex social dilemmas, and plan for the future (Tyler 1990, 2006; Ostrom 1998). Some scholars suggest that, in order for environmental law to enable more flexible decision-making without jeopardizing security or stability, legal systems affected by dynamic stressors like climate change should emphasize legal standards and guiding principles, which guide decision-making without specifying exact solutions that could become outdated or too rigid when social-ecological conditions change (Garmestani and Benson 2013). For example, in the western United States, specific amounts of water have sometimes been allocated to particular users based on historic water supply; in many cases, these are outdated and difficult to modify, posing a barrier to climate adaptation and cooperation, as illustrated in the Middle Rio Grande (Chap. 7). Proportional allocation rules may be more robust (Schlager and Heikkila 2011), especially if reasonable legal sunsets are applied, so that potentially outdated or problematic rules can be periodically updated and renegotiated (Clarvis et al. 2014b). Governance systems can (a) develop inertia, preventing change, and (b) become paralyzed by excessive scrutiny, as illustrated in the Florida Everglades (Chap. 5). Hence, planned sunsets may also provide a balanced, structured, and predictable way to assess governance arrangements, without becoming too flexible or simultaneously sacrificing responsiveness (Craig et al. 2017; Chap. 11).

These approaches embody aspects of reflexive law. In reflexive law, legal authorities establish goals, standards, and ground rules for engagement but leave open final solutions, so that decision-makers have both the legal guidance and flexibility they need to respond to complex societal dilemmas (Orts 1995; Teubner 1983). Reflexivity can apply to government agencies (Craig and Ruhl 2014), transboundary compacts and treaties (Schlager and Heikkila 2011), and community-based governance systems (Ostrom 1990, 2014) and is an important aspect of iterative decision-making, which is an essential characteristic of AEG (Armitage 2007).

Legally Binding Authority

The stakeholders within each center of activity need some authority to make decisions, carry out plans, and otherwise self-govern. Lack of authority to self-organize or contribute to important governance processes is a common barrier to AEG (Marshall 2007; Ostrom 2010), and a major source of conflict, as illustrated in the Anacostia (Chap. 3). Therefore, traditional centers of authority must have both the will and capacity to legitimize such self-organization (Shivakumar 2005; Sarker 2013).

Adaptive governance systems also tend to organize into a polycentric but coherent whole, composed of multiple stakeholders, organizations, and structures (e.g., partnerships), supported in some way by laws and institutions that legitimize their activity (Wheeler 2000; Sarker 2013). Dynamic federalism (Engel 2006; Schlager et al. 2011) and trans-governmental network theory (Craik and DiMento 2008) may provide legal guidance for these polycentric governance systems. Under dynamic federalism, loosely coordinated centers of authority (e.g., state and federal government agencies) carry and exchange various roles, powers, and responsibilities according to demand (Engel 2006). This aspect of US government may have allowed state agencies to address aspects of climate change adaptation before federal government entered this policy realm (Thomson and Arroyo 2011). Trans-governmental networks refer to formal and informal collaborative networks of governmental and nongovernmental stakeholders, which sometimes emerge to address shared social-ecological dilemmas. These governance structures and processes may be more effectively directed toward adaptive ends (Ruhl 2011) with formal incorporation in environmental governance (Craik and DiMento 2008).

Legally Binding Responsibility

Legally binding responsibility for environmental dilemmas at different governance scales may be a necessary component of effective AEG. Major stakeholders may be motivated to help resolve an environmental dilemma if they are formally vested with responsibility or perceive themselves as having a shared stake (Ostrom 1990). Legal action can clarify and assign this responsibility, triggering novel stewardship activity (Cosens et al. 2017; Chap. 10). For example, after the State of Oregon required state agencies and local governments to align their development plans with its 1970 Statewide Planning Goals, governments in Portland, Oregon, shared resources and cooperated to solve key environmental issues, creating novel intergovernmental agencies and other collaborative arrangements (Wheeler 2000).

Tangible Support

Devolved responsibility may be overwhelming without genuine decision-making authority and proper technical and financial support. Traditional centers of government may need regulatory tools like grants, tax incentives, and penalties (Wheeler 2000); new administrative standards and procedures (Bingham et al. 2005; Craig and Ruhl 2014); and stronger commitment to democratic norms like subsidiarity (Cosens et al. 2017; Chap. 10) and self-governance (Shivakumar 2005) to effectively support meaningful devolution of responsibility. In addition to reluctance to devolve responsibility and inadequate legal authority to self-organize (see previous principle), insufficient administrative and technical support are major sources of adaptive and cooperative failures in environmental governance (Marshall 2007; Clement 2010; Chap. 10).

16.3 Institutional Design Principles

The following institutional design principles come from Vincent and Elinor Ostrom's examination of self-organization in federal states (Ostrom 1971, 1999) and community-based environmental governance (Ostrom 1998, 2010). The Ostrom School of Institutional Analysis studies how governance systems influence societal capacity to adaptively solve complex problems (Shivakumar 2005), paying special attention to factors that encourage otherwise competitive, self-interested actors such as political factions or citizens to cooperatively solve shared social dilemmas (Cole and McGinnis 2015). Rather than look solely to centralized government for answers to societal problems, the Ostrom School emphasizes citizen self-organization (or self-governance) in communities (Ostrom 1990), municipalities (Ostrom 2000), and federations (Ostrom 1971, 1994; Schlager et al. 2011).

Ostrom's (1990) principles contributed to early articulations of adaptive governance (Dietz et al. 2003). The principles increase capacity for adaptive decision-making and cooperation (Ostrom 2014) as well as facilitate self-organization at smaller scales (Ostrom 1965, 1992). They include (a) well-defined political and geographic boundaries for resource management, (b) inclusive and deliberate participatory decision-making, (c) self-governed monitoring and enforcement, and (d) internal conflict resolution mechanisms (Cox et al. 2010). Access to official legal courts further supports self-governance when disputes cannot be settled internally, or informally (Ostrom 1965). These principles create a governance context of transparency, credibility, and enforceability which builds trust, solidifies environmental agreements, and helps maintain productive relationships (DeCaro et al. 2017a, b).

16.3.1 Informal Norms and Practical Solutions

Formal and informal norms and institutions are also important in AEG. Self-organizing agents frequently devise operational rules, norms, procedures, and technical methods to address significant management issues (Ostrom 1990; Folke et al. 2005). These solutions are often influenced by place-based knowledge, including cultural and ecological knowledge, and adapted over time, improving their fit and effectiveness (Ostrom 2014).

Reflexive law, such as legal standards, supports flexibility in decision-making. Legally binding devolution of authority gives stakeholders legal authority to make and act upon those decisions. Technical, financial, material, and informational assistance give decision-makers necessary resources to implement decisions and learn through iterative, experimental processes. Factors like well-defined sociopolitical and geographic boundaries, however difficult to achieve, make management of those systems more feasible, logistically and politically. Institutional design principles, such as internal mechanisms for dispute resolution, monitoring, and enforcement, coupled with participatory decision-making, support stakeholder self-organization and simultaneously generate conditions for procedural fairness, legitimacy, and security needed to cultivate trust and promote cooperation in complex social-ecological dilemmas. Finally, legal sunsets and quasi-legislative and judicial processes provide windows of opportunity for adaptation, and latitude for informal social norms and processes to emerge, within semirigid governance systems. If generalized more broadly throughout society, and linked across centers of activity, sectors, and scales, these design principles may support AEG.

Governance scholars are exploring legal mechanisms to support emergence and enable state-reinforced self-governance. For example, new legislation could clarify and enhance participatory decision-making (NCL 2013), expand quasi-legislative and judicial processes (e.g., citizen juries, participatory budgeting) to support community-based governance (Bingham et al. 2005), and give citizens more authority to collaboratively manage important aspects of ecosystems and environmental policy (Sarker 2013).

16.3.2 Demonstration of Design Principles

The design principles are illustrated using examples from the case studies in this volume. These examples cover a fairly broad range of factors, including scale, legal domains, and institutional activities. However, they are not comprehensive and should not be interpreted as ideal. These are presented for definitional purposes and to point out gaps in current AEG within each basin. A more complete demonstration of the principles within a single case study may be seen in recent analyses of Japan's complex irrigation governance system (Sarker 2013; Sarker et al. 2014) and the governance of Rhone's upper river basin (Clarvis et al. 2014b).

16.3.3 Examples of Institutional Design Principles

Clearly defined boundaries for legal and political jurisdiction, and biophysical systems, are a fundamental principle (Table 16.1) for successful management of complex systems, like river basins (Ostrom 1990). Stakeholders often spend considerable effort attempting to clarify these boundaries. For example, in the Everglades, the jurisdiction of the state water management institution is defined by hydrologic basin, but the legitimacy of this demarcation has been questioned on several occasions, hindering cooperation (Gunderson et al. 2014; Chap. 5). Formalized inter-jurisdictional river agreements, such as the Platte River Recovery and Implementation Program (Birge et al. 2014; Chap. 8) or the Columbia River Treaty (Cosens and Fremier 2014; Chap. 4), also attempt to clarify political, legal boundaries for water management, with varying degrees of success. Water management agreements such as the New York Watershed Memorandum of Agreement of 1997 (Galusha 2002) not only clarify boundaries but also refine sectoral or subsystem jurisdiction (Schlager and Heikkila 2011). The success of these attempts appears to depend on the extent to which boundaries are accurate, known, and accepted by each stakeholder.

Participatory decision-making is another Ostrom (1990) design principle: a variety of participatory methods and processes should be used so that affected stakeholders can influence the design and implementation of important governance systems . The adequacy of these methods can be questioned for every case study in this volume (Chap.3). For example, stakeholder groups, such as the Everglades Coalition (Gunderson et al. 2014; Chap. 5), are comprised of environmental groups, stakeholders, and nongovernmental agencies. Such groups meet annually to provide input into formal decision processes, such as Everglades restoration program (Gunderson et al. 2014; Chap. 5). However, according to a study conducted by the Clean Water Fund (2010), a significant number (74%) of public stakeholders feel such public engagement is inadequate. Furthermore, implementation of broad-scale collaboration has encountered many obstacles, which continue to shape cooperation (Gerlak and Heikkila 2011). The National Civic League (NCL 2013) and Bingham (2009, 2010) have proposed amendments to US Administrative Procedure Act, which may clarify standards for adequate public participation, creating more robust legal support for collaborative governance.

Compliance and enforcement of rules, regulations, and laws is done through a variety of mechanism throughout the case studies. All of the US study areas use federal agencies, such as the USGS, to monitor water and stream flow conditions in the systems. In addition to external monitoring and enforcement systems, organizations and collectives have internal mechanisms to monitor and enforce compliance with existing laws and regulation. The South Platte River Compact, in the context of the Platte River Recovery and Implementation Program (Birge et al. 2014; Chap. 8), and traditional community-based irrigation governance (Sarker 2013; Cox 2014) have external and internal mechanisms to monitor and regulate stakeholder compliance for items such as water usage and deliveries (Heikkila et al. 2011). These enforcement mechanisms were a frequent point of conflict in all the case studies, with government agencies responsible for their implementation often being unable or unwilling to carry out their responsibilities. To address these issues, it has been proposed that agencies become more proactive, transparent, and collaborative in their enforcement (Schlager and Heikkila 2011).

Many resource management systems have established transparent internal conflict resolution mechanisms. Traditional, community-based irrigation institutions use internal courts for this purpose, turning to outside courts only when needed (Ostrom 1994; Cox 2014). Irrigation commons in Japan are contemporary polycentric, multi-stakeholder farmer irrigation collectives that use communication and quasi-formal internal courts to resolve disputes before going to external venues (Sarker 2013). The large, bureaucratic water management cases in this volume, such as the Everglades (Chap. 5), rely on lawsuits in federal and state courts to resolve conflicts over changes in water management. More research is needed to determine the long-term benefits and costs of these different arrangements for AEG (Schlager and Heikkila 2011).

16.3.4 Administrative Law

The previous example broadly illustrated multiple legal instruments and structures. The following section describes some more targeted amendments to administrative law, illustrating how specific instruments might be modified to facilitate governmental adaptation and support broader public participation, in particular.

The US federal procedural statutes (APA, NEPA) sometimes hinder adaptability and decision-making authority of government agencies (Craig and Ruhl 2014). Before modifying their actions, government agencies typically complete comprehensive pre-decision assessments and minimum public participation; decisions can also be challenged and judicially reviewed. These requirements are meant to improve legitimacy and prevent serious mistakes. However, they can cripple environmental agencies, slowing their adaptation and miring them in protracted lawsuits as illustrated by the Everglades case study (Gunderson et al. 2014).

Craig and Ruhl (2014) outline amendments intended to alleviate these concerns. Instead of continuous review, agency activities for specific programs could be reviewed periodically so that agencies can test alternative environmental practices without excessive scrutiny, thereby providing a legal sunset. Moreover, only specific projects, with high social-ecological uncertainty, rapid change, and human controllability, would be eligible for these more flexible procedures, helping to identify appropriate applications of reflexive law. Such is the case with Platte River's Platte River Recovery and Implementation Program, which seems to be beneficial (Birge et al. 2014; Chap. 8). Nevertheless, this type of experimental adaptive management failed at the river basin scale in the Columbia River basin as a result of the lack of an inclusive governance mechanism to make decisions on trade-offs and potentially due to the high risk of failure in experimentation at this scale (Volkman and McConnaha 1993; Cosens and Williams 2012). Using nested governance to allow innovation and experimentation at small scales while providing resources from larger scales may be one way to address this problem (Cosens et al. 2017; Chap. 10).

Public participation is important, but is routinely undermined by government agencies who fear losing control (DeCaro et al. 2017a; Chap. 17). Hence, agency discretion must be balanced against participatory democracy. Bingham (2009, 2010) and colleagues (Bingham et al. 2005) outline amendments to the APA, which may complement Craig and Ruhl's (2014) amendments. These adjustments are designed to enhance public participation without paralyzing government agencies. Bingham (2009) argues that government agencies are able to avoid meaningful public participation partly because of ambiguities in the APA. Government agencies routinely use superficial forms of public engagement (e.g., public hearings) that give citizens virtually no decision-making influence or circumvent public participation entirely (Arnstein 1969; NCL 2013). Inadequate public participation, including marginalization of minority groups, is a major problem in US river basins, such as the Anacostia basin (Arnold et al. 2014; Chap. 3).

The National Civic League (NCL 2013) proposed model provisions to state APAs, municipal ordinances, and city charters, to address shortfalls in public engagement by city governments. The amendments clarify the standards for adequate public engagement by carefully defining participation to encompass broader democratic ideals, identifying types of participation that satisfy those ideals, and outlining best practices to implement them (Reed 2008). The amendments adhere to the principles of reflexive law, legally binding authority and responsibility, and tangible support in the following ways. First, agencies are authorized to select from these participatory tools without external judicial review, supporting agency discretion. Second, agencies are granted more authority to collaborate with governmental and nongovernmental stakeholders. Third, agencies are responsible to create and fund a staffed participation specialist to coordinate community engagement and train their personnel. City governments are also required to establish Citizen Advisory Bodies, give them powers to set agendas and recommend policy, and provide them "adequate administrative support," including training, space, and staff.

16.4 Summary and Future Research

This chapter synthesized multiple literatures and cases to propose legal and institutional design principles that may foster the emergence of adaptive governance among various governance centers. Future research should test the principles and explore their effects in different SESs and governance contexts (DeCaro et al. 2017a, b).

16.4.1 Describing Emergence Frontiers

It is not clear how emergence frontiers are shaped by the constellation of legal, institutional, and social-ecological factors that comprise a given SES. This information is needed to fully understand the dynamics of AEG and anticipate appropriate policies. Specifically, additional case studies should describe how emergence frontiers expand and contract, as well as how the proposed design principles (and other factors) influence the emergence and resilience of governance systems. Scholars may target enabling conditions for emergence, focusing on a few key governance centers in particular policy issue to understand how the factors influence their self-organization, cooperation, and resilience (Wheeler 2000; Sarker et al. 2014). These data could be supplemented by interviewing key actors in one or more focal governance centers to better describe how existing conditions shape decisions (DeCaro et al. 2017a; Chap. 17). Interview questions would target each design principle: for example, sources and sufficiency of federal and state funding, supportiveness of existing legal frameworks for the center's semiautonomous decision-making, and clarity of political and geographic boundaries. Comparative case studies would also be beneficial to triangulate common factors and contrast diverse governance settings (Poteete et al. 2010; Yin 2014), as illustrated by the case studies in this book: for example, the Anacostia's metropolitan basin in the eastern United States (Chap. 3) compared to the less metropolitan Klamath in the west (Chap. 6).

16.4.2 Understanding Law and Adaptive Environmental Governance

This chapter has brought together literature from law, institutional analysis, adaptive environmental governance, ecological resilience, and collaborative environmental problem management. As such, it presents a novel intersection of these disparate areas of scholarship and highlights areas for future research, as described in the following paragraphs.

A few aspects of law that directly support self-organization in AEG were introduced previously. However, many other aspects of law influence AEG indirectly, such as common law (Green et al. 2014), property law (Doremus and Hanemann 2008), and law governing judicial decision-making (Arnold and Gunderson 2013). Furthermore, legal instruments interact in ways that are poorly understood. For instance, grassroots watershed organizations may derive their legal authority to monitor watersheds from the Clean Water Act (Arnold et al. 2014; Chap. 3). In essence, this creates indirect legal authority for self-governance by grassroots organizations. Investigating these issues may identify important areas for refinement and better inform AEG.

Some scholars have adopted principles of resilience to understand patterns of robustness, resistance, and adaptability of governance systems in complex SESs (Chaffin et al. 2014b). Ruhl (2011) argues that AEG requires a balance between engineered resilience (specialization, robustness, resistance) and ecological resilience (diversity, flexibility) to provide both beneficial rigidity and flexibility (Craig et al. 2017; Chap. 11). Adaptation to significant stressors may entail any combination of rigid and flexible legal and institutional components within and across scales, governance centers, and domains of law. Additional research is needed to understand these and other interactions that drive complex dynamics in SESs (Cosens et al. 2017; Craig et al. 2017; Chap. 10; Chap. 11). For example, stakeholders sometimes use rigid aspects of governance systems to trigger change, such as the use of inflexibilities in the Endangered Species Act to create a crisis that may lead to transformation (Freeman 2010; Chaffin et al. 2014a; Chap. 6).

Governance scholars must learn how to integrate legal instruments to enable AEG for transboundary social-ecological problems spanning jurisdictions, sectors, and subsystems (Green et al. 2014, 2015). It is hypothesized that the design principles (Table 16.1) facilitate effective transboundary governance by creating favorable conditions for integrative governance systems such as bridging organizations, partnerships, and transboundary agreements to emerge. However, this may not be a complete explanation. For example, even though Japan's irrigated water is governed by state-reinforced self-governance, rice production is still governed centrally (Sarker 2013). Future research should determine whether compartmentalized state-reinforced self-governance is the best societies can do at such broad scales (Prager 2010), or if not, what role particular design features may play in establishing a broader capacity for AEG (Berkes 2007; Stern 2011). Finally, the function of any governance system depends on the decisions people make. Therefore, it is important to also understand how human decisions influence legal systems and contribute to societal cooperation and adaptation (Moser and Ekstrom 2010, DeCaro et al. 2017a; Chap. 17).

The ideas presented in this chapter are intended to inform a broad audience about potential legal and institutional foundations of AEG and start a robust dialogue about societal adaptation, which may lead to useful principles and research for AEG. Governance scholars and policymakers are therefore encouraged to explore the potential implications of state-reinforced self-governance and the legal and institutional principles outlined in this chapter.

Acknowledgments This work was developed in part under the Adaptive Water Governance Project, funded by the US National Socio-Environmental Synthesis Center (SESYNC) under funding from the US National Science Foundation, NSF DBI-1052875. The views set forth by contributors to this volume represent their own and do not represent the views of any public or private entity the contributor is affiliated with.

References

Allen CR, Garmestani AS (2015) Adaptive management of social-ecological systems. Springer, Dordrecht

Armitage D (2007) Governance and the commons in a multi-level world. Int J Commons 2(1):7–32

Arnold CA, Gunderson LH (2013) Adaptive law and resilience. Environ Law Report 43:10426–10443

Arnold CA, Green OO, DeCaro DA, Chase A, Ewa JG (2014) The social-ecological resilience of an eastern urban-suburban watershed: the Anacostia River Basin Idaho Law Rev 51(1):29–90

Arnstein SR (1969) A ladder of citizen participation. J Am Inst Plann 35:216–224

Berkes F (2007) Community-based conservation in a globalized world. Proc Natl Acad Sci U S A 104(39):15188–15193

Bingham LB (2009) Collaborative governance: emerging practices and the incomplete legal framework for public and stakeholder voice. J Disput Resolut 2(2):1–58

Bingham LB (2010) The next generation of administrative law: building the legal infrastructure for collaborative governance. Wisconsin Law Rev 10(2):297–356

Bingham LB, Nabatchi T, O'Leary R (2005) The new governance: practices and processes for stakeholder and citizen participation in the work of government. Public Adm Rev 65(5):547–558

Birge HE, Allen CR, Craig RK, Garmestani AS, Hamm JA, Babbitt C, Nemec K, Schlager E (2014) Social-ecological resilience and law in the Platte River Basin Idaho Law Rev 51(1):229–327

Chaffin BC, Gunderson LH (2016) Emergence, institutionalization and renewal: rhythms of adaptive governance in complex social-ecological systems. J Environ Manag 165:81–87

Chaffin BC, Craig RK, Gosnell H (2014a) Resilience, adaptation, and transformation in the Klamath River Basin social-ecological system. Idaho Law Rev 51(1):157–183

Chaffin BC, Gosnell H, Cosens BA (2014b) A decade of adaptive governance scholarship: synthesis and future directions. Ecol Soc 19(3):56. https://doi.org/10.5751/ES-06824-190356

Chaffin BC, Garmestani AS, Gunderson LH, Benson MH, Angeler DG, Arnold CA, Cosens BA, Craig RK, Ruhl JB Allen CR (2016) Transformative environmental governance. Annu Rev Environ Resour 41:399–423

Clarvis MH, Allan A, Hannah DM (2014a) Water, resilience and the law: from general concepts and governance design principles to actionable mechanisms. Environ Sci Pol 43:98–110. https://doi.org/10.1016/j.envsci.2013.10.005

Clarvis MH, Fatichi S, Allan A, Fuhrer J, Stoffel M, Romerio F, Gaudard L, Burlando P, Beniston M, Xoplaki E, Toreti A (2014b) Governing and managing water resources under changing hydro-climatic contexts: the case of the upper Rhone basin. Environ Sci Pol 43:56–67

Clean Water Fund (2010) Everglades for all? An in-depth look at public perception and public participation in the world largest ecosystem restoration project. http://www.cleanwater.org/files/publications/fl/everglades-for-all_survey_final_201003.pdf. Accessed 15 July 2017

Clement F (2010) Analysing decentralised natural resource governance: proposition for a "politicised" institutional analysis and development framework. Policy Sci 43(2):129–156

Cole DH, McGinnis MD (eds) (2015) Elinor Ostrom and the Bloomington school of political economy, Polycentricity in public administration and political science, vol 1. Lexington Books, Lexington

Cosens BA (2013) Legitimacy, adaptation, and resilience in ecosystem management. Ecol Soc 18(1):3. https://doi.org/10.5751/es-05093-180103

Cosens B, Fremier A (2014) Assessing system resilience and ecosystem services in large river basins: a case study of the Columbia River Basin. Idaho Law Rev 51(1):91–125

Cosens BA, Williams MK (2012) Resilience and water governance: adaptive governance in the Columbia River Basin. Ecol Soc 17(4):3. https://doi.org/10.5751/es-04986-170403

Cosens BA, Gunderson L, Chaffin B (2014) The Adaptive Water Governance Project: assessing law, resilience and governance in regional social-ecological water systems facing changing climate. Idaho Law Rev 51(1):1–27

Cosens BA, Craig RK, Hirsch S, Arnold CA, Benson MH, DeCaro DA, Garmestani AS, Gosnell H, Ruhl JB, Schlager E (2017) The role of law in adaptive governance. Ecol Soc 22(1):30. https://doi.org/10.5751/es-08731-22130
Cox M (2014) Applying a social-ecological system framework to the study of the Taos Valley irrigation system. Hum Ecol 42(2):311–324
Cox M, Arnold G, Tomás SV (2010) A review of design principles for community-based natural resource management. Ecol Soc 15(4):38. http://www.ecologyandsociety.org/vol15/iss4/art38/. Accessed 31 July 2017
Craig RK (2010) Stationarity is dead: long live transformation: five principles for climate change adaptation law. Harv Environ Law Rev 31:9–75
Craig RK, Ruhl JB (2014) Designing administrative law for adaptive management. Vanderbilt Law Rev 67(1):1–87
Craig RK, Garmestani AS, Allen CR, Arnold CA, Birgé H, DeCaro DH, Fremier AK, Gosnell H, aSchlager E. (2017) Balancing stability and flexibility in adaptive governance: an analysis of tools available in U.S. environmental law. Ecol Soc 22(2):3. https://doi.org/10.5751/ES-08983-220203
Craik AN, DiMento JF (2008) Environmental cooperation in the (partially) disaggregated state: lessons from the Security and Prosperity Partnership of North America. Chicago J Int Law 8:479
DeCaro DA, Stokes MK (2013) Public participation and institutional fit: a social–psychological perspective. Ecol Soc 18(4):40. https://doi.org/10.5751/es-05837-180440
DeCaro DA, Arnold CA, Boamah EF, Garmestani AS (2017a) Understanding and applying principles of social cognition and decision making in adaptive environmental governance. Ecol Soc 22(1):33. https://doi.org/10.5751/ES-09154-220133
DeCaro DA, Chaffin BC, Schlager E, Garmestani AS, Ruhl JB (2017b) Legal and institutional foundations of adaptive environmental governance. Ecol Soc 22(1):32. https://doi.org/10.5751/ES-09036-220132
Dietz T, Ostrom E, Stern PC (2003) The struggle to govern the commons. Science 302:1907–1912
Doremus H, Hanemann M (2008) The challenges of dynamic water management in the American West. UCLA J Environ Law & Policy 26(2):55–75
Engel KH (2006) Harnessing the benefits of dynamic federalism in environmental law. Emory Law J 56:159
Folke C, Hahn T, Olsson P, Norberg J (2005) Adaptive governance of social–ecological systems. Annu Rev Environ Resour 30:441–473
Freeman D (2010) Implementing the Endangered Species Act on the Platte Basin Water Commons. University Press of Colorado, Boulder
Galusha D (2002) Liquid assets: a history of New York City's water system. Purple Mountain Press, New York
Garmestani AS, Allen CR (2014) Social-ecological resilience and law. Columbia University Press, New York
Garmestani AS, Benson MH (2013) A framework for resilience-based governance of social-ecological systems. Ecol Soc 18(1):9. https://doi.org/10.5751/ES-05180-180109
Gerlak AK, Heikkila T (2011) Building a theory of learning in collaboratives: evidence from the Everglades restoration program. J Public Adm Res 21:619–644
Green OO, Garmestani AS, Hopton ME, Heberling MT (2014) A multi-scalar examination of law for sustainable ecosystems. Sustain For 6(6):3534–3551
Green OO, Garmestani AS, Allen CR, Gunderson LH, Ruhl JB, Arnold CA, Graham NA, Cosens BA, Angeler DG, Chaffin BC, Holling CS (2015) Barriers and bridges to the integration of social–ecological resilience and law. Front Ecol Environ 13(6):332–337
Gunderson LH, Garmestani A, Rizzardi KW, Ruhl JB, Light A (2014) Escaping a rigidity trap: governance and adaptive capacity to climate change in the everglades social ecological system Idaho Law Rev 51:127–156

Heikkila T, Schlager E, Davis MW (2011) The role of cross-scale institutional linkages in common pool resource management: assessing interstate river compacts. Pol Stud J 39(1):121–145
Holling CS (ed) (1978) Adaptive environmental assessment and management. Wiley, New York
Holling CS (1986) The resilience of terrestrial ecosystems: local surprise and global change. In: Clark WC, Munn RE (eds) Sustainable development of the biosphere. Cambridge University Press, Cambridge, pp 292–317
Holling CS (2001) Understanding the complexity of economic, ecological, and social systems. Ecosystems 4(5):390–405
International Union for Conservation of Nature and Natural Resources [IUCN] (1980) World conservation strategy: living resource conservation for sustainable development. IUCN, Gland
International Union for Conservation of Nature and Natural Resources [IUCN] (2008) Shaping a sustainable future: the IUCN programme 2009–2012. https://cmsdata.iucn.org/downloads/iucn_programme_2009_2012_dfc.pdf. Accessed 31 July 2017
Larson KL, Wiek A, Keeler LW (2013) A comprehensive sustainability appraisal of water governance in Phoenix, AZ. J Environ Manag 116:58–71
Marshall G (2007) Nesting, subsidiarity, and community-based environmental governance beyond the local scale. Int J Commons 2(1):75–97
Moser SC, Ekstrom JA (2010) A framework to diagnose barriers to climate change adaptation. Proc Natl Acad Sci USA 107(51):22026–22031
National Civic League [NCL] (2013) Making public participation legal. http://ncdd.org/rc/wp-content/uploads/MakingP2Legal.pdf. Accessed 27 July 2017
Olsson P, Gunderson LH, Carpenter SR, Ryan P, Lebel L, Folke C, Holling CS (2006) Shooting the rapids: navigating transitions to adaptive governance of social-ecological systems. Ecol Soc 11(1):18. http://www.ecologyandsociety.org/vol11/iss1/art18/
Orts EW (1995) A reflexive model of environmental regulation. Bus Ethics Q 5(4):779–794
Ostrom E (1965) Public entrepreneurship: a case study in ground water management. Dissertation, University of California at Los Angeles
Ostrom V (1971) The political theory of a compound republic: designing the American experiment, 2nd edn. Institute for Contemporary Studies Press, California
Ostrom E (1990) Governing the commons: evolution of institutions for collective action. Cambridge University Press, Cambridge
Ostrom E (1992) Crafting institutions for self-governing irrigation systems. Institute for Contemporary Studies Press, California
Ostrom V (1994) The meaning of American Federalism: constituting a self-governing society. Institute for Contemporary Studies Press, California
Ostrom E (1998) A behavioral approach to the rational choice theory of collective action: presidential address, American Political Science Association, 1997. Am Polit Sci Rev 91:1–22
Ostrom V (1999) Polycentricity (Parts 1 and 2). In: McGinnis MD (ed) Polycentricity and local public economies. University of Michigan Press, pp 52–74, 119–138
Ostrom E (2000) The danger of self-evident truths. PS Polit Sci Polit 33(01):33–46
Ostrom E (2005) Understanding institutional diversity. Princeton University Press, Princeton
Ostrom E (2007) A diagnostic approach for going beyond panaceas. Proc Natl Acad Sci USA 104(39):15181–15187
Ostrom E (2010) Beyond markets and states: polycentric governance of complex economic systems. Am Econ Rev 100:1–33
Ostrom E (2014) Do institutions for collective action evolve? J Bioecon 16:3–30
Pahl-Wostl C (2009) A conceptual framework for analysing adaptive capacity and multi-level learning processes in resource governance regimes. Glob Environ Chang 19(3):354–365
Poteete AR, Janssen MA, Ostrom E (eds) (2010) Broadly comparative field-based research. In: Working together: collective action, the commons, and multiple methods in practice. Princeton University Press, Princeton, pp 64–88
Prager K (2010) Local and regional partnerships in natural resource management: the challenge of bridging institutional levels. J Environ Manag 46:711–724

Reed M (2008) Stakeholder participation for environmental management: a literature review. Biol Conserv 141:2417–2431

Ruhl JB (2011) General design principles for resilience and adaptive capacity in legal systems: applications to climate change adaptation law. North Carolina Law Rev 89:1374–1401

Sabatier PA, Focht W, Lubell M, Trachtenberg Z, Vedlitz A, Matlock M (2005) Swimming upstream: collaborative approaches to watershed management. The MIT Press, Cambridge, MA

Sarker A (2013) The role of state-reinforced self-governance in averting the tragedy of the irrigation commons in Japan. Public Adm 91(3):727–743

Sarker A, Itoh T, Kada R, Nakashima M, Herath G (2014) User self-governance in a complex policy design for managing water commons in Japan. J Hydrol 510:246–258

Schlager E, Heikkila T (2011) Left high and dry? Climate change, common-pool resource theory, and adaptability of Western water compacts. Public Adm Rev 71(3):461–470

Schlager E, Engel K, Rider S (eds) (2011) Navigating climate change policy: the opportunities of federalism. The University of Arizona Press

Schlager E, Heikkila T, Case C (2012) The costs of compliance with interstate agreements: lessons from water compacts in the western United States. Publius, J Federalism 42(3):494–515

Shivakumar S (2005) The constitution of development: crafting capabilities for self-governance. Palgrave Macmillan, New York

Stern PC (2011) Design principles for global commons: natural resources and emerging technologies International. Int J Commons 5(2):213–232

Teubner G (1983) Substantive and reflexive elements in modern law. Law Soc Rev 17(2):239–286

Thomson VE, Arroyo V (2011) Upside-down cooperative federalism: climate change policymaking and the states. Virginia Environ Law J 29(1):1–61

Tyler TR (1990) Why people obey the law. Yale University Press, New Haven

Tyler TR (2006) Psychological perspectives on legitimacy and legitimization. Annu Rev Psychol 57:375–400

U. S. State Department (2017) Non-Governmental Organizations (NGOs) in the United States. https://www.state.gov/j/drl/rls/fs/2017/266904.htm#_ftn1. Accessed June 20 2017

Volkman JM, McConnaha WE (1993) Through a glass, darkly: Columbia River salmon, the Endangered Species Act, and adaptive management. Environ Law 23:1249–1272

Wheeler SM (2000) Planning for metropolitan sustainability. J Plan Educ Res 20(2):133–145

Yin KR (2014) Case study research: design and methods, 5th edn. SAGE, Thousand Oaks

Chapter 17
Theory and Research to Study Principles of Social Cognition and Decision-Making in Adaptive Environmental Governance

Daniel A. DeCaro, Craig Anthony (Tony) Arnold, Emmanuel Frimpong Boamah, and Ahjond S. Garmestani

Abstract Environmental governance systems must adapt to address increased uncertainty and new social-ecological conditions posed by stressors like climate change. This chapter presents several principles of social cognition and decision-making that influence adaptive governance. The principles are illustrated with examples from six US river basins. Future research opportunities are also outlined.

Keywords Adaptive governance · Cognition · Cooperation · Environmental law · Legitimacy · Social decision making

17.1 Introduction

Environmental governance systems establish a framework for societal cooperation and stewardship of earth's vital ecosystems. When governance systems perform poorly, vital resource systems may be depleted or destroyed, increasing conflict and undermining human welfare (IUCN 2008; UN 2015). Many of society's core environmental governance systems may not be suited to address contemporary

D. A. DeCaro (✉)
Department of Urban and Public Affairs, Department of Psychological and Brain Sciences, University of Louisville, Louisville, KY, USA
e-mail: daniel.decaro@louisville.edu

C. A. Arnold
Brandeis School of Law, Department of Urban and Public Affairs, University of Louisville, Louisville, KY, USA
e-mail: tony.arnold@louisville.edu

E. Frimpong Boamah
Department of Urban and Regional Planning, University of Buffalo, Bufallo, NY, USA
e-mail: efrimpon@buffalo.edu

A. S. Garmestani
School of Natural Resources, University of Nebraska, Lincoln, NE, USA

B. Cosens, L. Gunderson (eds.), *Practical Panarchy for Adaptive Water Governance*, https://doi.org/10.1007/978-3-319-72472-0_17

problems and will need to become more adaptive to cope with novel stressors such as climate change (Clarvis et al. 2014). Social-ecological systems (SESs) will change in complex and unpredictable ways as the global environment changes. For example, the Syrian Civil War may have been triggered by a combination of sustained drought caused by climate change, global economic and political shifts, and poor governance (Kelley et al. 2015; Vidal 2016). These same factors are involved in water conflict and resource system degradation in the USA, as illustrated in the Platte Basin (Chap. 8; Parker 2016).

Society's capacity to adapt to broad-scale change is manifest in both the social (governance) and ecological components of water resource systems. The adaptive capacity of governance systems and society are strongly influenced by social cognition and other aspects of decision making. This chapter presents several principles of social cognition—social aspects of motivation, perception, and decision-making (Lee and Harris 2013)—which may be especially relevant to adaptive governance and the law (DeCaro et al. 2017a). This chapter illustrates the principles using examples presented in Part I of this volume.

Adaptive environmental governance (AEG) is difficult because changes to governance systems are often resisted by society (Weber and Stern 2011) or poorly designed due to inadequate knowledge of human behavior and complex systems (Moser and Ekstrom 2010). Environmental decision-making is a widely studied topic (Kollmuss and Agyeman 2002; Cornforth 2009), and adaptation to climate change has received some attention (Moser and Ekstrom 2010; Swim et al. 2011). However, few psychological analyses have specifically examined environmental law and AEG, specifically from a social cognitive standpoint (Castro 2012). Failure to grasp this aspect of AEG contributes to governance failures and decreases societal support, hindering adaptation (Castro 2012; DeCaro et al. 2015). To develop more effective governance systems, scholars and policymakers must understand how inherently social actors perceive legal innovations and interact with one another in a highly social and political societal system.

The following section clarifies AEG and presents some conceptual lenses, such as that of societal cooperation, which may be used to better understand social cognitive processes involved in AEG. Research priorities will be discussed later.

17.2 Conceptual Lenses

17.2.1 Adaptive Environmental Governance

Environmental governance refers to all formal and informal actions societal stakeholders take to decide, design, implement, and enforce rule systems to oversee SESs (Chaffin et al. 2014b). Actions may be formal, such as official governmental policies or legal frameworks, or informal activities such as everyday practice and social norms (Pahl-Wostl 2009). Anyone with a stake in a particular social-ecological problem can be considered an important decision-maker, whether they are a government official or member of the general public, because societal outcomes are determined jointly by everyone's actions in that system or subsystem (Larson et al. 2013). Stakeholders fulfill many roles, ranging from rule-makers and enforcers to interpreters and compliers. For example, in the Anacostia River (Chap. 3), outcomes are jointly determined by different grassroots organizers who work together to influence state and local government officials, who are themselves connected to federal government actors and additional actors. It is important to consider each actor's role in the system to understand the social cognitions that drive their decisions (Ostrom 2005).

Current forms of formal environmental governance, such as federal laws, treat SESs as if they are linear and relatively unchanging (Green et al. 2015). Many of these policies are outdated and overlook important social-ecological dynamics (Arnold and Gunderson 2013). Adaptive environmental governance seeks to embrace the complexity of SESs with more flexible environmental policies and decision-making procedures that support experimentation, innovation, social learning, participatory democracy, and diversity (Armitage 2007). In theory, these elements enable society to respond to stressors more effectively (DeCaro et al. 2017b; Chap. 16).

In practice, however, many factors determine society's adaptive potential. This chapter discusses fundamental social cognitive features of adaptive governance, using societal cooperation as an interpretitive lens.

17.2.2 Social Cognition and Societal Cooperation

Governance systems are socially constructed. Therefore, these governance systems are strongly influenced by social cognition (Castro 2012). From a social cognitive standpoint, the core problem faced in AEG is cooperation and coordination in a complex social dilemma (Dietz et al. 2003), situations where individuals are tempted to pursue their narrow self-interest instead of the collective interest or common good (Parks et al. 2013). For society to adapt effectively, stakeholders must work together to develop new solutions to complex problems and encourage policy compliance.

AEG poses at least three cooperative challenges: (1) encouraging collaborative problem-solving despite inherent conflict, (2) garnering social acceptance and commitment, and (3) cultivating a culture of tolerance for change and uncertainty (DeCaro et al. 2017a). First, AEG is a contested social dilemma because stakeholders with different beliefs and desires compete for resources and social influence. Governance systems must encourage these stakeholders to work together (Ostrom 1990, 2014). Second, AEG requires social acceptance and commitment, lest new policies will be rejected or ignored (Castro 2012). Legal systems should explore ways to promote acceptance directly, through fair processes (Cosens 2013), but realize that solutions that empower the public to solve problems are equally important (DeCaro et al. 2017b; Chap. 16). Third, change tends to be perceived as a threat, because it is uncertain, alters existing power relationships, and disrupts people's lifestyles (Winter 2000). AEG systems must help fearful stakeholders cope with this stress by maintaining a sense of security, even while societal systems change (Craig

Table 17.1 Description of cognitive principles, social cognitive principles and social-psychological needs

Cognitive principles	
Mental models	Simplistic mental representations of complex problems can lead to inaccurate conclusions that adversely affect policy design, cooperation, and compliance
Loss aversion	People's natural aversion to losses can cause them to resist change, especially if the change is perceived (or framed) as a loss of power, endowments, or lifestyle
Social cognitive principles	
Social framing	Interest groups may portray important issues (e.g., alternative energy) as a threat or loss (e.g., "The Liberal's War on Coal") to increase public resistance and mobilize particular social groups
Social representations	Governance systems are subjectively interpreted and reshaped through popular discourse and everyday practice. These social representations (e.g., Agreeing in Principle, Equivocation; Castro 2012) affect the nature and effects of new governance solutions
Symbolic policy	Actors (e.g., politicians, government agencies) may engage in symbolic policy gestures designed to placate particular social groups, deceive, or mobilize constituents, often for political or strategic gain
Social-psychological needs	
Legitimacy and trust	Legitimacy and trust may help society cope with uncertainty and change during adaptation. Fundamental needs influence perceived legitimacy
Participatory democracy	Stakeholder participation in design and/or implementation of governance supports fundamental needs for self-determination (SD) and procedural justice (PJ):
	(A): Subjective definitions of genuine participation vary across individuals, groups, and contexts. Objective levels of involvement do not translate directly into high levels of felt SD or PJ. Must assess perceptions of PJ and SD (subjective)
	(B): Use multiple forms of participation to satisfy different stakeholders; use different methods when needs/goals change (participatory fit)
	(C): Iteratively assess stakeholder needs and perceptions of SD and PJ to identify desirable participatory processes and make necessary adjustments (adaptive)
Regulatory systems	Stakeholders need reassurance that they will not be taken advantage of by others and that rules and cooperative agreements will be honored. Regulatory systems (e.g., enforcement, monitoring) may provide this security when they are:
	(A): Legitimized: participatory democracy may legitimize enforcement, without sacrificing SD, PJ, or voluntary motivation
	(B): Used to empower stakeholders, not just to punish rule violators

et al. 2017; Chap. 11). As described later, legitimacy and trust and governance systems that satisfy fundamental needs, such as procedural justice, play an important role (Tyler 2006; Cosens 2013; Chap. 10).

Cooperative dilemmas are well known in water governance (Schlager and Heikkila 2011). However, many scholars and policymakers involved in AEG, with backgrounds outside political or psychological science, are unaware of the social cognitive foundations of these dilemmas. Many problems in AEG arise because of shortcomings in human cognition. The psychological principles presented in this chapter (Table 17.1) influence AEG by constraining governance systems' ability to resolve these cooperative challenges. Hence, knowledge of the principles may improve legal and institutional design.

17.3 Cognitive Biases

Decision-makers are plagued by cognitive biases. These biases distort their judgment and yield flawed decisions. They occur because people have incomplete knowledge, time, and mental resources to carry out complex calculations (Kahneman 2003). When decision-makers deal with complex issues, they often use cognitive shortcuts or heuristics (Gigerenzer and Todd 1999). Such cognitive shortcuts and biases actually increase people's susceptibility to certain kinds of social cognitive biases and social manipulation, which powerful stakeholders in society use to undermine societal cooperation and undercut AEG. Flawed mental models and loss aversion are two cognitive biases or limitations that increase people's susceptibility to social manipulation and social bias in decision making (DeCaro et al. 2017a).

17.3.1 Mental Models

Mental models are people's conceptual representations and lay theories of complex problems (Johnson-Laird 2013). Mental models strongly influence people's decisions, because they make certain aspects of a decision situation more salient and constrain the types of information and choice options they consider (Beach 1998). Simplistic or inaccurate mental representations can contribute to poor policy design and undermine public support for AEG (Moser and Ekstrom 2010). For example, misconceptions about climate dynamics and the factors that contribute to climate change have contributed to widespread disbelief in both climate change and the necessity for human intervention, thereby eroding public support for climate change policy (Weber and Stern 2011).

Disparate mental models can also be a barrier to adaptation, because they are difficult to reconcile. Failure to reconcile contradictory frames can create conflict, hindering coordination (Ostrom 2005) and yielding suboptimal solutions (Vireudenhil et al. 2010). For example, watersheds can be framed in conflicting ways by different stakeholders (Arnold 2014). In the Anacostia watershed, city managers see the Anacostia River primarily as a development asset, encouraging gentrification and undercutting broader desires for social and environmental justice by concerned citizens and environmental groups (Arnold et al. 2014; Chap. 3). Reconciling contradictory mental models is important for AEG (Ostrom 2005). In the Klamath River Basin, a window of cooperation emerged when some stakeholders stopped seeing each other as self-interested competitors and began to see each other as potential collaborators with common interest in shared problems (Chaffin et al. 2014a; Chap. 6). However, competing frames still threaten their ongoing deliberations and could ultimately disrupt AEG (Arnold et al. 2014; Chap. 3).

17.3.2 Loss Aversion

People strongly dislike loss, and this can cause them to resist change (Kahneman 2003). Unfortunately, many adaptive behaviors, such as reducing water use, seem like a loss initially, because they represent a departure from the status quo or involve personal sacrifice (Winter 2000). For example, in the Middle Rio Grande Basin (Benson et al. 2014; Chap. 7) and Platte Basin (Birgé et al. 2014; Chap. 8), poor ecosystem management transformed a once dynamic stretch of river with seasonally renewed floods, trees, and banks into a more static system of old growth and invasive species. Even though this new ecological regime is unnatural and arguably maladaptive, policymakers are reluctant to alter it, to restore vital seasonal flows needed for protected species, because the public has become attached to this landscape and would view such alteration as a loss.

Loss aversion is also important because lifestyles and power relationships are deeply entrenched in society. Most adaptations disrupt existing social structures and may be seen as a loss of entitlements or power, especially to stakeholders and interest groups who currently benefit from existing relationships (Kemp et al. 1998; Cote and Nightingale 2012). For example, even though robust public participation in environmental governance is often required by law, public stakeholders are routinely denied such participation because government officials perceive public involvement as a loss of control (Castro and Batel 2008; Marshall 2008; DeCaro and Stokes 2013). Failure to properly involve key stakeholders in important decisions was an underlying problem for adaptation in many river basins in this volume, such as the Anacostia (Arnold et al. 2014; Chap. 3) and Columbia (Mashuda et al. 2016).

Loss aversion and shortcomings in people's mental models increase susceptibility to three social cognitive biases that hinder AEG: social framing, social representations, and symbolic policy.

17.4 Social Cognitive Biases

17.4.1 Social Framing

Social framing refers to the strategic ideological or politicized portrayal of information to sway public opinion and influence public policy (Edelman 1960). Three aspects of social cognition increase susceptibility to social framing: in-group bias, confirmation bias, and flawed risk assessment.

First, regarding in-group bias, people often look to their accepted social groups (in-groups) for important information (Levine 1999). Laypersons rely on popular media and figureheads (e.g., family members, faith leaders, politicians) in their ideological community to understand complex issues such as climate change (Weber 2010). Figureheads within some communities couch information in politically charged, self-interested ways, which deviate from science (McCright et al. 2014). Second, people feel threatened by information that contradicts their beliefs (Winter 2000) and often ignore or devalue different perspectives (Nickerson 1998). They also seek out people who share their beliefs, thereby insulating and reinforcing their beliefs (Abrams et al. 1990), making them more difficult to change (Stern and Easterling 1999). Third, laypeople use shortcuts, like personal experience, familiarity, and the perceived severity or intensity (e.g.,vividness) of examples, to estimate probabilities of life events, such as catastrophic environmental events. Such estimates are distorted by extraneous information, like emotional labels and how often events are shown in media (Lichtenstein et al. 1978; Loewenstein et al. 2001). Imbalanced exposure to particular perspectives, coupled with highly emotional labels, can distort people's perceptions and then be used by interest groups to undermine societal cooperation (Moskowitz 2014).

Weber and Stern (2011) describe how strategic social framing has undermined climate change policy in the USA. When news outlets cover climate change, they often focus on controversial issues, use emotional labels, and feature fringe guests who have extreme beliefs, in a misguided attempt to get stronger ratings and show all sides of an argument (Boykoff and Boykoff 2004; Entman 2007). Thus, issues are often presented in emotional terms with disproportionate emphasis on climate change skepticism. News outlets give interest groups who oppose climate change adaptation an ideal platform to undermine societal cooperation (Lever-Tracy 2010; Farrell 2016). For example, conservative political leaders routinely disparage climate change science and frame environmental policies negatively, preying on the public's cognitive biases. They pit environmental responsibility against economic security to portray AEG as a loss (Schlichting 2013). They also use emotional labels like, "the war on coal," to anger and mobilize their ideological supporters (McGinley 2011; Smith 2012).

Popular misconceptions about climate change, spread by the media and further amplified by people's social networks, have polarized US environmental politics (McCright et al. 2014). According to Gallup polls, the public agreed with the scientific community's concern for climate change in 1997 but began to diverge along conservative/liberal party lines in 2001, creating a large bipartisan gap (Dunlap and McCright 2010). This split coincides with documented political lobbying and financial support from opposition leaders in major oil and gas companies, religious groups, and powerful political conservatives (Schlichting 2013; Stoknes 2014; Farrell 2016). Now political affiliation (in-group bias) reliably predicts support for environmental policy innovation (Hardisty et al. 2010; Gromet et al. 2013).

17.4.2 Social Representations

Social representations refer to societal interpretations of important issues, similar to social frames but more systemic to everyday discourse and practice (Moscovici 1988). Castro (2012) describes how social representations can negatively affect legal innovation in AEG. For ecample, people can agree with legal innovations in principle but disagree with all concrete instances of their application (Table 17.1). This tactic allows people to feign support for the US Endangered Species Act and disguise non-compliance. Landowners can agree with the principle of protecting biodiversity, as required by the Act, but resist its enforcement on their own land (Opotow and Brook 2003; Castro and Mouro 2011). Similarly, government officials routinely avoid legal mandates for public participation by arguing that, while they agree with democratic participation in principle, their stakeholders are not knowledgeable, trustworthy, or interested enough to participate in a manner they deem correct. (Arnstein 1969; Castro and Batel 2008). Legal innovation can also be undercut by equivocation (Castro 2012). Legal provisions for public participation are often ambiguous because they do not clarify what constitutes genuine participation (NCL 2013). Government officials frequently equate minimal forms of public engagement such as public hearings that are designed to maintain their centralized control with more robust types of participation that truly empower other stakeholders (Arnstein 1969; Castro and Batel 2008; Clement 2010; NCL 2013). Hence, government officials can say that stakeholders participated, without devolving real control. Such equivocation is sometimes intentional, but can also be due to interpretative differences or mere accident. Regardless the outcome is the same—inadequate public participation. Equivocation, and other sources of poor engagement, generated conflict and maladaptive outcomes in all of the US basins in Part I.

17.4.3 Symbolic Policy

Symbolic policy activity is another barrier to AEG. Symbolic policy activity refers to actions undertaken primarily to appear as though one is contributing to change, without making substantive commitments (Edelman 1960). For example, many governmental and commercial actors in the US National Estuary Program participate in collaborative programs, without making substantive contribution to policy or adaptive ecosystem management (Lubell 2004). They participate presumably for strategic social and political gain (Lubell et al. 2010). Symbolic policy was raised by concerned citizens and environmental groups as a problem in the Anacostia River system (Arnold et al. 2014; Chap. 3). Several green urban development projects along the river were identified as merely symbolic gestures designed to hide traditional economic development behind a guise of sustainability. In complex governance systems, actors engage in both symbolic and substantive policy; moreover, and some symbolic activities have beneficial social effects such as building trust, or facilitating social learning (Berrone et al. 2009; Kim and Lyon 2013). It is therefore important to assess each policy carefully.

17.4.4 Policy Implications

Some social scientists recommend using cognitive biases to benefit society by framing AEG in a positive light, using loss aversion to mobilize pro-environmental actors or subtly changing key policies without the public's knowledge (Cornforth 2009). Policymakers should be cautious about such an approach. Social biases are inherent to social discourse, so social frames, symbolic policy activity, and social representations are always open to counter-debate and reframing (Castro 2012). Both sides of the debate have engaged in social framing for decades in the US, and the general public is more polarized than ever (Dunlap and McCright 2010). Furthermore, research suggests that such political posturing can backfire, enflaming conflict and undermining policy credibility and trust (Jung and Mellers 2016). Thus, regardless of how treacherous it may seem, there may be no substitute for genuine public discourse and open communication about proposed governance innovations. Public discourse is especially necessary if society values democracy (Ostrom 1971) and sees cooperative problem-solving as essential to AEG (Pahl-Wostl 2009).

17.5 Legitimacy, Trust, and Fundamental Social-Psychological Needs

Legitimacy is essential for AEG (Cosens 2013; Chap. 10). Governance systems that are perceived as legitimate build trust and encourage cooperation (Tyler and Degoey 1995; Leach and Sabatier 2005). Moreover, trust helps society endure crisis and transformation, maintaining a sense of order amidst change (Lind and Van den Bos 2002; Tyler 2006).

Participatory and regulatory systems are important foundations of legitimacy (Tyler 1990, 2006); however, they are complex and can easily backfire (Bowles 2008; Reed 2008). Consideration of fundamental social cognitions involved in these systems may clarify when these systems fail and guide future design. Hence, in this section, the psychological relationship among participatory democracy, regulation, and legitimacy is discussed.

17.5.1 Participatory Democracy

Participatory democracy includes stakeholders in the design, decision-making, or implementation of governance systems. Successful participation facilitates legitimacy and can promote cooperation psychologically by satisfying fundamental needs for procedural justice (Tyler 2006) and self-determination (Moller et al. 2006). Fair, transparent, and empowering participation can reassure societal stakeholders and help build a sense of trust and security needed to constructively cope with change (McComas et al. 2011).

However, if implemented poorly, participation can reinforce existing power inequities and generally undermine these benefits through misapplication (Irvin and Stansbury 2004) and outright manipulation (Arnstein 1969; Clement 2010). These problems are more likely when scholars and policymakers do not understand the psychological principles involved in participation (Chess and Purcell 1999).

DeCaro and Stokes (2008, 2013) and DeCaro et al. (2015, 2017a) reviewed case studies of environmental governance from a social cognitive standpoint. They argued that there are at least three popular misconceptions about participation that contribute to its misapplication. First, popular belief suggests that certain types of participation such as deliberative democracy are superior, always desired, and perceived as procedurally fair and self-determined, regardless of the context or stakeholders (Hunt and Haider 2001). Contrary to this belief, stakeholder preferences and subjective perceptions for participation often depend on context, meaning no panacea exists. Second, public participation is often treated as a one-time event, rather than an ongoing democratic process (Reed 2008). Third, different types of public engagement are often treated as if they are interchangeable, ignoring their essential differences (Castro and Batel 2008). Some actors also fear losing control (Bailey and Grossardt 2010). Hence, participatory processes are generally oversimplified, superficial, and exploited, undermining their legitimacy.

Structural inequities and popular misconceptions about participation are built into governance systems and implemented in everyday practice (Arnold 2004; NCL 2013). These problems were major barriers to AEG in every basin in Part I. Greater participation by historically marginalized groups, as coequal stakeholders, was a necessary change to address longstanding social inequity and improve capacity for AEG in the Anacostia, Columbia, Klamath, and Lake Eyre basins in particular (Chaps. 3, 4, 6, Chap. 9). In contrast, in the Everglades, unmitigated conflict between Native Americans and citizen groups, as well as policymakers and scientists, continues to prevent fuller forms of public engagement and collaboration (Chap. 5).

17.5.2 Policy Implications

Several factors influence perceptions of procedural justice and self-determination: transparency, neutrality, timeliness, honesty, voice and choice, rationales, and opportunity for review or appeal (Tyler 1988; Moller et al. 2006; Chap. 10). These factors are contextual, interact, and may change over time. Therefore, multiple types of participation may be needed to properly satisfy different stakeholders and address emergent needs (Howard 2010). This calls for an adaptive approach (von Korff et al. 2010; Munaretto et al. 2014) in which people's subjective perceptions (e.g., procedural justice) are measured and used to guide public engagement (DeCaro and Stokes 2013). Turner et al. (2014) provide a partial example. They interviewed 70 households from 12 Caribbean communities about their perceptions of participation and related this to the community's different governance situations and social-ecological contexts. Perceptions of legitimacy, and other factors, differed across the communities, depending on the context. The researchers used these observations to make targeted recommendations like increased devolution of control for particular communities, changes to enforcement, or more support from NGOs to facilitate collaborations (Druschke and Hychka 2015; Sarkki et al. 2015).

17.5.3 Regulatory Systems

Participatory processes and regulatory systems often influence legitimacy jointly (Cosens et al. 2017; Chap. 10). Regulatory systems include monitoring and enforcement and are used to ensure security and encourage cooperation: credible safeguards against uncooperative behavior are needed to enforce rules and protect conservation agreements (Hardin 1968; Rachlinski 1998). Regulatory systems can also facilitate trust by reassuring cooperators that they will not be exploited by defectors (Yamagishi 1986; Gächter 2007). Regulations are fundamental aspects of AEG (Dietz et al. 2003). Effective enforcement supports legitimacy of AEG systems (Tyler 2006, Cosens 2013).

However, if regulatory mechanisms are perceived as coercive, improper, or unfair, then regulatory systems may increase conflict and decrease cooperation (Peluso 1993; Tyler 1990), hindering policy acceptance and AEG. Intrinsic, voluntary forms of cooperation such as moral obligation and belief in the inherent importance of an environmental policy, which are necessary for long-term societal problem-solving and commitment, may also decrease (Bowles 2008).

Several psychological processes may be involved in counterproductive regulatory effects (Bowles 2008; DeCaro et al. 2015). Regulatory systems can (a) signal distrust encouraging stakeholders to behave uncooperatively (Mulder et al. 2006); (b) reframe formerly moral obligations (doing what is right) as economic transactions, making ethical behavior a purchasable service (Tenbrunsel and Messick 1999); (c) make stakeholders dependent on enforcement by convincing them that others are cooperating only because they fear punishment (Chen et al. 2009); and (d) undermine self-determination, which creates resistance and reduces intrinsic motivation (Brehm and Brehm 1981; Moller et al. 2006).

17.5.4 Policy Implications

The primary behavioral challenge in designing regulatory systems is to discourage defectors without undermining internal motivations or increasing conflict. Regulatory systems must therefore ensure security, without jeopardizing procedural justice and self-determination. DeCaro et al. (2015) suggest three guiding principles for policy implementation. First, participatory processes and enforcement are mutually reinforcing. Stakeholders should participate in the design of rules and their enforcement whenever possible (Ostrom 1990; Hilbe et al. 2014). Doing so justifies and legitimizes the regulations, safeguarding internal motivations. Second, regulatory systems should empower stakeholders by helping them reach shared goals—not simply force compliance. Having the ability to enforce rules that are chosen by stakeholders is usually empowering, whereas enforcing rules that are imposed externally, often feels unjustified (illegitimate) and oppressive (Tyler 1990), though there may be exceptions to this principle that are context specific (Agrawal and Ribot 2014). Third, effective regulatory systems often incorporate components of restorative justice (van Ness et al. 2013), which educate and rehabilitate rule violators, helping build community and sustain trust (Ostrom 1990).

Supporting these principles, Ostrom's (1990, 2010) analysis of community-based governance systems demonstrates that effective regulatory systems are often accompanied by participatory processes and rarely without stakeholders' endorsement. Such systems often use graduated sanctions, whereby punishments increased in severity with each violation, beginning with a warning, education, and open dialogue. Similar approaches have been successful at larger scales, with river compacts and treaties (Schlager et al. 2012) and complex public-private irrigation governance systems that span local, regional, and national scales (Sarker 2013). According to Schlager and Heikkila (2011), river compacts could benefit from increased multi-stakeholder decision-making, monitoring, and enforcement. Voluntary conflict resolution mechanisms, which may resolve disputes before they escalate, can also help (Ostrom 1965; Schlager et al. 2012). The settlement group provided this role in the in Klamath River (Chaffin et al. 2014a; Chap. 6). In short, laws targeting participatory democracy and enforcement must be based on sound principles of social cognition to ensure they facilitate AEG.

17.6 Broader Implications

All societal stakeholders play an important role in AEG. Public discourse, information, and media together influence societal cooperation. Hence, these systems are important elements of AEG. Societal stakeholders must be knowledgeable about AEG—what it is, why it is necessary, and its challenges—so that they can (a) better tolerate the uncertainty and conflict associated with adaptation and (b) constructively participate in adaptive governance. Failure to sufficiently grasp the core principles of adaptive governance (e.g., complexity, uncertainty, cooperative problem-solving) may be the single most important psychological barrier. According to O'Leary et al. (2010) and Ostrom (1998, 2010), civic education must include robust philosophical and practical training about many forms of democracy, including community-based governance (Ostrom 1990), quasi-legislative and quasi-judicial systems (Bingham et al. 2005, 2008), tactical demonstrations (Lydon et al. 2010), and collaboration (Bingham 2009, 2010). Public education in cognitive science and social psychology, as well as cooperative decision-making (i.e., social dilemmas), may also be crucial preconditions for AEG. Society cannot reliably solve complex cooperative problems without such basic knowledge of human behavior.

Scientific communities, politicians, and media have a crucial role to play. Distrust of science and government has grown among influential groups (Weber and Stern 2011; Chokshi 2015). This distrust is partly caused by clear violations of authority, honesty, fairness, and rule of law (Arnstein 1969; Bullard and Johnson 2000). News media has contributed by featuring provocative stories to increase ratings and by popularizing misinformation (Weber and Stern 2011).

At a minimum, governments and professional societies should frankly communicate the rationales and supporting evidence behind their decisions and abstain from deliberate social manipulation (Lockwood 2010). Important decisions should not be made unilaterally (Reed 2008). Popular news organizations should also recognize their role and adopt precautionary methods to minimize bias and accurately portray information (Entman 2007; Silverman 2007).

17.7 Future Research

The primary goal of this chapter was to inform governance scholars of important social cognitive processes that shape AEG in order to guide future research. These processes fundamentally influence societal capacity for cooperation and adaptation. Cognitive limitations create informational, motivational, and decision-making barriers to societal cooperation and adaptive problem solving. These shortcomings lead to suboptimal solutions and slow necessary change; they are amplified by social discourse (e.g., political framing, news), increasing susceptibility to certain social cognitive biases (e.g., in-group favortism, disparate mental models) that enflame conflict and decrease acceptance of governance innovations, further undermining cooperation. Legitimacy perceptions and fundamental needs for security, procedural justice, and self-determination influence the reception of AEG, affecting participatory, regulatory, and other governance processes. What design features might improve AEG? There is not enough existing research to adequately address this question. More research is needed to understand how these principles operate in particular social-ecological contexts and ultimately influenceAEG.

17.7.1 Stages of Legal Innovation

Castro's (2012) Stages of Legal Innovation framework may guide such research. Future research should investigate how each such stage is shaped by social cognitions and translate into governance outcomes. According to Castro (2012), governance adaptations are incorporated into the "existing fabric of society" through a process of emergence, institutionalization, generalization, and stabilization similar to the adaptive governance cycle (Chaffin and Gunderson 2016) in resilience theory. During emergence, an innovative governance concept or policy issue is introduced by a particular stakeholder group to address a perceived shortcoming in current governance. These innovations enter public discourse and are debated by different interest groups, who compete to reframe the topic to control it (Castro 2012). For example, climate change policy is strategically reframed by opposing interest groups, altering public discourse and law (Dunlap and McCright 2010). During institutionalization, particularly influential interpretations of these innovations are "translated into a set of legal, policy, and institutional changes" (Castro 2012). Social cognitive processes such as those described in this chapter are hypothesized to influence the governance arrangements that are created. Moreover, the participatory and regulatory processes used to decide and implement these changes are expected to affect their perceived legitimacy, acceptance, and subsequent cooperation and compliance (Cosens 2013; Chap. 10; DeCaro et al. 2015). During generalization, proponents of these legal innovations try to encourage widespread adoption. However, informal social, organizational, and institutional norms may remain unchanged or resistant, such as when government officials sidestep laws that require increased public involvement (Castro and Batel 2008). During stabilization, some of these legal innovations are sufficiently entrenched within society to become common practice. Entrenched governance systems remain open to opposition, which could trigger further cycles of public discourse and legal innovation.

For best results, it is recommended that researchers study how the psychological principles discussed in this chapter influence pivotal events and transition points in cases of AEG. This approach will generate vital descriptive information needed to develop more rigorous explanations of behavior in AEG. In doing so, it is important to consider (1) how a focal issue is being framed by different groups, and how those frames are influencing conflict, cooperation, and change, (2) the role that perceptions of regulatory systems and participatory processes are playing in ongoing adaptive processes, (3) the factors that have contributed to the perceived legitimacy or illegitimacy of particular adaptive processes among certain stakeholders, and (4) how each of these processes interacts with specific aspects of the law.

For example, in the Florida Everglades, there have been decades of litigation over proposals to manage the Everglades adaptively. This litigation has undermined cooperation and hindered experimental approaches to ecosystem management (Gunderson et al. 2014; Chap. 5). In contrast, in Oregon's Klamath River Basin, prolonged litigation led to a turning point, where some stakeholders reconciled their differences and developed new cooperative proposals to cooperatively manage the ecosystem (Chaffin et al. 2014a; Chap. 6). It is unclear what social and psychological factors led to these differing trajectories. By studying the social cognitions involved in such cases it may be possible to create facilitative governance frameworks that encourage better conflict resolution, cooperation, and AEG (e.g., Leach and Sabatier 2005).

17.8 Conclusion

Adaptive environmental governance faces many challenges that stem from the fallibility of human decision-makers struggling to make sense of complex social and environmental problems. The fundamental challenge from a social cognitive perspective is to reconcile different, often antagonistic, values and beliefs shared by insular social groups, in order to create broad-scale societal cooperation capable of producing beneficial change. Cognitive biases and limitations (e.g., faulty mental models, loss aversion) further complicate this problem and make it especially difficult for stakeholders to devise sensible solutions to the problems they face. People fundamentally want governance systems that address broader needs such as legitimacy, fairness, and security, in addition to their basic economic needs and livelihood. Governance scholars and policymakers may be better equipped to address the challenges posed by AEG if they heed the principles outlined in this chapter.

Acknowledgments This work was developed in part under the Adaptive Water Governance Project, funded by the US National Socio-Environmental Synthesis Center (SESYNC) under funding from the US National Science Foundation, NSF DBI-1052875. The views set forth by contributors to this volume represent their own and do not represent the views of any public or private entity the contributor is affiliated with.

References

Abrams D, Wetherell M, Cochrane S, Hogg MA, Turner JC (1990) Knowing what to think by knowing who you are: self-categorization and the nature of norm formation, conformity, and group polarization. Br J Soc Psychol 29:97–119

Agrawal A, Ribot J (2014) Are Ostrom's design principles sufficient for design? Policy Matters 19:111–115
Armitage D (2007) Governance and the commons in a multi-level world. Int J Commons 2(1):7–32
Arnold CA (2004) Working out an environmental ethic: anniversary lessons from Mono Lake. Wyoming Law Rev 4:1–55
Arnold CA (2014) Framing watersheds. In: Kirokawa K (ed) Environmental law and contrasting ideas of nature: a constructivist approach. Cambridge University Press, Cambridge, pp 271–302
Arnold CA, Gunderson LH (2013) Adaptive law and resilience. Environ Law Rep 43:10426–10443
Arnold CA, Green OO, DeCaro DA, Chase A, Ewa JG (2014) The social-ecological resilience of an easter urban-suburban watershed: the anacostia river basin. Idaho Law Rev 51(1):29–90
Arnstein SR (1969) A ladder of citizen participation. J Am Inst Plann 35:216–224
Bailey K, Grossardt T (2010) Toward structured public involvement: justice, geography and collaborative/geovisual decision support systems. Ann Assoc Am Geogr 100(1):57–86
Beach LR (1998) Image theory: theoretical and empirical foundations. Routledge, Mahwah
Benson MH, Llewellyn D, Morrison R, Stone M (2014) Water governance challenges in New Mexico's Middle Rio Grande Valley: a resilience assessment. Idaho Law Rev 51 (1):195–228. https://doi.org/10.2139/ssrn.2464387
Berrone P, Gelabert L, Fosfuri A (2009) The impact of symbolic and substantive actions on environmental legitimacy. IESE Business School working paper 778. http://papers.ssrn.com/sol3/papers.cfm?abstract_id=1349063. Accessed 13 Mar 2016
Bingham LB (2009) Collaborative governance: emerging practices and the incomplete legal framework for public and stakeholder voice. J Dispute Resolution 2(2):1–58
Bingham LB (2010) The next generation of administrative law: building the legal infrastructure for collaborative governance. Wisconsin Law Rev 2010(2):297–356
Bingham LB, Nabatchi T, O'Leary R (2005) The new governance: practices and processes for stakeholder and citizen participation in the work of government. Public Adm Rev 65(5):547–558
Bingham LB, Sandfort J, O'Leary R (2008) Learning to do and doing to learn: teaching managers to collaborate in networks. In: Bingham LB, O'Leary R (eds) Big ideas in collaborative public management. Taylor and Francis, New York, pp 270–286
Birge HE, Allen CR, Craig RK, Garmestani AS, Hamm JA, Babbitt C, Nemec K, Schlager E (2014) Social-ecological resilience and law in the platte river basin. Idaho Law Rev 51(1):229–327
Bowles S (2008) Policies designed for self-interested citizens may undermine "the moral sentiments": evidence from economic experiments. Science 320:1605–1609
Boykoff MT, Boykoff JM (2004) Balance as bias: global warming and the US prestige press. Glob Environ Chang 14(2):125–136
Brehm SS, Brehm JW (1981) Psychological reactance: a theory of freedom and control. Academic, New York
Bullard RD, Johnson GS (2000) Environmental justice: grassroots activism and its impact on public policy decision making. J Soc Issues 56(3):555–578
Castro P (2012) Legal innovation for social change: exploring change and resistance to different types of sustainability laws. Polit Psychol 33(1):105–121
Castro P, Batel S (2008) Social representation, change and resistance: on the difficulties of generalizing new norms. Culture & Psychol 14(4):475–497
Castro P, Mouro C (2011) Psycho-social processes in dealing with legal innovation in the community: insights from biodiversity conservation. Am J Commun Psychol 47(3-4):362–373
Chaffin BC, Gunderson LH (2016) Emergence, institutionalization and renewal: rhythms of adaptive governance in complex social-ecological systems. J Environ Manag **165**:81–87
Chaffin BC, Craig RK, Gosnell H (2014a) Resilience, adaptation, and transformation in the Klamath River Basin social-ecological system. Idaho Law Rev 51(1):157–183
Chaffin BC, Gosnell H, Cosens BA (2014b) A decade of adaptive governance scholarship: synthesis and future directions. Ecol Soc 19(3):56. https://doi.org/10.5751/es-06824-190356

Chen XP, Pillutla MM, Yao X (2009) Unintended consequences of cooperation inducing and maintaining mechanisms in public goods dilemmas: sanctions and moral appeals. Group Process Intergroup Relat 12(2):241–255

Chess C, Purcell K (1999) Public participation and the environment: do we know what works? Environ Sci Technol 33(16):2685–2692

Chokshi N (2015) "Turn off the computer," Oregon sheriff begs. Stop "tearing each other apart." Post Nation. In: The Washington Post. https://www.washingtonpost.com/news/post-nation/wp/2016/02/12/turn-off-the-computer-oregon-sheriff-begs-stop-tearing-each-other-apart/. Accessed 1 Apr 2016

Clarvis MH, Fatichi S, Allan A, Fuhrer J, Stoffel M, Romerio F, Gaudard L, Burlando P, Beniston M, Xoplaki E, Toreti A (2014) Governing and managing water resources under changing hydro-climatic contexts: the case of the upper Rhone basin. Environ Sci Pol 43:56–67

Clement F (2010) Analysing decentralised natural resource governance: proposition for a "politicised" institutional analysis and development framework. Policy Sci 43(2):129–156

Cornforth A (2009) Behaviour change: insights for environmental policy making from social psychology and behavioural economics. Policy Q 5(4):21–28

Cosens BA (2013) Legitimacy, adaptation, and resilience in ecosystem management. Ecol Soc 18(1):3. https://doi.org/10.5751/es-05093-180103

Cote M, Nightingale AJ (2012) Resilience thinking meets social theory situating social change in socio-ecological systems (SES) research. Prog Hum Geogr 36(4):475–489

Craig RK, Garmestani AS, Allen CR, Arnold CA, Birgé H, DeCaro DA, Fremier AK, Gosnell H, Schlager E (2017) Balancing stability and flexibility in adaptive governance: an analysis of tools available in U.S. environmental law. Ecol Soc 22(2):3. https://doi.org/10.5751/ES-08983-220203

DeCaro DA, Stokes M (2008) Social-psychological principles of community-based conservation and conservancy motivation: attaining goals within an autonomy-supportive environment. Conserv Biol 22(6):1443–1451

DeCaro DA, Stokes MK (2013) Public participation and institutional fit: a social–psychological perspective. Ecol Soc 18(4):40. https://doi.org/10.5751/es-05837-180440

DeCaro DA, Janssen MA, Lee A (2015) Synergistic effects of voting and enforcement on internalized motivation to cooperate in a resource dilemma. Judgm Decis Mak 10(6):511–537

DeCaro DA, Arnold CA, Boamah EF, Garmestani AS (2017a) Understanding and applying principles of social cognition and decision making in adaptive environmental governance. Ecol Soc 22(1):33. https://doi.org/10.5751/ES-09154-220133

DeCaro DA, Chaffin BC, Schlager E, Garmestani AS, Ruhl JB (2017b) Legal and institutional foundations of adaptive environmental governance. Ecol Soc 22(1):32. https://doi.org/10.5751/ES-09036-220132

Dietz T, Ostrom E, Stern PC (2003) The struggle to govern the commons. Science 302:1907–1912

Druschke CG, Hychka KC (2015) Manager perspectives on communication and public engagement in ecological restoration project success. Ecol Soc 20(1):58. https://doi.org/10.5751/es-07451-200158

Dunlap RE, McCright AM (2010) Climate change denial: sources, actors, strategies. In: Lever-Tracy C (ed) Routledge handbook of climate change and society. Routledge, London, pp 240–259

Edelman M (1960) Symbols and political quiescence. Am Polit Sci Rev 54(03):695–704

Entman RM (2007) Framing bias: media in the distribution of power. J Commun 57(1):163–173

Farrell J (2016) Corporate funding and ideological polarization about climate change. Proc Natl Acad Sci U S A 113(1):92–97

Gächter S (2007) Conditional cooperation. In: Frey BS, Stutzer A (eds) Economics and psychology: a promising new cross-disciplinary field. MIT Press, Cambridge, MA, pp 19–50

Gigerenzer G, Todd PM (1999) Fast and frugal heuristics: the adaptive toolbox. In: Gigerenzer G, Todd PM (eds) Simple heuristics that make us smart. Oxford University Press, New York, pp 2–34

Green OO, Garmestani AS, Allen CR, Gunderson LH, Ruhl JB, Arnold CA, Graham NA, Cosens BA, Angeler DG, Chaffin BC, Holling CS (2015) Barriers and bridges to the integration of social-ecological resilience and law. Front Ecol Environ 13(6):332–337

Gromet DM, Kunreuther H, Larrick RP (2013) Political ideology affects energy-efficiency attitudes and choices. Proc Natl Acad Sci U S A 110(23):9314–9319

Gunderson LH, Garmestani A, Rizzardi KW, Ruhl JB, Light A (2014) Escaping a rigidity trap: governance and adaptive capacity to climate change in the everglades social ecological system Idaho Law Rev 51:127–156

Hardin G (1968) Tragedy of the commons. Science 162(13):1243–1248

Hardisty DJ, Johnson EJ, Weber EU (2010) A dirty word or a dirty world? Attribute framing, political affiliation, and query theory. Psychol Sci 21(1):86–92

Hilbe C, Traulsen A, Röhl T, Milinski M (2014) Democratic decisions establish stable authorities that overcome the paradox of second-order punishment. Proc Natl Acad Sci U S A 111(2):752–756

Howard JL (2010) Managing for justice in community-based water planning: a conceptual framework. Environ Conserv 37(3):356–363

Hunt L, Haider W (2001) Fair and effective decision making in forest management planning. Soc Nat Resour 14:873–887

International Union for Conservation of Nature and Natural Resources [IUCN] (2008) Shaping a sustainable future: The IUCN programme 2009–2012. https://cmsdata.iucn.org/downloads/iucn_programme_2009_2012_dfc.pdf. Accessed 31 July 2017

Irvin RA, Stansbury J (2004) Citizen participation in decision making: is it worth the effort? Public Adm Rev 64(1):55–65

Johnson-Laird PN (2013) Mental models and cognitive change. J Cognit Psychol 25(2):131–138

Jung JY, Mellers BA (2016) American attitudes toward nudges. Judgm Decis Mak 11(1):62–74

Kahneman D (2003) A perspective on judgment and choice: mapping bounded rationality. Am Psychol 58(9):697–720

Kelley CP, Mohtadi S, Cane MA, Seager R, Kushnir Y (2015) Climate change in the fertile crescent and implications of the recent Syrian drought. Proc Natl Acad Sci U S A 112(11):3241–3246

Kemp R, Schot J, Hoogma R (1998) Regime shifts to sustainability through processes of niche formation: the approach of strategic niche management. Tech Anal Strat Manag 10(2):175–198

Kim E-H, Lyon TP (2013) Beyond the dichotomy of symbolic versus substantive actions. Acad Manag Proc 1:14398. https://doi.org/10.5465/AMBPP.2013.14398abstract

Kollmuss A, Agyeman J (2002) Mind the gap: why do people act environmentally and what are the barriers to pro-environmental behavior? Environ Educ Res 8(3):239–260

Larson KL, Wiek A, Keeler LW (2013) A comprehensive sustainability appraisal of water governance in Phoenix, AZ. J Environ Manag 116:58–71

Leach WD, Sabatier PA (2005) To trust an adversary: integrating rational and psychological models of collaborative policymaking. Am Polit Sci Rev 99(04):491–503

Lee VK, Harris LT (2013) How social cognition can inform social decision making. Front Neurosci 7(259):1–13

Lever-Tracy C (ed) (2010) Routledge handbook of climate change and society. Routledge, New York

Levine JM (1999) Solomon Asch's legacy for group research. Personal Soc Psychol Rev 3(4):358–364

Lichtenstein S, Slovic P, Fischhoff B, Layman M, Combs B (1978) Judged frequency of lethal events. J Exp Psychol Hum Learn 4:551–578

Lind EA, Van den Bos K (2002) When fairness works: toward a general theory of uncertainty management. Res Organ Behav 24:181–223

Lockwood M (2010) Good governance for terrestrial protected areas: a framework, principles and performance outcomes. J Environ Manag 91:754–766

Loewenstein GF, Weber EU, Hsee CK, Welch N (2001) Risk as feelings. Psychol Bull 127(2):267–286

Lubell M (2004) Collaborative environmental institutions: all talk and no action? J Policy Anal Manag 23(3):549–573
Lubell M, Henry AD, McCoy M (2010) Collaborative institutions in an ecology of games. Am J Polit Sci 54(2):287–300
Lydon M, Bartman D, Woudstra R, Khawarzad A (2010) Tactical urbanism: short term action: long term change, vol 1. Issuu. http://issuu.com/streetplanscollaborative/docs/tactical_urbanism_vol.1 . Accessed 1 Aug 2017
Marshall G (2008) Nesting, subsidiarity, and community-based environmental governance beyond the local scale. Int J Commons 2(1):75–97
Mashuda S, Good-Stefani G, France T et al (2016). Letter: Re: Federal agencies' NEPA process on Columbia-Snake River dam operations: process obscures facts, fails to invite meaningful public participation. EarthJustice. http://earthjustice.org/sites/default/files/files/NEPA%20Scoping%20Failure%20Ltr%20FINAL.pdf. Accessed 14 June 2017
McComas KA, Stedman R, Hart PS (2011) Community support for campus approaches to sustainable energy use: the role of "town–gown" relationships. Energy Policy 39:2310–2318
McCright AM, Xiao C, Dunlap RE (2014) Political polarization on support for government spending on environmental protection in the USA, 1974–2012. Soc Sci Res 48:251–260
McGinley PC (2011) Climate change and the war on coal: exploring the dark side. Vermont J EnviroN Law 13:255–331
Moller AC, Ryan RM, Deci EL (2006) Self-determination theory and public policy: improving the quality of consumer decisions without using coercion. J Publ Policy and Mark 25(1):104–116
Moscovici S (1988) Notes towards a description of social representations. Eur J Soc Psychol 18(3):211–250
Moser SC, Ekstrom JA (2010) A framework to diagnose barriers to climate change adaptation. Proc Nat Acad Sci USA 107(51):22026–22031
Moskowitz P (2014) Cold weather snap fuels misinformation over climate change. Al Jazeera America. http://america.aljazeera.com/articles/2014/1/5/cold-weather-snapbringsconfusionoverclimatechange.html. Accessed 16 Mar 2016
Mulder LB, van Dijk E, DeCremer D, Wilke HAM (2006) Undermining trust and cooperation: the paradox of sanctioning systems in social dilemmas. J Exp Soc Psychol 42:147–162
Munaretto S, Siciliano G, Turvani ME (2014) Integrating adaptive governance and participatory multicriteria methods: a framework for climate adaptation governance. Ecol Soc 19(2):74. https://doi.org/10.5751/ES-06381-190274
National Civic League [NCL] (2013) Making public participation legal. http://ncdd.org/rc/wp-content/uploads/MakingP2Legal.pdf. Accessed 31 July 2017
Nickerson RS (1998) Confirmation bias: a ubiquitous phenomenon in many guises. Rev Gen Psychol 2(2):175–220
O'Leary R, Bingham LB, Choi Y (2010) Teaching collaborative leadership: ideas and lessons for the field. J Public Affairs Edu 16(4):565–592
Opotow S, Brook A (2003) Identity and exclusion in rangeland conflict. In: Clayton S, Opotow S (eds) Identity and natural environment: the psychological significance of nature. MIT Press, Cambridge, MA, pp 249–272
Ostrom E (1965) Public entrepreneurship: a case study in ground water management. Dissertation, University of California at Los Angeles
Ostrom V (1971) The political theory of a compound republic: designing the American experiment, 2nd edn. Institute for Contemporary Studies Press, San Francisco
Ostrom E (1990) Governing the commons: evolution of institutions for collective action. Cambridge University Press, Cambridge
Ostrom E (1998) A behavioral approach to the rational choice theory of collective action: presidential address, American Political Science Association, 1997. Am Polit Sci Rev 91:1–22
Ostrom E (2005) Animating institutional analysis. In: Ostrom E (ed) Understanding institutional diversity. Princeton University Press, Princeton, pp 99–133

Ostrom E (2010) Beyond markets and states: polycentric governance of complex economic systems. Am Econ Rev 100:1–33
Ostrom E (2014) Do institutions for collective action evolve? J Bioecon 16:3–30
Pahl-Wostl C (2009) A conceptual framework for analysing adaptive capacity and multi-level learning processes in resource governance regimes. Glob Environ Chang 19(3):354–365
Parker L (2016) What happens to the U.S. Midwest when the water's gone? National Geographic, August 2016. http://www.nationalgeographic.com/magazine/2016/08/vanishing-midwest-ogallala-aquifer-drought/. Accessed 15 June 2017
Parks C, Joireman J, Lange P (2013) Cooperation, trust, and antagonism: how public goods are promoted. Psychol Sci Public Interest 14(3):119–165
Peluso NL (1993) Coercing conservation? The politics of state resource control. Glob Environ Chang 3(2):199–217
Rachlinski JJ (1998) Limits of social norms. Chicago-Kent Law Rev 74:1537–1567
Reed M (2008) Stakeholder participation for environmental management: a literature review. Biol Conserv 141:2417–2431
Sarker A (2013) The role of state-reinforced self-governance in averting the tragedy of the irrigation commons in Japan. Public Adm 91(3):727–743
Sarkki S, Rantala L, Karjalainen TP (2015) Fit between conservation instruments and local social systems: cases of co-management and payments for ecosystem services. Change and Adapt Socio-Ecol Syst 2(1):59–78
Schlager E, Heikkila T (2011) Left high and dry? Climate change, common-pool resource theory, and adaptability of Western water compacts. Public Adm Rev 71(3):461–470
Schlager E, Heikkila T, Case C (2012) The costs of compliance with interstate agreements: lessons from water compacts in the western United States. Publius, J Federalism 42(3):494–515
Schlichting I (2013) Strategic framing of climate change by industry actors: a meta-analysis. Environ Commun, J Nat Culture 7(4):493–511
Silverman C (2007) Regret the error: how media mistakes pollute the press and imperil free speech. Union Square Press, New York
Smith V (2012) Analysis: 'war on coal' label belies realities. Lexington Herald-Leader. http://www.kentucky.com/news/business/article44384148.html. Accessed 3 Mar 2016
Stern PC, Easterling WE (eds) (1999) Making climate forecasts matter. National Research Council, National Academy Press, Washington, DC
Stoknes PE (2014) Rethinking climate communications and the "psychological climate paradox". Energy Res Soc Sci 1:161–170
Swim JK, Stern PC, Doherty TJ, Clayton S, Reser JP, Weber EU, Gifford R, Howard GS (2011) Psychology's contributions to understanding and addressing global climate change. Am Psychol 66(4):241–250
Tenbrunsel AE, Messick DM (1999) Sanctioning systems, decision frames, and cooperation. Adm Sci Q 44(4):684–707
Turner RA, Fitzsimmons C, Forster J, Mahon R, Peterson A, Stead SM (2014) Measuring good governance for complex ecosystems: perceptions of coral reef-dependent communities in the Caribbean. Glob Environ Chang 29:105–117
Tyler TR (1988) What is procedural justice? Law and Soc Rev 22:301–355
Tyler TR (1990) Why people obey the law. Yale University Press, New Haven
Tyler TR (2006) Psychological perspectives on legitimacy and legitimization. Annu Rev Psychol 57:375–400
Tyler TR, Degoey P (1995) Collective restraint in social dilemmas: procedural justice and social identification effects on support for authorities. J Pers Soc Psychol 69(3):482–497
United Nations [UN] (2015) The United Nations world water development report 2015: water for a sustainable world. World Water Assessment Programme, Paris. http://unesdoc.unesco.org/images/0023/002318/231823E.pdf. Accessed 16 Feb 2017
Van Ness DW, Strong KH (2013) Restoring justice: an introduction to restorative justice. Bender, New Jersey

Vidal J (2016) Water supplies in Syria deteriorating fast due to conflict, experts warn. The Guardian. https://www.theguardian.com/environment/2016/sep/07/water-supplies-in-syria-deteriorating-fast-due-to-conflict-experts-warn. Accessed 15 June 2017
Von Korff Y, d'Aquino P, Daniell KA, Bijlsma R (2010) Designing participation processes for water management and beyond. Ecol Soc 15(3):1. http://www.ecologyandsociety.org/vol15/iss3/art1/. Accessed 1 August 2017
Vreudenhil H, Slinger J, Kater E, Thissen W (2010) The influence of scale preferences on the design of a water innovation: a case in Dutch River management. Environ Manag 46(1):29–43
Weber EU (2010) What shapes perceptions of climate change? Wiley Interdiscip Rev Clim Chang 1(3):332–342
Weber EU, Stern PC (2011) Public understanding of climate change in the United States. Am Psychol 66(4):315–328
Winter DD (2000) Some big ideas for some big problems. Am Psychol 55(5):516–522
Yamagishi T (1986) The provision of a sanctioning system as a public good. J Pers Soc Psychol 51(1):110–116

Part IV
Summary and Conclusions

Chapter 18
Adaptive Water Governance: Summary and Synthesis

Barbara Cosens and Lance H. Gunderson

Abstract The Adaptive Water Governance project was set up to examine how law, policy, and ecological dynamics influence the governance of regional-scale water-based social-ecological systems in the United States and Australia. With the onset of the Anthropocene, global and regional changes in biophysical inputs to these systems will challenge their capacity to respond while maintaining functions of water supply, flood control, hydropower production, water quality, and biodiversity in a time of aging infrastructure. At the heart of the capacity of these systems to respond to these challenges is their governance. Governance of these complex and dynamic social-environmental systems has moved beyond state-centric, legally bounded control to a complex mix of public/private self-organization. The resilience assessments of case studies and integrative scholarship led to this synthesis, which is presented in the form of three challenges to adaptive water governance. The first is to consider the role of government in removing barriers to adaptive governance by facilitating actions that take advantage of windows of opportunity and institutionalize the results of innovative solutions. The second challenge is to assure that in embracing these new approaches, society must continue to evaluates trade-offs. Such governance should assure that engagement of private and economic actors and the increase in governmental flexibility do not destabilize nor come at the expense of legitimacy, accountability, equity, and justice. Law in its role of establishing the structure and process of government and in placing bounds on the interaction of governmental entities with private actors is a key component in striking the balance between stability in government and adaptability of governance. The third challenge is to identify whether government might be given the authority to facilitate and participate in adaptive governance. Meeting these challenges will increase the capacity of these social-ecological systems to innovate, adapt, and learn their way into an uncertain future, by increasing participation in adaptive governance in ways that are legitimate, transparent, and just.

B. Cosens (✉)
College of Law and Waters of the West Program, University of Idaho, Moscow, ID, USA
e-mail: bcosens@uidaho.edu

L. H. Gunderson
Department of Environmental Sciences, Emory University, Atlanta, GA, USA
e-mail: lgunder@emory.edu

B. Cosens, L. Gunderson (eds.), *Practical Panarchy for Adaptive Water Governance*, https://doi.org/10.1007/978-3-319-72472-0_18

Keywords Resilience and law · Adaptive law · Adaptive governance · Water management · Social-ecological systems · Windows of opportunity · Good governance · Panarchy · Adaptive capacity · Participatory capacity

18.1 Introduction

With the onset of the Anthropocene, the age in which the biophysical imprint of humans is being recorded in earth's geologic record (Steffen et al. 2007, 2011), society is faced with both the challenge and the obligation of human dominance of the planet. No longer can the relation between humans and nature be thought of as separate – a state in which action flows from the social to the ecological and regulation of human interaction with ecosystems can assume a linear response. Humans and nature are one social-ecological system (Barnosky et al. 2017). The observations of ecosystems that led to the development of resilience theory illustrate that systems are self-organizing, adaptive, and nonlinear (Holling 1973; Gunderson and Holling 2002; Walker et al. 2004; Walker and Salt 2006, 2012). Environmental governance must respond accordingly (Ostrom 1990; Dietz et al. 2003; Delmas and Young 2009).

While the imprint of human dominance on the planet plays out on a geologic time scale, society is also experiencing accelerating change at the much more immediate human timescale. The Great Acceleration (Steffen et al. 2007, 2015) is described in the literature as the intersection of the Anthropocene with the digital revolution characterized by economic globalization and social connectivity, both of which have been changing at increasingly exponential rates since the 1950s (Friedman 2016). Just as this presents a challenge to higher education to move from learning as the accumulation of substantive knowledge to a focus on learning to learn, governance must be nimble and have the capacity to learn and evolve. Governance must be adaptive.

Defining adaptive governance as emergent, a definition followed in this volume, masks the reality that emergence is occurring in places with highly developed systems of government. The authority, structure, and process of environmental government is defined by law and government acts intentionally under the mandates provided by law. Defining adaptive governance as a mix of state-centric government with private and public self-organization aimed at the challenge of collective choice requires the ability of government to participate in and even facilitate adaptive governance. Thus, for adaptive governance to succeed, the law related to environmental government requires reform. The Adaptive Water Governance project brought together scholars of law, institutions, and human geography with ecological resilience scholars to consider the role of government in adaptive governance and how legal reform might improve the climate for its emergence and sustainability. This chapter synthesizes lessons from the Adaptive Water Governance project.

18.2 Synthesis: The Adaptive Water Governance Project

Governmental facilitation and participation in adaptive governance faces three major challenges: (1) the lack of a clear legal mandate and appropriate structure for agency action to facilitate the preparation for, emergence of, and institutionalization of the results of adaptive governance; (2) the lack of legal mechanisms to assure that a move from state-centric governmental control of environmental issues to state-enabled adaptive governance is not destabilizing and does not occur at the expense of legitimacy, accountability, equity, and justice; and (3) the lack of legal tools and a governmental structure necessary to empower governmental agencies to facilitate and participate while enhancing the participatory capacity of local actors to engage in adaptive governance. The following sections synthesize the lessons from the Adaptive Water Governance project relevant to these three challenges.

18.2.1 The Role of Government in Emergent Adaptive Governance: Preparation, Institutionalization, and Panarchy

A window of opportunity in which adaptive governance may emerge is observed to be the result of a rapid to intermediate rate crisis or disturbance triggered by change in the biophysical, social, or economic aspects of the system (Kingdon 1995; Olsson et al. 2006). Olsson et al. (2006) explored the questions of when is a window of opportunity taken advantage of and when does it lead to transformation to a more desirable state for the social-ecological system. Importantly, they identified the need for preparation prior to the opening of a window. One need go no further than the daily news to see this play out when a crisis occurs. The response to a devastating weather event such as a flood or hurricane is to rebuild. Having spent no time engaging those in the zone of risk in envisioning a different future, the empathy engendered by the shock of the occurrence and the sense of local loss creates a strong momentum toward return to the status quo.

Natural disasters are not the only events that trigger the opening of a window of opportunity. The basin studies illustrate the creation of disturbances through legal action (Chaps. 4 (Columbia), 6 (Klamath), and 7 (Middle Rio Grande)) and suggest that the inadequacy of solutions under strict application of laws such as the Endangered Species Act (Chap. 8 (Platte)), or the redistribution of power relations brought about by assertion of rights by marginalized communities (Chaps. 4 (Columbia) and 6 (Klamath)), may be sufficient to trigger emergence of adaptive governance. The basin studies also illustrate situations in which, despite preparation and the opening of a window, the status quo path was taken (Chap. 5 (Everglades)) and situations in which adaptive governance emerged, but failed (or may have failed) to achieve the results of collaboration (Chap. 6 (Klamath).

In both situations, natural disaster or social disturbance, it is clear that preparation prior to the occurrence and institutionalization of the results of any emergence of adaptive governance are important points for intentional intervention. Thus, government can play an important role on both the front and back ends of adaptive governance processes.

Government facilitation of and participation in local processes using tools for envisioning multiple possible futures and for changing perceptions of risk and uncertainty such as scenario (Schwartz 1996) or adaptive planning (Arnold 2010), and resilience assessment (Resilience Alliance 2007), can help in managing expectations (Chap. 17) and even generating excitement around transformation. Importantly, these types of processes can lay the ground work for the development of informal networks across interests and levels of government with a stake in the landscape or system of interest. Relationships built in a relatively calm planning setting may then form the basis for emergence when a crisis does occur.

Because adaptive governance is observed to emerge in response to a particular problem, the informal networks that lead to its emergence may disintegrate when the problem subsides. The basin studies illustrate that even when the problem persists, the engagement of new actors at higher and more powerful levels may disturb fragile coalitions (Chaps. 4 (Columbia), 5 (Everglades), and 6 (Klamath)). The mere authority for government at various levels to participate in emergent processes is insufficient to prevent this result. Where government is engaged in the front end with planning, the emergent stage with resources and facilitation, it must have an avenue to institutionalize the results on the back end. Water settlements among Native American Tribes, States, and the United States have this type of process (Cosens and Chaffin 2016), and it has proven successful in situations where the three sovereigns involved plan and agree from the outset to use their resources and influence to see the process through to institutionalization should a settlement be reached (Cosens 1998, 2003). Institutionalization must include not only the formal recognition of a new vision for the future, but the resources from higher levels of government to achieve that future and the continued governmental facilitation of implementation of any necessary transformation.

18.2.2 The Role of Law in Balancing Stability and Flexibility: Good Governance and Panarchy

As touched on in Chaps. 10 and 17 and the topic of an entire Chap. 11 by Craig et al., the social need and expectation for stability in government and law is one of the primary barriers to governmental participation in adaptive governance. One of the primary services society expects from government is the provision of stability within which individuals and private entities may innovate, invest, thrive, and simply live a life free from conflict and disruption. As governance of environmental systems moves beyond state-centric control bounded by law, to a complex mix of public/private self-organization, society must rethink the role that law and government play in that mix. Importantly, in trading state-centric governmental control for innovative adaptive governance, it is essential not only that the law allows governmental participation in flexible governance, but that the engagement of private and economic actors and the increase in governmental flexibility is not done at the expense of legitimacy, accountability, equity, and justice. Failure to do so risks either rejection of reform to enhance environmental governance by democratic societies or destabilization of those very societies. Of even greater importance, resilience theory illustrates that the reliance on stability through maintenance of the status quo may provide a false sense of security as systems respond in surprising and nonlinear ways as climate change unfolds. In fact, a system of government that is unable to participate in and facilitate adaptation and transformation may, in the face of change, be unstable. The question becomes, how does society achieve both flexibility in governance and stability in government?

Two primary lessons can be drawn from the Adaptive Water Governance project. First, the process elements that have been developed in administrative law to assure attention to accountability, legitimacy, equity, and justice in actions by governmental entities should be modified to assure these attributes remain in focus as greater flexibility in governmental action is introduced and as public/private networks emerge as adaptive governance. The process of governmental agency interaction with the public, whether in the context of public participation or the enforcement of regulation, is governed by administrative law. This is where requirements for transparency, public participation, and review of agency action can be found, and it is understanding the purpose and design of administrative law that will lead to the reforms necessary to assure that government participation in flexible adaptive governance remains legitimate (Chap. 10). Importantly, tools for flexible management, such as the process for adaptive management (Craig and Ruhl 2014), will have extremely limited applicability in complex systems with high uncertainty without the parallel use of process elements of administrative law that ensure legitimacy, accountability, equity, and justice (Cosens 2010, 2013; Chap. 10). The basin assessments indicate that robust, government-facilitated participatory processes and parallel capacity building to ensure equal ability to participate should be a primary focus of government reform in the process elements of water management.

In addition to changes in the process elements of administrative law, balancing stability and flexibility will require changes to substantive environmental law. Through rulemaking that assures that similar situations across a vast country will be treated in a similar manner, agencies ensure the legitimacy and equity of their implementation of environmental laws, but also limit the ability to tailor solutions to local systems. Standards to be achieved must replace rules pertaining to the manner of achieving them, opening up space for local innovation and tailoring (Chap. 11). This leads to the second lesson for balancing stability and flexibility through the structure of government.

The structure of government and the choice of the level at which authority to act and innovate is placed should be informed by the concept of panarchy (Gunderson and Holling 2002; Chaps. 10 and 11). The nested structure of ecosystems, referred to as panarchy, provides a metaphor for the structure of governance systems. Higher levels of government may provide stability through establishing standards, maintaining oversight, and providing resources for facilitation and capacity building. At lower levels, government or governmental networks at the scale of the social-ecological system may engage with local actors to innovate in the face of change. This move to subsidiarity while retaining higher level checks requires changes to substantive environmental laws.

The application of federal level solutions to, for example, nonpoint source pollution at the local watershed scale under the US Clean Water Act has led to failure in the headwaters of the Columbia River basin to achieve standards, unnecessary costs, and loss of legitimacy when failure to account for the physical aspects of the watershed rendered the chosen solution inadequate (Rittenburg et al. 2015). Modifying this type of program to allow local experimentation while maintaining a federal level standard would allow adaptation to local physical and social conditions as well as the inclusion of local knowledge. Implementation of federal environmental laws such as the Endangered Species Act have secondary consequences for those economic pursuits that are highly dependent on water, such as the production of hydropower (Columbia, Chap. 4) or irrigation (Klamath, Everglades, Middle Rio Grande, Platte, Chaps. 5, 6, 7 and 8). The basin assessments illustrate that in the face of these consequences, the ESA may lead basin communities to seek comprehensive solutions that benefit or transform existing economies while recovering species. Yet, as illustrated in the Klamath and Everglades (Chaps. 5, 6, and 12), the absence of a mechanism for government facilitation, flexibility in solution design, capacity building, and institutionalization of results within the ESA results in ad hoc processes that may require substantial resources, may be captured by more powerful interests and may, in the end, fail.

While intentional efforts to modify existing environmental law may provide solutions, it is also important to note that change is possible even without the political will at the federal level. Local government may step up during a gap in environmental leadership on the national stage. This may build local capacity and set the stage for a better local/national balance in the future. While arguably a messy process, the potential for innovation when structure is rebuilt from the bottom-up may be greater. In the end, better solutions may emerge. This points to the need to consider local capacity.

18.2.3 The Role of Government in Building Adaptive and Participatory Capacity: Authority, Resources, and Panarchy

The term adaptive capacity has been used to describe the ability of ecosystems to respond to disturbance while maintaining structure and function and to navigate ecological regime shifts (Gunderson 2000). The concept is equally relevant in application to social-ecological systems (Norris et al. 2008; Walker and Salt 2012). Government can play a role in two important aspects of adaptive capacity. First, by being adaptive itself and second by building the capacity of local governmental and nongovernmental actors to respond, participate, and adapt.

For government itself to be adaptive, it must have the legal authority to respond to change. This authority should be reflected in the statutes, regulations, and practices of agencies (Clarvis et al. 2014). Legal sources effectively define, shape, and limit this authority in water management by governmental entities by defining the authority of government to (1) adjust water allocations and water quality requirements in response to change, (2) implement adaptive management in appropriate circumstances (Craig and Ruhl 2014), (3) facilitate collaborative adaptive planning (Arnold 2010, 2014), (4) provide resources and flexibility for local innovation, and (5) facilitate the ability of individuals and private entities to adapt through water markets (Cosens 2016). The basin assessments reveal that water management agencies currently lack legal mechanisms that allow for alternatives to traditional environmental enforcement—alternatives that could enhance local innovation while maintaining stability and accountability toward achieving the goals that traditional enforcement mechanisms seek to achieve. As a result, when federal legal mechanisms—such as the Endangered Species Act and Clean Water Act—apply in a given basin, the absence of state and local legal avenues to develop solutions, the lack of federal level discretion to allow local coalitions authority to create new and more comprehensive solutions, and the lack of capacity of local actors to participate in efforts to tailor solutions may result in unsatisfactory or incomplete outcomes (Bingham 2009, 2010). Introducing legal mechanisms into federal environmental law for local development of comprehensive solutions and providing the resources to implement those solutions is a necessary legal reform. This approach is an illustration of panarchy. Resources from higher levels of government provide the capacity and stability for local levels of governance to navigate change.

The role of government in building local capacity to recover from, adapt to, and transform in the face of change is likely the single most important factor in building resilient communities. In this time of accelerating change at the intersection of the Anthropocene, population growth and the digital revolution, surprise in the form of disturbance to social-ecological systems is likely to be unpredictable in source, scale, location, and magnitude. It would be impossible to design a bioregional-scale governmental entity in anticipation of the type and scale of every disturbance. Instead, sufficient local capacity must be built for response to emerge through formal and informal networks in relation to the nature and scale of the disturbance. The basin assessments suggest that when the law facilitates increased local capacity, adaptive governance has a better chance of emerging.

Local actors often lack the knowledge, time, and resources to engage in the substance of decision-making, and even when capacity exists, it is often unevenly distributed with certain sectors of society marginalized in decision-making processes. The role of law in promoting local capacity is to provide the requirement for public participation, judicial forums that can recognize and enforce the rights of those without power, and, where appropriate, provide an avenue for capacity building through the legislative allocation of resources and authority to facilitate local response (Olsson et al. 2004; Bingham 2009; National Civic League 2013), but also to build the knowledge and access to resources of local actors. The substantial increase in participatory capacity of formerly marginalized communities as a result of recognition of rights through judicial processes coincides with the emergence of local collaborative processes in the Anacostia, Rio Grande, Platte, Klamath, and Columbia basin assessments (Chaps. 3, 4, 6, 7, and 8).

18.3 Conclusion

When an individual organism is faced with a changing environment, it has three choices: to ignore, to control, or to adapt to those changes. Collectively, humans face the same set of options when confronted by the challenges of the Anthropocene. Certainly, there are some groups that willfully ignore global changes because of vested political and economic interests. The second option characterizes all of the cases studied in this volume. All are in wealthy, developed countries, and all of the cases reflect a system constructed to optimally control water and water movement for a variety of societal purposes. Infrastructure, institutions, laws, and policies have worked together to develop large-scale systems that successfully command and control waters for agricultural, economic, urban, and other interests. One of the costs of this development has been degradation of a suite of environmental indicators, from endangered species and populations to declines in water quantity and quality. While managers, policy makers, and others attempt to reverse these environmental trends, they are faced with new sets of issues related in part to global changes. Increased variation in the biophysical inputs to these systems will challenge their capacity to respond while maintaining functions of water supply, flood control, hydropower production, water quality, and biodiversity. Other issues such as aging infrastructure, rigid bureaucracies, and other uncertainties associated with managing complex systems of people and nature, will challenge their capacity to respond and adapt to global change.

At the heart of the capacity of these systems to respond lies their governance. The Adaptive Water Governance project was set up to examine how law, policy, and ecological dynamics influence the governance of a set of seven regional-scale water-based social-ecological systems in the United States and Australia. Governance of these complex and dynamic social-environmental systems has moved beyond state-centric, legally bounded control to a complex mix of public/private self-organization. The resilience assessments of case studies and integrative scholarship have articulated at least three challenges to adaptive water governance. One is to consider the role of government in removing barriers to adaptive governance by facilitating actions that take advantage of windows of opportunity and institutionalize the results of innovative solutions. Another challenge is to assure that in embracing these new approaches, society must continue to evaluate trade-offs. Such governance should assure that engagement of private and economic actors and the increase in governmental flexibility do not destabilize nor come at the expense of legitimacy, accountability, equity, and justice. Law in its role of establishing the structure and process of government and in placing bounds on the interaction of governmental entities with private actors is a key component in striking the balance between stability in government and adaptability of governance. The third challenge is to identify whether government might be given the authority to facilitate and participate in adaptive governance. Meeting these challenges will increase the capacity of these systems to innovate, adapt, and learn while increasing participation in adaptive governance in ways that are legitimate, transparent, and just. The hope is that this volume will help guide such a trajectory into the Anthropocene.

Acknowledgments This work was developed in part under the Adaptive Water Governance project, funded by the US National Socio-Environmental Synthesis Center (SESYNC) under funding from the US National Science Foundation, NSF DBI-1052875.

References

Arnold CA (2010) Adaptive watershed planning and climate change. Environ Energy Law Policy J 5:417–488

Arnold CA (2014) Adaptive water law. Kansas Law Rev 62:101–149

Barnosky AD, Hadly EA, Gonzalez P, Head J, Polly PD, Lawing AM, Eronen JT, Ackerly DD, Alex K, Biber E, Blois J, Brashares J, Ceballos G, Davis E, Dietl GP, Dirzo R, Doremus H, Fortelius M, Greene HW, Hellmann J, Hickler T, Jackson ST, Kemp M, Koch PL, Kremen C, Lindsey EL, Looy C, Marshall CR, Mendenhall C, Mulch A, Mychajliw AM, Nowak C, Ramakrishnan U, Schnitzler J, Shrestha KD, Solari K, Stegner L, Stegner MA, Stenseth NC, Wake MH, Zhang Z (2017) Merging paleobiology with conservation biology to guide the future of terrestrial ecosystems. Science 355:594

Bingham LB (2009) Collaborative governance: emerging practices and the incomplete legal framework for public and stakeholder voice. J Disput Resolut 2(2):1–58

Bingham LB (2010) The next generation of administrative law: building the legal infrastructure for collaborative governance. Wisconsin Law Rev 2010(2):297–356

Clarvis MH, Allan A, Hannah DM (2014) Water, resilience and the law: from general concepts and governance design principles to actionable mechanisms. Environ Sci Policy 43:98–110

Cosens B (1998) The 1997 water rights settlement between the State of Montana and the Chippewa Cree Tribe of the Rocky Boy's reservation – the role of community and of the trustee. UCLA J Environ Law Policy 16:255–295

Cosens B (2003) Water dispute resolution in the west: process elements for the modern era in basin-wide problem solving. Environ Law 33:949–1018

Cosens B (2010) Transboundary river governance in the face of uncertainty: resilience theory and the Columbia River Treaty. Univ Utah J Land Resour Environ Law 30:229

Cosens B (2013) Legitimacy, adaptation, and resilience in ecosystem management. Ecol Soc 18(1):3. http://www.ecologyandsociety.org/vol18/iss1/art3/

Cosens B (2016) Water law reform in the face of climate change: learning from drought in Australia and the Western United States. Environ Plan Law J 33:372

Cosens B, Chaffin BC (2016) Adaptive governance of water resources shared with indigenous peoples: the role of law. Special Issue: "Water Governance, Stakeholder Engagement, and Sustainable Water Resources Management." Water 8(3):97. Eds Sharon BM, Eylon S, Susanna E. URL:http://www.mdpi.com/2073-4441/8/3/97/html

Craig RK, Ruhl JB (2014) Designing administrative law for adaptive management. Vanderbilt Law Rev 67:1–87

Holling CS (1973) Resilience and stability of ecological systems. Annu Rev Ecol Syst 4:1–24. https://doi.org/10.1146/annurev.es.04.110173.000245

Delmas MA, Young OR (eds) (2009) Governance for the environment: new perspectives. Cambridge University Press, Cambridge

Dietz T, Ostrom E, Stern PC (2003) The struggle to govern the commons. Science 302:1907–1912

Friedman TL (2016) Thank you for being late: an optimist's guide to thriving in the age of accelerations. Farrar, Straus and Giroux, New York

Gunderson L (2000) Ecological resilience—in theory and application. Annu Rev Ecol Syst 31:425–439. https://doi.org/10.1146/annurev.ecolsys.31.1.425

Gunderson LH, Holling CS (eds) (2002) Panarchy: understanding transformations in human and natural systems. Island Press, Washington, DC

Kingdon J (1995) Agendas, Alternatives, and Public Policies, 2nd edn. Harper Collins

National Civic League (2013) Making public participation legal. http://ncdd.org/rc/wp-content/uploads/MakingP2Legal.pdf

Norris FH, Stevens SP, Pfefferbaum B, Wyche KF, Pfefferbaum RL (2008) Community resilience as a metaphor, theory, set of capacities, and strategy for disaster readiness. Am J Commun Psychol 41:127–150

Olsson P, Folke C, Berkes F (2004) Adaptive comanagement for building resilience in social-ecological systems. Environ Manag 34:75–90. https://doi.org/10.1007/s00267-003-0101-7

Olsson P, Gunderson LH, Carpenter SR, Ryan P, Lebel L, Folke C, Holling CS (2006) Shooting the rapids: navigating transitions to adaptive governance of social-ecological systems. Ecol Soc 11(1):18. [online] URL: http://www.ecologyandsociety.org/vol11/iss1/art18/

Ostrom E (1990) Governing the commons: the evolution of institutions for collective action. Cambridge University Press, New York

Resilience Alliance (2007) Assessing resilience in social-ecological systems: a workbook for scientists., http://www.resalliance.org/index.php/resilience_assessment

Rittenburg RA, Squires AL, Boll J, Brooks ES, Easton ZM, Steenhuis TS (2015) Agricultural BMP effectiveness and dominant hydrological flow paths: concepts and a review. JAWRA J Am Water Resour Assoc 51(2):305–329

Schwartz P (1996) The art of the long view: planning for the future in an uncertain world. Doubleday, New York

Steffen W, Crutzen PJ, McNeill JR (2007) The Anthropocene: are humans now overwhelming the great forces of nature? Ambio 36:614–621

Steffen W, Persson Å, Deutsch L, Zalasiewicz J, Williams M, Richardson K, Crumley C, Crutzen P, Folke C, Gordon L, Molina M, Ramanathan V, Rockström J, Scheffer M, Schellnhuber HJ, Svedin U (2011) The anthropocene: from global change to planetary stewardship. Ambio 40:739–761

Steffen W, Broadgate W, Deutsch L, Gaffney O, Ludwig C (2015) The trajectory of the Anthropocene: the great acceleration. The Anthropocene Rev 2(1):81–98

Walker BH, Salt D (2012) Resilience practice: building capacity to absorb disturbance and maintain function. Island Press, Washington, DC

Walker B, Salt D (2006) Resilience thinking: sustaining ecosystems and people in a changing world. Island Press, Washington, DC

Walker BH, Holling CS, Carpenter SR, Kinzig AP (2004) Resilience, adaptability and transformability in social-ecological systems. Ecol Soc 9 (2):5. http://www.ecologyandsociety.org/vol9/iss2/art5/

MIX
Papier aus verantwortungsvollen Quellen
Paper from responsible sources
FSC® C105338

If you have any concerns about our products,
you can contact us on
ProductSafety@springernature.com

In case Publisher is established outside the EU,
the EU authorized representative is:
Springer Nature Customer Service Center GmbH
Europaplatz 3, 69115 Heidelberg, Germany

Printed by Libri Plureos GmbH
in Hamburg, Germany